# 2022 International Symposium on Semiconductor Manufacturing (ISSM 2022)

Tokyo, Japan
12-13 December 2022

IEEE Catalog Number:   CFP22SSM-POD
ISBN:   978-1-6654-7134-3

**Copyright © 2022 by the Institute of Electrical and Electronics Engineers, Inc.
All Rights Reserved**

*Copyright and Reprint Permissions*: Abstracting is permitted with credit to the source. Libraries are permitted to photocopy beyond the limit of U.S. copyright law for private use of patrons those articles in this volume that carry a code at the bottom of the first page, provided the per-copy fee indicated in the code is paid through Copyright Clearance Center, 222 Rosewood Drive, Danvers, MA 01923.

For other copying, reprint or republication permission, write to IEEE Copyrights Manager, IEEE Service Center, 445 Hoes Lane, Piscataway, NJ 08854. All rights reserved.

***\*\*\* This is a print representation of what appears in the IEEE Digital Library. Some format issues inherent in the e-media version may also appear in this print version.***

IEEE Catalog Number:      CFP22SSM-POD
ISBN (Print-On-Demand):   978-1-6654-7134-3
ISBN (Online):            978-1-6654-7133-6
ISSN:                     1523-553X

**Additional Copies of This Publication Are Available From:**

Curran Associates, Inc
57 Morehouse Lane
Red Hook, NY 12571 USA
Phone:    (845) 758-0400
Fax:      (845) 758-2633
E-mail:   curran@proceedings.com
Web:      www.proceedings.com

# TABLE OF CONTENTS

Self-Tuning Optimization to Compatible the Delivery and Low Energy Consumption .................................... 1
*Chending Mao, Jia Lin, Sumika Arima*

Maintenance Content Reduction and Digitalization for Performance Optimization ......................................... 5
*Christopher Bode*

Application of Natural Language Processing in Semiconductor Manufacturing ............................................... 9
*Daisuke Kobayashi, Shunsaku Yasuda, Takashi Iuti, Shiho Ito*

Advanced Process Control Model for Trench Shape of Power Devices .......................................................... 12
*Takumi Ito, Wang Xueting, Yasuhisa Oomuro, Kazutaka Nagashima*

Principal Component Analysis Based GaN Transistor Live Health Monitoring .............................................. 16
*Florian Chalvin, Yoshinori Miyamae, Yoshiaki Oku, Ken Nakahara*

Dynamic AI Computation Tasks with SECS/GEM in Semiconductor Smart Manufacturing .......................... 19
*Hung H Nguyen*

Application of Big Data Science in High Reliability Automotive Wafer Yield Management System
and Failure Analysis .................................................................................................................................... 23
*Chia-Cheng Kuo, Po-Chih Chen, Chang-Tsun Tseng*

Equipment Sensor Data Cleansing Algorithm Design for ML-Based Anomaly Detection ............................... 26
*Yun-Che Hsieh, Chieh-Yu Chen, Da-Yin Liao, Peter B. Luh, Shi-Chung Chang*

Secure and Reliable Power Monitoring for Low Consumption Factory Equipment via
Programmable IoT Devices .......................................................................................................................... 30
*Sergio Garnica, Robert Wieland*

Systematic Search for Stabilizing Dopants in $ZrO_2$ and $HfO_2$ using First-Principles Calculations ................... 34
*Yosuke Harashima, Hiroaki Koga, Zeyuan Ni, Takehiro Yonehara, Michio Katouda, Akira
Notake, Hidefumi Matsui, Tsuyoshi Moriya, Mrinal Kanti Si, Ryu Hasunuma, Akira Uedono,
Yasuteru Shigeta*

A Novel Approach to Dynamic Line Balance Control and Scheduling with a Digital Twin
Production ..................................................................................................................................................... 37
*Hirofumi Tsuchiyama, Holland Smith*

Data-Driven Modeling for Production Dynamics ......................................................................................... 41
*Sumika Arima, Yu Sasaki, Sho Morie, Yuto Kataoka, Chending Mao, Jia Lin*

Recent Status of EUV Lithography, What is the Stochastic Issues? ............................................................ 45
*Toru Fujimori*

Technology Trends and Characteristics of Patent Information Disclosure in Advanced
Semiconductor Photoresist ........................................................................................................................... 47
*Kosuke Watahiki, Yoshihiro Midoh, Kazuya Okamoto*

Practical Load Impedance Monitoring System Externally Installed in Plasma Etching Equipment .................. 51
*Yuji Kasashima, Shinji Kuniie, Toshiyuki Sayama, Tatsuo Tabaru*

Advanced Process Monitoring Through Fault Detection and Classification for Robust Statistical Process Control of Tantalum Nitride Reactive Sputtering ............... 54

Stephanie Y Chang, Shiban Tiku, Lam Luu

Characterization of Light Propagation Loss in Photonics Devices using High-Resolution CDSEM Metrology ............... 59

S. Levi, R. Le Tiec, C. Dupre, C. Vannuffel, T. Dewolf, S. Garcia, K. Millard, B. Meynard, Y. Lee, M. Colard, H. Al Dujaili, J. Faugier-Tovar, Shinsuke Mizuno

Plasma Process Classification using Causal Discovery Technique ............... 63

Dai Kobayashi, Masaki Kitsunezuka, Yuki Kataoka, Jun Shinagawa

Plasma Diagnostics and Characteristics of Hydrofluorocarbon Films in Capacitively Coupled $CF_4/H_2$ Plasmas ............... 66

Shih-Nan Hsiao, Yusuke Imai, Nikolay Britrun, Takayoshi Tsutsumi, Kenji Ishikawa, Makoto Sekine, Masaru Hori

In Situ Measurement and Analysis of Low Pressure Gas Concentration Distribution using 70-DB SNR 1,000 Frame-Per-Second Absorption Imaging System ............... 69

Yushi Sakai, Yoshinobu Shiba, Takafumi Inada, Tetsuya Goto, Tomoyuki Suwa, Akihito Sutoh, Tatsuo Morimoto, Yasuyuki Shirai, Shigetoshi Sugawa, Tetsu Oikawa, Aoi Hamaya, Rihito Kuroda

A Study on Robust Noninteracting Control System Design with Disturbance Feedforward for 6-DoF Active Vibration Isolation Platform ............... 73

Thinh Huynh, Dong-Hun Lee, Young-Bok Kim

Impact of Cation Vacancies on Leakage Current on $TiN/ZrO_2/TiN$ Capacitors Studied by Positron Annihilation ............... 79

Akira Uedono, Naomichi Takahashi, Ryu Hasunuma, Yosuke Harashima, Yasuteru Shigeta, Zeyuan Ni, Hidefumi Matsui, Akira Notake, Atsushi Kubo, Tsuyoshi Moriya, Koji Michishio, Nagayasu Oshima, Shoji Ishibashi

Optimization of RF Frequencies in Dual-Frequency Capacitively Coupled Plasma Apparatus using Genetic Algorithm (GA) and Plasma Simulation ............... 83

Shigeyuki Takagi, Tatsuhiro Nakaegawa, Shih-Nan Hsiao, Makoto Sekine

Deposition Rate Dependence of the 5 Nm-Thick Ferroelectric Nondoped $HfO_2$ on MFSFET Characteristics ............... 86

Masakazu Tanuma, Joong-Won Shin, Shun-Ichiro Ohmi

Ultra-Fast Etching of Photoresist by Reactive Atmospheric-Pressure Thermal Plasma Jet ............... 90

Hibiki Kato, Hiroaki Hanafusa, Takuma Sato, Seiichiro Higashi

Obtaining of Carbon Nanowalls with a Specified Morphology ............... 93

Yerassyl Yerlanuly, Maratbek T. Gabdullin, Renata R. Nemkayeva, Rakhymzhan Zhumadilov, Balaussa Ye. Alpysbayeva, Tlekkabul S. Ramazanov

Process Optimizations for Ge-On-Si Depletion Mode Transistors using Mesa Architecture ............... 97

Sumit Choudhary, Daniel Schwarz, Hannes S. Funk, Kumar Palit Sharma, Satinder K. Sharma, Jöorg Schulze

Experimentally Study on the Effect of RIE Etching Power on Etching Rate of $\beta$-$Ga_2O_3$ Thin Film ............... 101

Wang Xu, Ran Jing Yang, Yang Lai, Yang Fa Shun, Ma Kui

Nanoimprint Lithography with $CO_2$ Ambient .................................................................................... 105
  *Toshiki Ito, Yuto Ito, Isao Kawata, Ken-Ichi Ueyama, Kouhei Nagane, Weijun Liu, Timothy Stachowiak, Wei Zhang, Teresa Estrada*

Hydrogen Diffusion Behavior in $CH_4N$-Molecularion-Implanted Wafers for Three-Dimensional Stacked CMOS Image Sensors .................................................................................... 110
  *Ryosuke Okuyama, Takeshi Kadono, Ayumi Masada, Akihiro Suzuki, Koji Kobayashi, Satoshi Shigematsu, Ryo Hirose, Yoshihiro Koga, Kazunari Kurita*

Preparation of Uniform $SiO_2$ Insulating Layer on the Inner Wall of TSV by Thermal Oxidation .................. 114
  *Guo Fengjie, Ran Jing Yang, Wang Shuo, Ma Kui, Yang Fa Shun*

Yield Prediction with Machine Learning and Parameter Limits in Semiconductor Production ...................... 117
  *Rebecca Busch, Michael Wahl, Peter Czerner, Bhaskar Choubey*

Positive/Negative Decision via Outlier Detection Towards Automatic Performance Evaluation for Defect Detector .................................................................................... 121
  *Toshinori Yamauchi, Kentaro Ohira, Takefumi Kakinuma*

A Study on Detection Method using 2-Class Classifiers for Defective Wafer Maps ...................... 125
  *Seima Sakaguchi, Yasushi Arimura, Takayuki Yamauchi, Yuichi Tokuyama, Tomoya Kawai, Hidetaka Eguchi, Hiroyuki Morinaga, Hiroharu Kawanaka, Tetsushi Wakabayashi*

Influence of High Temperature $N_2$ Annealing on Photoluminescence of SiC and Si Quantum Dots in $SiO_2$ Layer .................................................................................... 129
  *Kohki Murakawa, Norihito Mayama, Tomohisa Mizuno*

Noise Reduction in SEM Images using Deep Learning .................................................................................... 133
  *Yuki Sato, Masato Kazui, Shinji Kobayashi*

Automatic Classification of C-SAM Voids for Root Cause Identification of Bonding Yield Degradation .................................................................................... 137
  *Julien Baderot, Solange Garrais, Sergio Martinez, Johann Foucher, Ryuji Eto, Kazumasa Tanida, Takatoshi Yasui, Tomoya Tanaka*

**Author Index**

# 2022 International Symposium on Semiconductor Manufacturing (ISSM)

## PROCEEDINGS OF TECHNICAL PAPAERS

December 12-13, 2022

Hybrid

# Message from ISSM 2022 Committee

**Mr. Shozo Saito**
**Chairman of Organizing Committee of ISSM 2022**

Chairman & CEO
Device & System Platform Development Center Co., Ltd.

On behalf of the organizing committee, it is my great pleasure to extend to you all a very warm welcome to the International Symposium on Semiconductor Manufacturing (ISSM) 2022. ISSM is an international semiconductor manufacturing forum where the most updated and advanced manufacturing technologies in solid-state and semiconductor fields will be presented. Started in 1992, ISSM 2022 will be the 29th edition of the symposium.

Semiconductor, as an indispensable commodity for economic security, has contributed to the development of the digital society in the past and will continue to be a key technology that will provide the foundation for future industrial development.

Distinct types of semiconductors are installed in a wide variety of IT equipment. With the spread of IoT technologies, the amount of information processing in IT devices is dramatically increasing. In the future, all kinds of things will be connected to the network, and the transition from the era of storing data to the era of utilizing data will progress, creating newer services.

Domestic manufacturing is crucial for supply chain resilience and contributes significantly to the development of the Japanese economy. Growth of the semiconductor industry requires revitalization of the domestic IT and digital infrastructure industries. Securing and fostering human resources is necessary for both R&D and semiconductor manufacturing. Japan is one of the few countries that has a complete semiconductor manufacturing ecosystem, with strength in the equipment and materials fields specially.

Our goal is not only to strengthen our technological competitiveness. We have a mission for semiconductors from the perspective of protecting the earth. Aiming to be a game changer in greening semiconductor industry and supply chain, it is necessary to promote integrated research on systems, circuits, devices, processes, equipment, and materials, as well as human resource development, to realize green semiconductors with low environmental impact and other green features.

Today we are together here at ISSM to reaffirm the key role that semiconductor manufacturing and its environment throughout the supply chain, in ensuring our industry's ongoing progress.

Besides the series of technical papers from all over the world, we are honor to have eight marvelous keynote and invited speakers, and two tutorial speakers sharing the view of the distinguished scholars, engineering, and management on the vital semiconductor value chain which are cornerstones for the future digital society.

ISSM has been collaborating with global affiliation including Taiwan Semiconductor Industry Association (TSIA), SEMI, SEAJ, Minimal Fab Promotion Organization (MINIMAL). I would like to express our special appreciation to our partner, TSIA, in Taiwan for their support to ISSM.

ISSM is an excellent opportunity to make connections and explore new collaborations, as we share ideas and perspectives on challenges and opportunities in both technology and manufacturing. I would like to express my deepest gratitude to the sponsored companies, and to all the committee members involved in organizing this symposium. Finally, I offer my best wishes for highly productive information exchange among everyone at ISSM 2022.

**Dr. Ayako Shimazaki**

**Chairman of Executive Committee of ISSM 2022**

Technology Executive
Toshiba Nanoanalysis Corporation

On behalf of the ISSM 2022 Executive Committee, I would like to thank all of you for your participation in the ISSM 2022.

The digital transformation has been accelerated by 5G, AI, and high-performance computing and we expect to see the change even in a faster pace. 5G and AI will continue to be a driving factor for the global semiconductor industry to grow. It is expected that the global semiconductor industry will grow double to over $1 trillion US dollars by 2030.

The ISSM executive committee invited global notable keynote speakers to address their deep insights for both application perspectives and manufacturing aspects for the rapidly changing semiconductor technologies and industries. *Prof. Masayoshi Yamamoto of Nagoya University* will cover the demand technique for EV using teardown report of Model 3 (TESLA), Mustang Mach-E (Ford), ID.3 (VW), Taycan (Porsche) and HongGuang Mini EV (Wuling). *Mr. Yutaka Emoto of TSMC Japan 3DIC R&D Center*, Inc. will present chiplet integration through 2.5D and 3D stacking technologies to increase circuit densities through the collaboration with Japanese partners in addition to SoC scaling to follow Moore's Law. *Dr. Nelson Felix of IBM* will cover the strategic goals for the next three years regarding semiconductor research, which suggests the main scaling paths starting to look upward, from stacked FETs to advanced chiplet technology. *Dr. James Moyne of University of Michigan and Applied Material* will present background on the current state of the Digital Twin (DT) space in semiconductor manufacturing. *Mr. Timothy Lee of IEEE and The Boeing Company* will address immense need for the Heterogeneous Integration technology roadmap addressing an Ecosystem Agenda comprising of future vision, difficult challenges and potential solutions to pave the way for Microelectronics Resurgence. *Mr. Michitaka Tokeiji of Zeroboard* will cover the intention behind the disclosure of $CO_2$ emissions, including Scope 3, and its impact on the entire supply chain, so that it can be used for strategic planning for decarbonization management.

This year, we will also receive talks from presentations at EDTM 2022 and IITC 2022 that have been recommended by their respective committees as outstanding papers. We are looking forward to deepening the cooperation among the societies related to the field of ISSM.

We invited two tutorial speakers. *Prof. Masaru Hori of Nagoya University* will introduce the possibility of environmental innovation in semiconductor manufacturing through the challenge of advanced plasma science and technology for green semiconductor manufacturing. *Prof. Satoshi Hamaguchi of Osaka University* will review the latest development of etching and deposition technologies with atomic-scale accuracy, i.e., atomic-layer etching (ALE) and atomic-layer deposition (ALD)

This year's ISSM will be held as a full hybrid format of in-person and online participation. Through oral presentations, authors' interviews, interactive poster session, and exhibits, we are pleased to offer participants the opportunity to engage in lively communication in a venue. We hope that you will take advantage of the benefits of face-to-face discussions among the participants.

I hope that the participants of ISSM 2022 evaluate the outlook of ISSM committee for preparation of change in this industry.

**Dr. Shin-ichi Imai**
**Chairman**
**Program Committee of ISSM 2022**
Hitachi High-Tech Corporation

Welcome to the 29th International Symposium on Semiconductor Manufacturing (ISSM) 2022.

ISSM is a premiere conference focusing on manufacturing technologies for semiconductor which is a vital core driver to make our society more comfortable.

Semiconductor devices are manufactured by performing various microfabrication processes using huge amounts of tuning parameters to achieve nanometer-order precision. Digital enabler technologies such as Artificial Intelligence (AI), machine learning, cloud computing, and digital twin have been developed one after another and applied to semiconductor manufacturing. Data is the base of machine learning and AI. These results in increasing expectation for higher level of data acquisition and analysis. The mission of ISSM, *"Converting Know-How into Science"* is being put into practice.

ISSM Program Committee consisted by 26 members from 24 affiliations including semiconductor device, equipment and materials manufacturers, revised the areas of the interests and highlighted themes for ISSM 2022 through intimate discussion among members. ISSM 2022 highlights on IoT and AI Solution, Production Innovation in 200-mm Fab, High Reliability Device Process Technology for Automotive and Medical Applications, and Game-Changing Manufacturing Technologies with Heterogeneous Integration.

ISSM 2022 has received an abundance of high-quality abstracts full of insight and useful know-hows. Through tough review process by Program Committee members, 37 papers were selected including 8 papers for Process/Material Optimization (PO), followed by 7 papers for Intelligent Data Management (ID) and Process Monitoring & Control Method (PM), 6 papers for Yield & Defect Control (YD) and 2 papers for Manufacturing Strategy (MS), New Gas, New Liquid, and New Resist Technologies (NM), and Manufacturing Technology for Variety Devices (VD). And 1paper for Fab Operation Method (FO), Environment, Safety and Health, Carbon Neutral (ES)and Material Informatics (MI). We will also feature 6 keynote speeches by world-renowned industry experts, focusing on topics relevant for today's economic and manufacturing climate. Also 2 tutorial sessions focused on covering a well-defined topic will take place in the morning of the first day. And also 5 invited talks from cooperative conferences, EDTM&IITC.

ISSM 2022 will hold its 2nd "ISSM AI Solution Contests to Revolutionize Semiconductor Manufacturing" aiming to accelerate to adopt AI technologies in the field of semiconductor manufacturing. ISSM committee hope to discover human resources for artificial intelligence and improve motivation for learning and research that spreads from the excellent technologies and ideas among participants at the contests. This is an excellent opportunity for students who want to play an active role by applying the latest information technologies to solve various challenges at semiconductor manufacturing fabs. At the two divisions of the contest, there were 10 entries for ISSM AI Algorithm Contest Smart Metrology Challenge Using Semiconductor Actual Tool Data and 8 entries for Semiconductor Manufacturing Fab Data AI utilization Idea Contest. I would like to encourage all participants to view the presentations for the Fab Data AI utilization Idea Contest and to poll your vote. The winner announcement and the Award Ceremony for both 2 divisions start at 13:00, December 13.

I would like to extend my sincere appreciation to all authors and speakers, sponsors, committee members, moderators, and, last but not least, all the participants. I hope you will enjoy the technical program of ISSM 2022, take this opportunity to network with experts around the world, and bring back good memories with you.

# ISSM 2022 Committee

## Japan Organizing Committee

**Chairman:** Shozo Saito, Device & System Platform Development Center Co., Ltd.
**Vice Chairman:** Shuichi Inoue, ATONARP INC.
**Treasurer:** Hiroaki Kato, TOSHIBA ELECTRONIC DEVICES & STORAGE Corp.
**Member:**

Tadahiro Suhara, JSR Corporation
Michihiro Inoue, National Institute of Advanced Industrial Science and Technology (AIST)
Hiroyuki Umimoto, Panasonic Corporation
Yasutoshi Okuno, SCREEN Semiconductor Solutions Co., Ltd.
Takahito Matsuzawa, Tokyo Electron Ltd.
Tomoyuki Sasaki, Tower Partners Semiconductor Co., Ltd.
Atsuyoshi Koike, Rapidus Corporation

## Japan Executive Committee

**Chairman:** Ayako Shimazaki, Toshiba Nanoanalysis Corp.
**Member:**

Kensuke Uriga, Dura Systems, Inc.
Naoyuki Ishiwata, Fujitsu Semiconductor Limited
Takeshi Hattori, Hattori Consulting International
Hiroshi Akahori, KIOXIA Corporation
Toshiyuki Uchino, Kokusai Electric Corporation
Katsutoshi Ozawa, OMRON Corp.
Hiroyuki Umimoto, Panasonic Corporation
Hiroyuki Mori, Renesas Electronics Corporation
Takashi Shimane, Rohm Co., Ltd.

Yasutoshi Okuno, SCREEN Semiconductor Solutions Co., Ltd.
Hiroyuki Chuma, Nisshinbo Holdings Inc.
Masahiko Hamajima, SEMI
Kiyoshi Watanabe, Semiconductor Equipment Association of Japan (SEAJ)
Takahito Matsuzawa, Tokyo Electron, Ltd.
Kazuya Okamoto, Yamaguchi University Graduate School

## Japan Program Committee

**Chairman:** Shin-ichi Imai, Hitachi High-Tech Corporation
**Executive Vice Chairman:** Ayako Shimazaki, Toshiba Nanoanalysis Corp.
**Committee leader:**
Kazunori Kato, Advanced Interface Technology Corp.
Isamu Namose, OMRON Corporation
Kazuhito Matsukawa, SUMCO Corporation
Tomio Otsuki, JX Nippon Mining & Metals Corporation
**ISSM Conference Editor for IEEE TSM:**
Tsuyoshi Moriya, Tokyo Electron Ltd.
**Member:**

Shinsuke Mizuno, Applied Materials Japan, Inc
Kazunori Nemoto, Hitachi High Technologies America, Inc.
Takanori Kawakami, JSR Corporation
Yuji Yamada, KIOXIA Corp.
Masami Aoki, KLA-Tencor Japan Ltd.
Minoru Akaishi, ON Semiconductor
Masahiro Shimbo, ON Semiconductor
Kenji Miyake, PMT Corporation
Kazuki Yokota, Renesas Electronics Corporation
Shoji Takei, Rohm Co., Ltd.

Kenji Watanabe, Western Digital
Takayuki Hisamatsu, Sony Semiconductor Manufacturing Corporation
Hiroyuki Inoue, Texas Instruments Japan Limited
Shun-ichiro Ohmi, Tokyo Institute of Technology
Toshio Konishi, Toppan Photomask Co., Ltd.
Takatoshi Yasui, Tower Partners Semiconductor Co., Ltd.
Takayuki Matsumoto, United Semiconductor Japan Co., Ltd.
Takahiro Tsuchiya, United Semiconductor Japan Co., Ltd.
Sumika Arima, University of Tsukuba

## e-Manufacturing & Design Collaboration Conference (eMCD) Advisory Committee

Dr. Nicky C.C. Lu, Etron Technology, Inc.
Dr. TY Wu, TSIA
Ms. Celia Shih, TSIA
Mr. Thomas Chen, tsmc
Mr. Robert Chien, tsmc
Dr. C. Hsu, tsmc
Prof. SC Chang, National Taiwan University

# Symposium Schedule (Day-1)

## ISSM2022
## Monday, December 12th, 2022

### Room: KFC Hall

| Time | Content |
|------|---------|
| 8:45- 8:50 | **Opening Remarks**<br>Shozo Saito, Chairman of Organizing Committee of ISSM2022 / Device & System Platform Development Center |
| 8:50- 8:55 | **ISSM 2022**<br>Ayako Shimazaki, Chairman of Executive Committee of ISSM2022 / Toshiba Nanoanalysis |
| 8:55- 9:00 | **ISSM 2022 Program Outline**<br>Shin-ichi Imai, Chairman of Program Committee of ISSM2022 / Hitachi High-Tech Corporation |
| 9:00- 10:30 | **Tutorial Session**      *Session Chair: Ayako Shimazaki, Toshiba Nanoanalysis / Tsuyoshi Moriya, Tokyo Electron* |
| 9:00- 9:40 | **"Challenges of Plasma Science and Technology for Green Semiconductor Manufacturing"**<br>Prof. Masaru Hori, Professor, Center for Low-temperature Plasma Sciences (cLPS), Nagoya University |
| 9:40- 9:50 | Break |
| 9:50-10:30 | **"Atomic-level control of plasma processing toward sub-nm node technologies"**<br>Prof. Satoshi Hamaguchi, Center for Atomic and Molecular Technologies, Osaka University |
| 10:30-10:40 | Break |
| 10:40-12:00 | **Technical Session A-1 & B-1** |
| 12:00-12:50 | Sponsor Exhibition (on-site) & Video (Sponsor: Silver-Platinum) / Lunch Break |
| 12:50-13:30 | **Keynote Speech: "Cutting Edge Technologies of Power Semiconductor Device Applications for Electric Vehicle's Power Electronics Systems in Japan, US, Europe and China Case"**<br>Prof. Masayoshi Yamamoto, Professor, Institute of Materials and Systems for Sustainability (IMaSS), Graduate School of Engineering and School of Engineering, Department of Electrical Engineering, Nagoya University<br>*Session Chair: Kazunori Kato, Advanced Interface Technology* |
| 13:30-13:40 | Break |
| 13:40-14:20 | **Keynote Speech: "The Contribution and Expectation of TSMC Japan 3DIC R&D Center to Japan Industry"**<br>Mr. Yutaka Emoto, Vice President, TSMC Japan 3DIC R&D Center, Inc.<br>*Session Chair: Kazuhito Matsukawa, SUMCO* |
| 14:20-14:30 | Break |

## Room1: KFC Hall

**Session A-1: Manufacturing Strategy (MS) & Environment, Safety and Health, Carbon Neutral (ES) & Fab Operation Method (FO)**
Session Co-Chairs: Katsutoshi Ozawa, OMRON /
Takahiro Tsuchiya, United Semiconductor Japan

| | |
|---|---|
| 10:40 | MS-44 : A Novel Approach to Dynamic Line Balance Control and Scheduling with a Digital Twin Production System<br>Hirofumi Tsuchiyama, INFICON |
| 11:00 | ES-62 : Self-tuning optimization to compatible the delivery and low energy consumption<br>Chending MAO, University of Tsukuba |
| 11:20 | ONLINE FO-43 : Maintenance Content Reduction and Digitalization for Performance Optimization<br>Christopher Bode, INFICON |
| 11:40 | Author's Interview (On-Site ONLY) & Break |
| 12:00 | Sponsor Exhibition (on-site) & Video (Sponosr: Silver-Platinum) / Lunch Break |
| 12:50 | Plenary Sessions |

**Session A-2: Invited & Highlight AI**
Session Co-Chairs: Isamu Namose, OMRON /
Takatoshi Yasui, Tower Partners Semiconductor

| | |
|---|---|
| 14:30 | INVITED<br>EDTM-01 : Inter Spike Interval and Stochasticity Engineering of Floating Gate Technology-based Neurons for Spiking Neural Network Hardware<br>Akira Goda, University of Tokyo |
| 14:50 | YD-21 : Noise Reduction in SEM Images using Deep Learning<br>Yuki Sato, Tokyo Electron |
| 15:10 | ID-6 : Application of Natural Language Processing in Semiconductor Manufacturing<br>Daisuke Kobayashi, Sony Semiconductor Manufacturing Corporation |
| 15:30 | Author's Interview & Exhibition (On-Site ONLY) & Break |

**Session A-3: Highlight AI**
Session Co-Chairs: Takatoshi Yasui, Tower Partners Semiconductor /
Masami Aoki, KLA-Tencor Japan

| | |
|---|---|
| 16:00 | YD-10 : Positive/Negative Decision via Outlier Detection Towards Automatic Performance Evaluation for Defect Detector<br>Toshinori Yamauchi, Hitachi High-Tech |
| 16:20 | YD-15 : A Study on Detection Method Using 2-Class Classifiers for Defective Wafer Maps<br>Seima Sakaguchi, Mie University |
| 16:40 | ONLINE<br>YD-5 : Yield prediction with Machine Learning and parameter limits in semiconductor production<br>Rebecca Busch, University of Siegen |
| 17:00 | Author's Interview (On-Site ONLY) & Break |

**Session A-4: Invited & Yield & Defect Control (YD) & Manufacturing Technology for Variety Devices (VD)**
Session Co-Chairs: Kenji Watanabe, Western Digital /
Tomio Otsuki, JX Nippon Mining & Metals

| | |
|---|---|
| 17:20 | INVITED ONLINE<br>IITC-01 : Dual Damascene 28nm-Pitch Single Exposure EUV Design Rules Evaluation by Voltage Contrast Characterization<br>Victor M. Carballo, IMEC |
| 17:40 | ONLINE YD-35 : Automatic classification of C-SAM voids for root cause identification of bonding yield degradation<br>Julien Baderot, Pollen Metrology |
| 18:00 | VD-33 : Hydrogen diffusion behavior of CH4N-molecular-ion-implanted wafers for 3D-stacked CMOS image sensors<br>Ryosuke Okuyama, SUMCO |
| 18:20 | Author's Interview & Exhibition (On-Site ONLY) & Break |

## Room 2: KFC Hall Annex

**Session B-1: Invited & Process/Material Optimization (PO)**
Session Co-Chairs: Tsuyoshi Moriya, Tokyo Electron /
Hiroyuki Inoue, Texas Instruments Japan

| | |
|---|---|
| 10:40 | INVITED<br>EDTM-02 : Experimental Analysis of Process Impacts on Fluorine Incorporated Gate Oxide Film Properties Near Gate Edge Region<br>Shuntaro Fujii, Asahi Kasei |
| 11:00 | PO-63 : Nanoimprint Lithography with CO2 Ambient<br>Toshiki ITO, Canon |
| 11:20 | PO-11 : Impact of cation vacancies on leakage current on TiN/ZrO2/TiN capacitors studied by positron annihilation<br>Akira Uedono, University of Tsukuba |
| 11:40 | Author's Interview (On-Site ONLY) & Break |
| 12:00 | Sponsor Exhibition (on-site) & Video (Sponosr: Silver-Platinum) / Lunch Break |
| 12:50 | Plenary Sessions |

**Session B-2: Invited & Process/Material Optimization (PO)**
Session Co-Chairs: Kazunori Kato, Advanced Interface Technology /
Shun-ichiro Ohmi, Tokyo Institute of Technology

| | |
|---|---|
| 14:30 | INVITED ONLINE<br>IITC-02 : TSV fabrication technology using direct electroplating of Cu on the electroless plated barrier metal<br>Shoso Shingubara, Kansai University |
| 14:50 | PO-42 : Ultra-fast Etching of Photoresist by Reactive Atmospheric-pressure Micro-Thermal Plasma Jet<br>Hibiki Kato, Hiroshima University |
| 15:10 | PO-28 : Optimization of RF frequencies in dual-frequency capacitively coupled plasma apparatus using genetic algorithm (GA) and plasma simulation<br>Shigeyuki Takagi, Tokyo University of Technology |
| 15:30 | Author's Interview & Exhibition (On-Site ONLY) & Break |

**Session B-3: Invited & Process/Material Optimization (PO)**
Session Co-Chairs: Shun-ichiro Ohmi, Tokyo Institute of Technology /
Yuji Yamada, KIOXIA

| | |
|---|---|
| 16:00 | INVITED<br>EDTM-03 : Temperature Dependence of Current-Voltage Characteristics of Ionic Liquid Type Intelligent Connection Device<br>Masakazu Kobayashi, Nagase |
| 16:20 | PO-40 : Deposition rate dependence of the 5 nm-thick ferroelectric nondoped HfO2 on MFSFET characteristics<br>Masakazu Tanuma, Tokyo Institute of Technology |
| 16:40 | ONLINE<br>PO-55 : Process Optimization for Ge-on-Si depletion mode transistors using mesa architecture<br>Sumit Choudhary, Indian Institute of Technology, (IIT), Mandi |
| 17:00 | Author's Interview (On-Site ONLY) & Break |

**Session B-4: Material Informatics (MI) & New Gas, New Liquid, and New Resist Technologies (NM)**
Session Co-Chairs: Shinsuke Mizuno, Applied Materials Japan /
Takanori Kawakami, JSR

| | |
|---|---|
| 17:20 | ONLINE<br>MI-25 : Systematic search for stabilizing dopants in ZrO2 and HfO2 using first-principles calculations<br>Yosuke Harashima, Nara Institute of Science and Technology |
| 17:40 | NM-23 : Recent status of EUV lithography, what is the stochastic issues ?<br>Toru Fujimori, FUJIFILM |
| 18:00 | NM-45 : Technology Trends and Characteristics of Patent Information Disclosure in Advanced Semiconductor Photoresist<br>Kosuke Watahiki, Yamaguchi University |
| 18:20 | Author's Interview & Exhibition (On-Site ONLY) & Break |

# Symposium Schedule (Day-2)

## ISSM2022
## Tuesday, December 13th, 2022

| 8:30- | Registration |
|---|---|

### Room: KFC Hall

| | |
|---|---|
| 8:50 | **Introduction of Day 2 program**<br>Shin-ichi Imai, Chairman of Program Committee of ISSM2022 / Hitachi High-Tech Corporation |
| 9:00- 9:40 | ONLINE **Keynote Speech: "The Path to 100 Billion Goes Upward – IBM Research Semiconductors Technology Atlas"**<br>Dr. Nelson Felix, Director, Process Technology, IBM<br>*Session Chair: Yasutoshi Okuno, SCREEN Semiconductor Solutions* |
| 9:40- 9:50 | Break |
| 9:50-10:30 | ONLINE **Keynote Speech: "A Requirements Driven Digital Twin Framework to Support Semiconductor Manufacturing: Specifications and Opportunities"**<br>Dr. James Moyne, Associate Research Scientist, Mechanical Engineering Department, University of Michigan<br>Consultant for Standards and Technology, Advanced Services Engineering, Applied Global Services, Applied Materials<br>*Session Chair: Shuichi Inoue, ATONARP* |
| 10:30-10:40 | Break |
| 10:40-11:20 | ONLINE **Keynote Speech: "Heterogeneous Integration Paving the way for Microelectronics Resurgence"**<br>Mr. Timothy Lee<br>IEEE Board of Director, IEEE HIR Chair for 5G Technical Working Group<br>*Session Chair: Kenji Miyake, PMT* |
| 11:20-12:10 | Sponsor Exhibition & Video (Sponosr: Silver-Platinum) / Lunch Break |
| 12:10-12:50 | **Keynote Speech: "Decarbonization Management to Improve Corporate Value"**<br>Mr. Michitaka Tokeiji, Founder and CEO, Zeroboard. Inc<br>*Session Chair: Hiroshi Akahori, KIOXIA* |
| 12:50-13:00 | Break |
| 13:00-14:00 | **AI Contest Award**<br>Session Co-Chairs: *Shin-ichi Imai, Hitachi High-Tech / Isamu Namose, OMRON* |
| 14:00-14:10 | Break |
| 14:10-14:40 | **3miniutes Summary presentation by Interactive Poster Speakers**<br>Session Co-Chairs: *Takanori Kawakami, JSR / Toshio Konishi, Toppan Photomask* |
| 14:40-15:00 | Sponsor Exhibition (On-site) & Break |

---

### Room1: KFC Hall

**Session A-5 : Intelligent Data Management (ID)**
Session Co-Chairs: Masami Aoki, KLA-Tencor Japan /
Hiroyuki Inoue, Texas Instruments Japan

| | |
|---|---|
| 15:00 | **ID-13 : Advanced Process Control Model for Trench Shape of Power Devices**<br>Takumi Ito, TOSHIBA DEVICE & STORAGE |
| 15:20 | **ID-17 : Principal Component Analysis based GaN transistor live health monitoring**<br>Florian Chalvin, Rohm |
| 15:40 | ONLINE<br>**ID-27 : Application of Big Data Science in High Reliability Automotive Wafer Yield Management System and Failure Analysis**<br>Chia-Cheng Kuo, Taiwan Semiconductor Co., Ltd. |
| 16:00 | Author's Interview (ON-Site ONLY) & Break |

**Session A-6 : Intelligent Data Management (ID)**
Session Co-chairs: Yuji Yamada, KIOXIA /
Takayuki Hisamatsu, Sony Semiconductor Manufacturing

| | |
|---|---|
| 16:20 | **ID-32 : Equipment Sensor Data Cleansing Algorithm Design for ML-Based Anomaly Detection**<br>Shi-Chung Chang, National Taiwan University |
| 16:40 | **ID-26 : Dynamic AI Computation Tasks with SECS/GEM in Semiconductor Smart Manufacturing**<br>Hung H Nguyen, Yield Engineering Systems |
| 17:00 | **ID-36 : Secure and Reliable Power Monitoring for Low Consumption Factory Equipment via Programmable IoT Devices**<br>Sergio Garnica, Fraunhofer Research Institution for Microsystems and Solid State Technologies EMFT |
| 17:20 | - |
| 17:40 | Author's Interview (ON-Site ONLY) & Break |

---

### Room 2: KFC Hall Annex

**Session B-5 : Process Monitoring & Control Method (PM)**
Session Co-Chairs: Takahiro Tsuchiya, United Semiconductor Japan /
Takayuki Matsumoto, United Semiconductor Japan

| | |
|---|---|
| 15:00 | **PM-22 : Characterization of light propagation loss in Si Photonics using High-Resolution CDSEM metrology**<br>Shimon Halevi, Applied Materials |
| 15:20 | **PM-41 : In Situ Measurement and Analysis of Low Pressure Gas Concentration Distribution Using 70-dB SNR 1,000 Frames-per-second Absorption Imaging System**<br>Yushi Sakai, Tohoku University |
| 15:40 | **PM-18 : Advanced Process Monitoring through Fault Detection and Classification for Robust Statistical Process Control of Tantalum Nitride Reactive Sputtering**<br>Stephanie Y Chang, Skyworks Solutions |
| 16:00 | Author's Interview (ON-Site ONLY) & Break |

**Session B-6 : Process Monitoring & Control Method (PM)**
Session Co-chairs: Shin-ichi Imai, Hitachi High-Tech /
Tsuyoshi Moriya, Tokyo Electron

| | |
|---|---|
| 16:20 | **PM-30 : Plasma Process Classification using Causal Discovery Technique**<br>Dai Kobayashi, Tokyo Electron |
| 16:40 | **PM-14 : Practical load impedance monitoring system externally installed in plasma etching equipment**<br>Yuji Kasashima, National Institute of Advanced Industrial Science and Technology |
| 17:00 | **PM-31 : Plasma diagnostics and characteristics of hydrofluorocarbon films in capacitively coupled CF4/H2 plasmas**<br>Shih-Nan Hsiao, Nagoya University |
| 17:20 | **PM-49 : A Study on Robust Noninteracting Control System Design with Disturbance Feedforward for 6-DoF AVIS**<br>Thinh Huynh, Pukyong National University |
| 17:40 | Author's Interview (ON-Site ONLY) & Break |

---

| 18:00 | Poster Session (On-Site) @ Foyer   and   ISSM2022 Awards @ Room1 |
|---|---|

# Symposium Schedule (Day-2)

## ISSM2022
## Tuesday, December 13th, 2022
## Interactive Poster Session

**14:10-14:40   3-min Flash Presentation for Interactive Poster Session @ Room 1 : KFC Hall**

| Room1 : KFC Hall | |
|---|---|
| Session Co-Chairs: Takanori Kawakami, JSR / Toshio Konishi, Toppan Photomask | |
| **MS-57** | **Data-driven Modeling for Production Dynamics**<br>Yu Sasaki, University of Tsukuba |
| **PO-46** | **Obtaining carbon nanowalls with a specified morphology**<br>Yerassyl Yerlanuly, Kazakh-British Technical University |
| **YD-20** | **Influence of High Temperature N2 Annealing on Photoluminescence of SiC and Si Quantum Dots in SiO2 Layer**<br>Koki Murakawa, Kanagawa University |
| ONLINE<br>**PO-59** | **Experimentally study on the effect of RIE etching power on etching rate of β-Ga2O3 thin film**<br>Wang Xu, Guizhou University |
| ONLINE<br>**VD-60** | **Preparation of Uniform SiO2 Insulating Layer on the Inner Wall of TSV by Thermal Oxidation**<br>GuoFengJie, Guizhou University |

**18:00-18:25     Poster Session @ Foyer**

## Poster Layout

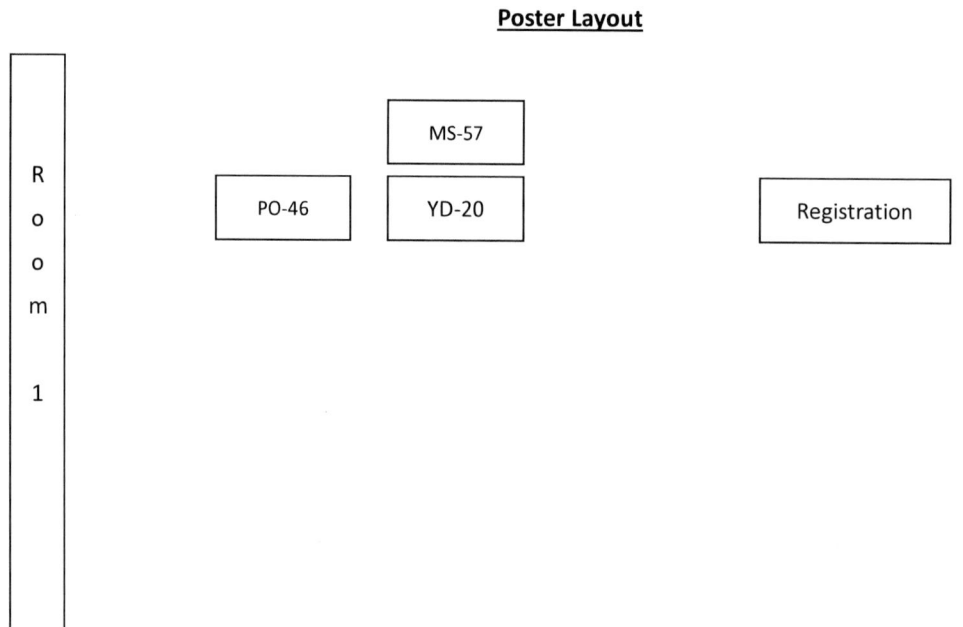

# Self-tuning Optimization to Compatible the Delivery and Low Energy Consumption

Chending MAO
*University of Tsukuba*
Tsukuba, Japan
s2020480@u.tsukuba.ac.jp

Jia LIN
*University of Tsukuba*
Tsukuba, Japan
s1930142@u.tsukuba.ac.jp

Sumika ARIMA
*University of Tsukuba*
Tsukuba, Japan
arima@sk.tsukuba.ac.jp

*Abstract*— *This paper introduced n-step hybrid flow-shop scheduling (nHFS) with batch process to consider a trade-offs between power consumption and productivity. Wider optimization scope with power conscious have been advanced to increase efficiency of algorithm and tune the parameter automatically based on our past research. Concretely, Energy consumption and due date are considered in Self-tuning Optimization. Actual companies' data are used to evaluate the performance of proposed and past methods.*

*Keywords— Hybrid flow-shop, Batch process, Energy consumption, Scheduling, Bayesian optimization*

## I. INTRODUCTION

High-Mix Low-Volume (HMLV) manufacturing is characterized by the diversity of product-mix and resource sharing processes. In this condition, it isn't easy to simultaneously improve multi-criteria objectives such as time-to-market, production efficiency, eco-friendliness, and so on. This research focuses on the optimization of loading condition of batch processes considering energy consumption in an n-step product-mix hybrid flow-shop scheduling (HFS).

In previous research, n-Gupta method [1] of the n-step HFS outperformed other scheduling methods due to its objective function (1), which contains the trade-off structures between the losses of all jobs' earliness and lateness ($E_i$ and $T_i$ for each job $i$), the due date ($d_i$), and the complete time ($C_i$). Tanaka et.al. proposed the extended n-Gupta method (n-GuptaEx) with the loading and work-leveling methods for MTO and MTO-MTS-mixed production systems [2]. Moreover, Bayesian optimization (BO) is applied to tune multi-criteria parameter automatically for improving n-GuptaEx solutions. The efficiency and accuracy of the BO-based n-GuptaEx method (n-GuptaEx-BO) have been proved by verifying practical cases. In addition, Lin and Ohno, et.al. proposed a heuristic algorithm of batching which outperformed the ordinary batch loading rules to reduce # of batches (e.g. under the half) while keeping the due-dates [3]. The batching rule can be flexibly combined to n-GuptaEx-BO.

$$F = \sum_{i=1}^{n}(u_i E_i + v_i T_i + w_i C_i + z_i d_i) \qquad (1)$$

Now, it is also an important issue to decide the upper/lower bound of the time window interval in batching algorithm when various orders of different arrival rates and due dates are dealt with. From the perspective of eco-friendliness, the operational methods have been proved to be feasible and effective to reduce the energy consumption of manufacturing companies [4]. From ecology aspect, this study will introduce an eco-friendly condition of batch-loading considering with the trade-offs between the due-date satisfaction (thus production speed) and total energy consumption of idle and in-process times in batch and post-batch process steps.

Such a case the number of parameters increases to be tuned, and the computational efficiency of n-GuptaEx-BO still has room for improvement. In this study, we firstly refined the local search with the tabu search method and pooling mechanism as well as applied Self-Tuning Portfolio-based BO (SETUP-BO) [5] in parameter tuning to improve more n-GuptaEx's computational efficiency as a baseline.

## II. PROBLEM DESCRIPTION AND PAST METHODS

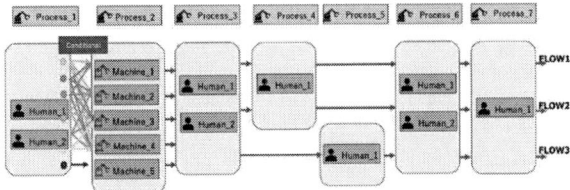

Figure1. Actual problem settings

In this research, a 7-step hybrid flow-shop with batch progress as shown in Figure1 is evaluated. In each process, a given number of machines or human laborers have settled under the specification condition. Process 1 and Process 2 are treated as batch-type processes and the others are individual lot-type processes. Process 2 is energy-extensive, due to the pressure processor. The scheduling problem is complicated because different products have different flows (processing steps and usable machines). Each product can use only a given range of machines or labor pre-allocated to each process step. Since both machines and labor have different capacities or conditions in addition to a variance of product-mix demand, one machine/labor can be shared with different products.

In order to explore optimal or near-optimal solutions for the nHFS with the batch process, we decomposed the problem into two subproblems which are job sequencing and job batching. To solve these subproblems, n-GuptaEx-BO method and a variable time-window batching heuristic are developed to decide the order of lots or batches and create batches with lots in our past research.

### A. n-Gupta Method and n-GuptaEx-BO method

To solve job sequencing problems, the n-Gupta method [1] can assign arbitrary weights to the multi-objective function for the actual process. The objective function of original n-Gupta method is defined in (1). As the description in Section 1, $E_i$, $T_i$ respectively are earliness penalty and tardiness penalty of job $i$. $C_i$ is the complete time of job $i$, and $d_i$ is due date decided by factory of job $i$. They are defined by the following formula.

$$E_i = \max\{0, d_i - C_i\} \qquad (2)$$

$$T_i = \max\{0, C_i - d_i\} \qquad (3)$$

$$d_i = min\{r_i + TI \cdot \sum_{j=1}^{n} t_{ij}, d_i^0\} \tag{4}$$

Here, $r_i$ is release time of job $i$. $t_{ij}$ is processing time of the $j$'s process of job $i$. $d_i^0$ is due date from the order receipt data and $TI$ is a tightness parameter for determining the process' due date based on the number of required hours. The smaller $TI$, the more difficult it is to complete the job by due date.

The n-Gupta method uses iterative algorithms based on local search that applied some neighborhood structure to improve the current initial solution iteratively. Some basic moves (Insertion and Exchange operations, Tightness parameter shifts) are used in n-Gupta iterative algorithm (Algorithm1) for generating a neighbor. Note that we use $B = 10$ and $\Delta TI = 0.05$ in this study, consistent with the settings of the original n-Gupta method.

---

**Algorithm1**: n-Gupta method

**Step1:** Determine the deadline $d_i$ based on the tightness parameter $TI$. The insertion order is IO: $= i_1, i_2, ... i_n$ as $i_1, i_2, ... i_n$ from the smallest $i$ of $d_i$. Initialize with partial order PO: $= \{\}$, partial order length $l = 0$.

**Step2:** While IO$\neq \varphi$

I.  Select the top $i$ of IO and remove it from IO.

II.  $F^* := \infty, k^* := min\{k, l+1\}$

III.  For $j$: $= 1$ to $k^*$

    a.  $u := l - k^* + j + 1$

    b.  Let PO' be job $i$ is inserted partial order **PO**.

    c.  Determine the allocation to the machine according to **PO'**. In the process with a plurality of machines, jobs are assigned to the machine where the previous processing ends the earliest. In each process, **PO'** is not ignored and the process is not started.

    d.  Calculate the evaluation value F(**PO'**).

    e.  If F (**PO'**) < F* 
      F*: = **F (PO')**, **PO''**: = **PO'**

IV.  **PO**: = **PO''**, $l = l + 1$

**Step3:** Processing order O: = **PO**

**Step4:** In the schedule based on the processing order O obtained in Step3, if $C_i - d_i > B$ is satisfied in any job $i$, check whether $TI$ improves the evaluation value F. $TI^0 := TI$, continuously considering values such as $TI^k = TI^0 + \Delta TI \cdot k$, $k$ is a natural number, and using a minimum $TI^k$ with a new $TI$. Even if $k$ is increased, $d$ does not increase, or $k$ changes in a range satisfying $C_i - d_i > B$.

**Step5:** The processing order $O$, and the schedule obtained from the final tightness parameter $TI$ are used as an initial solution.

---

In n-GuptaEx-BO method, we described improvement of solution by iterative method (local search method). The solution is improved by changing the job processing order $O$ and the tightness parameter $TI$. Jobs to be replaced are performed between $s$ or less adjacent jobs. The extended n-Gupta method(n-GuptaEx) is as shown in Algorithm2.

---

**Algorithm2**: extended n-Gupta method

**Step1:** From the range of $[TI-\delta1, TI]$ with probability *prdec* ($\leqq 1$) $TI'$is determined randomly from the range of $[TI - \delta1, TI + \delta2]$ with probability 1-*prdec*.

**Step2:** Calculate the deadline $d_i$ based on $TI'$. Choose a job $i$ of $C_i \neq d_i$.

**Step3:** Select job $j$ from a job with earlier processing order if $C_i > d_i$ (which means the completion is later than deadline), or from a job with later processing order if $C_i < d_i$ (which means the completion is earlier

---

than deadline). Job $j$ is selected from jobs in the adjacent order within $i$ to $s$.

**Step4:** Swap the positions of job $i$ and job $j$ in processing order O.

**Step5:** Change $TI$ in the same way as in Step4.

---

We also improved the weighting parameter settings of objective function by applying four parameters $(u, v, w, z)$ to discuss its characteristic details and optimization at each flow instead of each job (in (5)).

$$minimize\ F = \sum_{i=1}^{n}(uE_i + vT_i + wC_i + zd_i) \tag{5}$$

To optimize these interfering parameters, we applied Bayesian optimization (BO) which can tune the parameters automatically without domain knowledge implementation.

### B. Batch Heuristics

To deal with batch processes, we proposed batch heuristics as shown in Algorithm3 to form batches, and (6) to calculate the release date of jobs.

$$R_i = D_i - k_i \sum_{i,j=1}^{n} p_{ij} \tag{6}$$

Here, $R_i$ is the latest release date and $D_i$ is the due date of job $i$. $k_i$ means the slack coefficient to adjust the suitable release date. $p_{ij}$ is the processing time of job $i$ in process $j$.

---

**Algorithm3**: Batch Heuristic

**Step1:** Sort all the lots in descending order based on the due date.

**Step2:** Select the top one as lot A according to the order that determined in Step1 and calculate the lot A's latest release date $R_i$ with (6).

**Step3:** According the lot A's pressure condition, determine the machine which has the largest capacity among the available machines as the batch size. Also, only the lot with the same pressure condition as lot A has can be selected in the next steps.

**Step4:** Choose the lot that satisfies the batch composition conditions from the remaining lots that are determined in Step3. If the lot's $R_i$ is within a range of $x_1$ days before and after the lot A's $R_i$ and the lot's $D_i$ is within a range of $x_2$ days before and after lot A's $D_i$, the lot can be forming a batch with the lot A selected in Step2.

**Step5:** Put all the lots selected in Step4 in a batch with lot A in Step2 until reach the upper limit of the batch size.

**Step6:** Adjust the machine limit. If the batch size that formed in Step5 is equal to the upper limit of the batch size in Step3, use the machine that has been determined in Step3. If the batch size is smaller than the upper limit and smaller than the machine which have the second largest capacity, use the second largest. The same as the other machines. If the batch size is just smaller than the upper limit while larger than the second largest, use the machine selected in Step3.

**Step7:** Remove the lot that already in the batch from the pre-selected queue.

**Step8:** Repeat Step2~Step7 until all lots are in the batches.

---

## III. PROPOSED METHODS

Although the effectiveness of n-GuptaEx-BO and batch heuristic have been proved in past research, there are still some issues to be solved.

One important issue is that n-GuptaEx-BO method involves massive calculation, and it is time-consuming. This issue becomes more non-negligible when we try to perform experiments on parameter tuning. The other issue is the appropriate parameter setting of batching heuristic, which decides the upper/lower bound of the time window interval of chosen lots. It's still difficult to deal with various orders with different arrival rates and due dates and reduce the energy consumption of process 2 simultaneously. Therefore, we proposed an improved n-GuptaEx algorithm with SETUP-BO (n-GuptaEx-SETUPBO) to improve compute efficiency and a tuning approach for batching heuristic which considered energy consumption and due-date satisfaction.

*A. n-GuptaEx-SETUPBO Method*

The improvements of n-GuptaEx-SETUPBO are as follows.
Based on Algorithm1, we apply an initial solution pool approach to avoid duplicate solutions and improve the diversity of solutions. Specifically, after generating the initial solution each time, calculate the objective function value $F$ and create an ascending-ordered set $SO$ of the $F$ values of the initial solutions. When performing Bayesian optimization, an initial solution is selected from the pool and computed. The selection criterion is that select the best initial solution with a probability of $0 < PM < 1$, and randomly select from the inferior solutions with a probability of $1 - PM$.

We also improved the iterative method to avoid reruns by applying the tabu search method. Furthermore, a new probabilistic-based chosen strategy is proposed to refine the diversity of solutions and efficiency of the iterative method. The procedure is shown in Algorithm4.

| **Algorithm4**: Tabu Search Based Iterative Method |
| --- |
| **Step1:** Determine the $TI'$ by choosing randomly a value from the interval $[TI - \delta_1, TI]$ with probability $0 \le prdec \le 1$. Calculate the due date $d_i$ based on $TI'$. Collect the initial solution set $S'$ from the pool, reorder $S'$ with ascending order of $C_i - d_i$ to form set $X$.<br>**Step2:** Create the tabu list $L$. Define the stopping number of iterations $Runtimes$, length of tabu list $Length_{tb}$ and rule probability parameter $RI$.<br>**Step3:** Collect the jobs which meet the condition $C_i \ne d_i$ from $X$, and form the candidate list $P$ with the collected jobs.<br>**Step4:** Choose randomly a value $p$ from the interval $[0,1]$. If $0 < p \le RI$, apply rule A. If $RI < p$, apply rule B.<br>Rule A: From the candidate list $P$, choose the job (job $i$) that has the biggest $C_i - d_i$, and the job (job $j$) that has the smallest $C_i - d_i$.<br>Rule B: Choose randomly two jobs (job $i$ and job $j$) from $k$ adjacent jobs in the candidate list $P$.<br>**Step5:** If the pair of job $i$ and job $j$ in $L$, move to Step6. Otherwise, move to Step7<br>**Step6:** Delete job $i$ and job $j$ from $P$, return to Step4.<br>**Step7:** Swap the order of job $i$ and job $j$. Calculate $F(S')$ with the new order $S'$. If $F(S') < F^*$, $F^* = F(S')$, upgrade $X, P, L$, return to Step4. Otherwise, return the previous $S'$ and return to Step4.<br>**Step10:** Repeat Step4~Step7 until number of iterations < $Runtimes$ and all jobs meet the condition $C_i - d_i < B$. |

In parameter tuning of weighting parameter in n-GuptaEx, we apply portfolio-based methods, which allows us to choose a better acquisition strategy in Bayesian optimization. The procedure is shown in Algorithm5.

| **Algorithm5**: Self-Tuning Portfolio-based BO |
| --- |
| **Step1:** Set $G_j(0) = 0$ for $j = 1, 2, ..., J$.<br>**Step2:** Sample the hyperparameter $\eta \sim$ Gamma $(\alpha, \beta)$ and $m \sim$ Beta $(a, b)$. Nominate points from each acquisition function $h_j$: $x_j(t) = arg \max_x h_j(x)$ |

**Step3:** Compute $r_{\min}(t-1) = \min_j \left(G_j(t-1)\right)$ and $r_{\max}(t-1) = \max_j \left(G_j(t-1)\right)$, then compute the normalized rewards: $r_j(t-1) = \dfrac{G_j(t-1) - r_{\min}(t-1)}{r_{\max}(t-1) - r_{\min}(t-1)}$

**Step4:** Select a nominee $\boldsymbol{x}(t) = x_j(t)$ with probability $p_j(t) = \dfrac{\exp(\eta r_j(t-1))}{\sum_{j'=1}^{J} \exp(\eta r_j(t-1))}$

**Step5:** Compute $y(t)$ by evaluating the objective on point $\boldsymbol{x}(t)$.

**Step6:** Augment the data $D_t$ with the new pair $(\boldsymbol{x}(t), y(t))$.

**Step7:** Update the surrogate GP model.

**Step8:** Update the rewards $G_j(t) = mG_j(t-1) - \mu(x_j(t))$ from the updated GP posterior.

**Step9:** Update the posteriors $a = a + 1$, if $y(t)$ is the best point evaluate so far, otherwise update $b = b + 1$.

**Step10:** Update the posterior $\alpha \leftarrow \alpha + 1$, and update the posterior $\beta \leftarrow \beta + x(t)$

**Step11:** Repeat Step3~Step13 until stopping criterion is reached.

Here, $h_j$ represents acquisition function of strategy $j$. In our research, the chosen strategies are Probability of Improvement (PI), Expected Improvement (EI) and Lower Confidence Bound (LCB). $y(t)$ represents objective function, and in this research, it can be calculated by (7). Here $FA$ is non-weighted performance except $d$ of the objective function (1). This is because $d$ depends on the parameter TI, and $FA$ is more important for the actual factory performance.

$$FA = \sum_{i=1}^{n}(E_i + T_i + C_i) \tag{7}$$

*B. Auto-tuning Batch Heuristic*

In Auto-tuning batch heuristic, two parameters $(x_1, x_2)$ of batching heuristic are treated as decision variables, and the normalized total energy consumption and tardiness penalty are considered in objective functions. To reduce the energy consumption of process 2 while ensuring productivity, we take not only the total energy consumption during the idle and in-process time of process 2 but the energy consumption of idle time of process 3 into account (As shown in (8)). This is because the size of batches becomes larger and the waiting time in process 3(post-batch process) becomes longer if we focus only on reducing the in-process time of process 2.

$$TEC = \sum_{m \in M_{p2}} t_{proc} \cdot n_m \cdot P_{proc} + t_{idle,m} \cdot P_{idle} + \sum_{m \in M_{p3}} t_{idle,m} \cdot P_{idle} \tag{8}$$

Here, $t_{proc}$ is the processing time of one batch process, $n_m$ is the number of processing on machine $m$, and $P_{proc}$ is the power of in-process machine in the batch process. $P_{idle}$ represents the power when the machine or person is idle, and $t_{idle,m}$ represents the idling time of the machine or person $m$.

(9) represents the sum of tardiness for each job (lot), and it's calculated basing on the results of n-GuptaEx-SETUPBO.

$$TT = \sum_{i=1}^{n} T_i \tag{9}$$

The proposed two-layer Bayesian optimization (BO) based Auto-tuning approach is shown in Algorithm6. The layer 1 BO with the objective function (9) is to optimize the parameter $\alpha$ which balances trade-offs between total energy consumption and tardiness penalty. The layer 2 BO with the objective function (10) is to find the best parameter settings $(x_1^*, x_2^*)$ of batching heuristic under the current $\alpha$ given by layer 1 BO. The number of Random initialization points are controlled by parameter $k_{layer1}$ and $k_{layer2}$. $Q_{layer1}$ and $Q_{layer2}$ controlled the number of Query points.

$$F_{layer1} = \frac{TEC_{x_1^*,x_2^*} - TEC_{min}}{TEC_{max} - TEC_{min}} + \frac{TT_{x_1^*,x_2^*} - TT_{min}}{TT_{max} - TT_{min}} \quad (10)$$

$$F_{layer2} = \alpha \frac{TEC_{x_1,x_2} - TEC_{min}}{TEC_{max} - TEC_{min}} + (1-\alpha) \frac{TT_{x_1,x_2} - TT_{min}}{TT_{max} - TT_{min}} \quad (11)$$

---

**Algorithm6:** Auto-tuning Batch Heuristic

---

**Step1:** Set $x_1$, $x_2$ randomly, apply batch heuristic.

**Step2:** Apply n-GuptaEx-SETUPBO method, calculate $TEC_{x_1 x_2}$ and $TT_{x_1 x_2}$.

**Step3:** Repeat Step1~Step2 for 10 times, then calculate $TEC_{min}$, $TEC_{max}$ and $TT_{min}$, $TT_{max}$.

**Step4:** Randomly generate $k_{layer1}$ initialization points of $\alpha$.( layer 1 BO)

**Step5:** Randomly generate $k_{layer2}$ initialization points of $x_1$, $x_2$.( layer 2 BO)

**Step6:** Apply batching heuristic and n-GuptaEx-SETUPBO method. Evaluate the results by objective function (11).

**Step7:** If iterations of layer 2 BO reached $Q_{layer2}$, move to Step9, otherwise move to Step8.

**Step8:** Create surrogate model, optimize acquisition function (EI), and generate Query point of $x_1$, $x_2$. Return to Steps6.

**Step9:** Output results $x_1^*$, $x_2^*$, Evaluate the results by evaluation function (10).

**Step10:** If iterations of layer 1 BO reached $Q_{layer1}$, end the algorithm, otherwise move to Step11.

**Step11:** Create surrogate model, optimize acquisition function (EI), and generate Query point of $\alpha$. Return to Steps5.

---

## IV. NUMERICAL STUDY

The numerical experiments were performed on a standard computer environment (MacOS, Apple M1, RAM 8.0GB). Python 3.8 is used both for the n-GuptaEx-SETUPBO and auto-tuning Batch Heuristic.

We made a benchmark test based on actual companies' data to compare the n-GuptaEx-SETUPBO and n-GuptaEx-BO. As Tables 1 and 2 show, the n-GuptaEx-SETUPBO outperformed n-GuptaEx-BO both in efficiency (by reducing 69.3% calculation time when handling 250 jobs) and accuracy (by finding a better value of FA), especially in the case of large numbers of jobs.

TABLE I.    COMPUTATION TIME COMPARISON

| Number of jobs [lots] | Calculation time of n-GuptaEx-BO for one iteration [seconds] | Calculation time of n-GuptaEx-SETUP-BO for one iteration [seconds] | Calculation time reduction rate [%] |
|---|---|---|---|
| 50 | 4039 | 2738 | 32.2% |
| 150 | 19407 | 8202 | 57.7% |
| 250 | 43841 | 13481 | 69.3% |

TABLE II.    COMPARISON OF EI-BO AND SETUP-BO

| Number of jobs = 250 | n-GuptaEx-BO | n-GuptaEx-SETUP-BO |
|---|---|---|
| Best value of FA | 13379.13 | 12711.28 |
| The percentage of iterations that acquired better FA | 5.38%~15.81% | 3.15%~20.01% |
| Number of iterations that acquired better FA | 50 | 50 |
| The percentage of iterations that acquired better T | 15.20%~24.49% | 7.84%~32.44% |
| The percentage of iterations that acquired better C | 0.89%~11.67% | 0.93%~14.11% |
| Number of iterations that acquired better E | 19 | 29 |

FA: Non-weighted performance except "d" of the objective function (1)
T: Tardiness of jobs    C: Cycle time of jobs    E: Earliness of jobs

TABLE III.    COMPARISON OF AUTO-TUNING BATCH HEURISTIC AND PREVIOUS BATCH HEURISTIC

| Method | Period | Flow | # of lots | Due-date satisfaction rate [%] | Average tardiness [days/lot] | Average earliness [days/lot] | Average cycle time [days] | Average load rate of batch process [%] | Total Energy Consumption [kWh] |
|---|---|---|---|---|---|---|---|---|---|
| n-GuptaEX-SETUPBO with previous batch heuristic | Apr.~May | f1 | 9 | 100.00% | 0.00 | 19.78 | 6.89 | 23.07 | 16497.59 |
| | | f2 | 28 | 100.00% | 0.00 | 20.43 | 5.82 | | |
| | | f3 | 102 | 100.00% | 0.00 | 19.39 | 5.55 | | |
| n-GuptaEX-SETUPBO with auto-tuning batch heuristic | Apr.~May | f1 | 9 | 100.00% | 0.00 | 20.89 | 7.11 | 31.20 | 12672.63 |
| | | f2 | 28 | 100.00% | 0.00 | 22.32 | 5.82 | | |
| | | f3 | 102 | 100.00% | 0.00 | 18.54 | 5.52 | | |
| n-GuptaEX-SETUPBO with previous batch heuristic | Jun.~Jul. | f1 | 12 | 100.00% | 0.00 | 10.00 | 8.50 | 23.16 | 21271.62 |
| | | f2 | 38 | 100.00% | 0.00 | 17.66 | 6.34 | | |
| | | f3 | 138 | 100.00% | 0.00 | 16.44 | 5.72 | | |
| n-GuptaEX-SETUPBO with auto-tuning batch heuristic | Jun.~Jul. | f1 | 12 | 100.00% | 0.00 | 19.08 | 8.17 | 32.20 | 15551.83 |
| | | f2 | 38 | 100.00% | 0.00 | 20.45 | 6.89 | | |
| | | f3 | 138 | 100.00% | 0.00 | 19.57 | 5.45 | | |

Actual companies' data in period [April ~ May 2021] and [June ~ July 2021] are used to compare the performance of auto-tuning batch heuristic and previous batch heuristic. For auto-tuning batch heuristic, we set $k_{layer1} = k_{layer2} = Q_{layer1} = Q_{layer2} = 10$. As Table 3 shows, n-GuptaEx-SETUPBO with proposed auto-tuning batch heuristic outperformed n-GuptaEx-SETUP-BO with previous batch heuristic by reducing 23% TEC and maintaining 100% OTD rate in period [April ~ May 2021] and reducing 27% TEC and maintaining 100% OTD rate in period [June ~ July 2021].

## V. CONCLUSION

This study introduced new algorithm to consider the trade-offs between power consumption and productivity for the n-step hybrid flow-shop scheduling (nHFS) problem with batch process. Two points have been progressed to promote the efficiency of the algorithm and to tune the parameter automatically based on our past research. Besides, a self-tuning optimization is proposed to batch the jobs considering the energy consumption and due date. Both the improvements and approach are effective based on the result of experiments.

## REFERENCES

[1] J.N.D. Gupta, K. Krüger, V. Lauff, F. Werner, Y. N. Sotskov, 2002, "Heuristics for hybrid flow shops with controllable processing times and assignable due dates," Computers & Operations Research, 29(10), pp.1417-1439, Elsevier Science Ltd

[2] M. TANAKA, K. NISHIZAWA, T. OHNO, Y. OGAWA, S. ARIMA, 2019, "Applications on hybrid flow-shop scheduling under dynamic constraints of queue time and capacities," Proceedings of Joint Symposium of e-Manufacturing and Design Collaboration Symposium and ISSM, IEEE.

[3] J. LIN, T. ONO, QX. ZHU, CD. MAO, H. TAKAHASHI, S. MORIE, S. ARIMA, 2022, "Multi-criteria optimization of n-step hybrid flow-shop scheduling," Proceedings of the 2022 International Symposium on Flexible Automation, ISFA:271-278.

[4] Liu, Ying, et al. "A multi-objective genetic algorithm for optimisation of energy consumption and shop floor production performance." International Journal of Production Economics 179 (2016): 259-272.

[5] T. P. Vasconcelos, D. Augusto R.M.A. Souza, Gustavo C. de M. Virgolino, C. L.C. Mattos, J. P.P. Gomes. "Self-tuning portfolio-based Bayesian optimization," Expert Systems With Applications, 188, 115847.(2021).

[6] J. Lin, T. Ohno, S. Morie, Q.X. Zhu, Y. Sasaki, S. Arima, 2020, "Optimization of Multi-objective Function of n-step Hybrid Flow shop Scheduling," Proceedings of the International Symposium on Semiconductor Manufacturing (ISSM2020), IEEE, WC-51, pp.1-4.

# Maintenance Content Reduction and Digitalization for Performance Optimization

Christopher Bode
*Intelligent Manufacturing Solutions*
*INFICON*
Austin, TX USA
christopher.bode@inficon.com

*Abstract—* One bedrock requirement within the semiconductor industry is the need for a comprehensive maintenance program to support reliable and predictable tool performance. While this is true, it is certainly not the case where "more is better," but rather an effort to define the truly necessary actions and best-known methods in maintaining tools to optimize availability, productivity, and product quality. The ongoing development and integration of Smart Manufacturing solutions across the factory systems landscape is now playing a role toward these objectives in providing a foundation for defining, deploying, and managing optimized maintenance task workflows. This paper will present both the efforts of companies using FabRecover, a novel maintenance management decision support framework, and the resulting demonstrable improvements that were achieved through such investments.

*Keywords—preventative maintenance, Smart Manufacturing, Waddington Effect, C4U-compliant maintenance specifications, Digital Twin*

## I. INTRODUCTION

Equipment maintenance is one of the more significant components of semiconductor maintenance in terms of both the importance and impact on overall operations productivity and profitability. Preventative Maintenance specifically is employed to minimize unplanned downtime, ensure product quality through correct equipment performance, improve equipment longevity, and minimize impact to productivity through effective maintenance planning. It has been estimated that 23.9% of the cost of manufacturing is associated with downtime when accounting for the direct cost of the activities and loss of productivity. Fully one-third of these costs are estimated to be unnecessary or improperly carried out, and therefore represent operational waste that may otherwise be prevented [1]. The inherent challenge within maintenance program development and optimization is to identify the correct actions to be taken and support them to be done in the correct manner to achieve maximum value from the efforts.

This challenge has been explored since the earlier days of modern Operations Research, most notably by C. H. Waddington in the context of military aircraft maintenance during WWII [2]. In this context it was observed that unscheduled maintenance events were most frequent directly after a scheduled maintenance event, and decreased in frequency over time to reach a steady-state failure rate. The conclusion drawn was that the maintenance activities within the program were either performed too frequently, or were carried out in such a way as to cause unscheduled maintenance rather than prevent it. The discovery of this so-called Waddington Effect then lead to efforts to improve not only the maintenance itself, but the methodology for documenting and analyzing their overall program for efficacy and efficiency.

Such learning and methods can be carried forward into present day semiconductor manufacturing rather directly in both structure and effect. One illustrative structure is C4U-compliant specification, a program and acronym that seeks to identify the attributes of effective maintenance specifications in such a way as to benefit their practice [3]. Such specifications are meant to be Clear, Complete, Concise, Correct and Unambiguous. Such methodologies can certainly be put into practice directly with the improvements of maintenance specifications as currently defined and utilized within fabs. There is evidence that maintenance activities can and do benefit further from their integration into Smart Manufacturing solutions to best support their indoctrination into practice and cycles of improvements over time. The balance of this discussion will focus on both the elements of effective maintenance definition and their inclusion into modern semiconductor factory automation.

## II. EFFECTIVE MAINTENANCE SPECIFICATION

While many early and ongoing efforts to improve maintenance activities center on the discussion of planned or scheduled maintenance, there is opportunity to include any related activity for which there a prescribed methodology with which to complete a task. This may include unscheduled events that require a root cause analysis approach to identify a defect source, corrective maintenance procedures that remedy each of those root causes, and correlated Out-of-Control Action Plans (OCAP) that originate from process and equipment control application that suggest the need to address an underlying tool control issue. In all cases there is a need to clearly specify the needed activities and support their execution in such a way as to achieve repeatable results. This discussion will encompass all these aspects generally, though specific requirements may be highlighted for the components as needed.

Returning to the elements of C4U-compliant specification, we may examine each in term to understand how they translate into a needs statements.

- Clear: it is easy to understand what is to be accomplished, both in terms of presentation and instruction. The former is achieved by communicating the work instructions in sufficient details as to support each task in turn, and the latter is in providing correlated information about how to perform the task as to be compliant.

- Complete: all the necessary information needed to perform tasks are provided at hand and in context, such that additional sources and locations are not required.

978-1-6654-7134-3/22 $31.00 © 2022 IEEE

- Concise: only that information that is needed is provided, which is best achieved in providing only that information that is needed for the specific tasks be undertaken at a given time.

- Correct: the steps, procedure and methods are demonstrably proven to provide the desired benefit.

- Unambiguous: there is clarity in both what and how things are to be accomplished, and all decisions are supporting with any needed information.

The corollary to these methods as presented are in the development of program business processes that help to achieve these goals. Subject Matter Experts (SME) are key to specification development, as they are most familiar with the tool sets and their requirements. These individuals should be put in a position to create the documentation directly, as they are best equipped to judge their quality. The procedures should be put through a number of "dry runs" to test and assess their C4U attributes and should as well undergo peer review from others to better understand that all knowledge and instruction, often called "tribal knowledge," is contained in the finished specifications. All issues at each step should be captured and used for iterative refinement. Finally, all documentation should be deployed when deemed safe, credible, and effective without further need for peer-to-peer communication.

## III. SMART MANUFACTURING DEPLOYMENT

Smart Manufacturing and similar concepts refer to the use of sensors and data collection, cyber-physical systems known as "Digital Twins," and predictive technologies to predict and optimize the future state of operations based on a comprehensive knowledge of historical, current and forecasted events. Solutions in this domain first seek to collect more information from a broader array of sources to increase awareness and information about the factory state. Digital Twins and similar technology then merge these sources of information to facilitate more holistic decision support across domains and factories. True Smart Manufacturing systems then employ predictive modeling capabilities to more accurately estimate and react to future operational state based on that richly developed foundation.

While many of the technologies in the growing field of Smart Manufacturing development are highly technical, computationally intensive and extraordinarily complex, even the relatively straightforward concerns of maintenance specification and execution can make use of such technology to great benefit. The vehicle for the benefit does not even need to significantly alter the specifications, but rather the manner in which the information is delivered, captured and reviewed over time. Each of the discussed C4U criteria may be improved by leveraging modern factory automation technology in these similar ways.

FabRecover™ is a maintenance execution decision support system from INFICON that is used by semiconductor manufacturers to support exactly this type of Smart Manufacturing improvement process. It performs several Smart Manufacturing system functions, first facilitating the collection of maintenance execution time, data, and results to enrich the understanding of the tools and events that support them (Sensing.) This data is then combined with tool state data and related factory operations data to model the efficacy and efficiency of the performed tasks (Connecting.) This data may

then be used in that larger manufacturing context to predict maintenance needs, assess workforce and timing requirements, and forecast the optimal way to execute future tasks (Predicting.)

### A. Clear

This primary concern is perhaps the most straightforward to benefit from the inclusion of modern capabilities such as computers and web-enabled devices, HTML5 browsers and Graphical User Interfaces (GUI) should be focused on clear presentation of information to the user. Paper specification, whether literally on paper or in electronic documents, are by their nature static and for reference. Translation of these documents into digital workflows has the first benefit of turning these into "living" documents that can be accessed at any area within the factory and used to guide maintenance activities in real time. Sparse instructional information may also be augmented with content-rich supporting documentation that can go beyond the original specification information. Normal text and pictures may be augmented with likes to other web sites, video, other embedded reference documents and the like from the feature set offered by modern browser technology.

The presentation of the information may further be enhanced by providing reactive display of the information based upon execution of the maintenance and the information collected. Tasks may be shown one at a time for clarity, with subsequent steps being hidden until preceding tasks are marked as complete. Such capabilities have the combined benefit of recording and tracking progress while making clear what steps need yet to be completed within a task. In terms of Digital Twin data aggregation, which also translates into information about who completed which tasks, how long it took to complete them, and what data might be available to document the state of the tool at the time the task(s) were performed. This in situ data collection is an important part of accurately modeling maintenance execution.

### B. Complete

Smart Manufacturing solutions are well positioned to aid in the aggregation of all necessary information to support proper maintenance execution. While any specification must be complete for it to be valid, the information required to execute will be dependent upon the tool and factory state at the time the work is to be carried out. The specification may, for example, detail which part(s) are required to perform the maintenance, but an interconnected system is able to identify whether those parts are on hand and where they might be located at present. Optional or extended maintenance tasks may also be documented and added to the scope of a planned maintenance activity, such as changing out pump or performing maintenance on a tool computer while the tool is down to maintenance. Other systems and information repositories may also be accessed from the same system, thereby virtually extending the specification by tying in other sources of information to complete the context of the work.

Tool and work history are also often a major component to performing even scheduled maintenance work properly. Decisions within the context of specified work may be subjective in nature, or allow for more than one approach or conclusion, such that the eventual outcome is dependent upon the judgment of the individual. Tool event history, recent maintenance events and data, comments entered by other technicians and similar historical records can be invaluable in informing decisions about the nature and extent of maintenance interdictions needed during a given event.

### C. Concise

Workflow management and dynamic decision support both play a significant role in the ability to provide concise information. Specifications often need to account for differences across the applicable toolset, and in doing so document all the information that pertains to each use case. Some of the information may also be conditional, following a sort of if-then-else decision pattern where the needed tasks are dependent upon the equipment condition and context at the time the event is executed. For the sake of completeness, static specification must document the decision, the criteria and each of the potential outcomes from the decision.

Workflow engines such as FabRecover can process and resolve such conditional logic behind the scenes, if you will, such that all the user is presented with is the applicable situation and needed actions. This can take many forms such as the proper part numbers of the given tool ID, whether a given step in a procedure is applicable in this instance, and whether entire parts of an overall workflow need to be carried out for this level of maintenance event. The logic for all of these is encoded and specified behind the presentation layer for completeness but are managed through the interface to also show they are concise.

### D. Correct

This element benefits the most from the larger, integrated context of Smart Manufacturing systems and the way content is managed within them. An initial level of judgment is made by the Subject Matter Experts that develop the content and their peers who review it during both publication and execution. This is the nominal and typical level of review that seeks to make sure the information is proper and complete. Added content from the larger factory context in terms of factory Key Performance Indicators, success rate, variability of availability, unscheduled downtime and other measures of efficacy help also to ensure that the specifications are also data-driven in review of their delivered benefit.

Modern application development lifecycle management practices directly benefit maintenance specification in the way that such a business process delivers peer review, testing and iterative development. Maintenance specifications may first be developed and then released for peer review, perhaps through several iterations, such that the information is complete and clear to a plurality of individuals. In the way that FabRecover and like systems deploy this information through a decision support framework, dry runs of the procure may be executed in a test environment to ensure that the information is clear, complete and concise. Once initially deemed correct, the specification may then be released to Production as a living document to support actual maintenance events.

Continuous improvement of the specifications follows similar methods from application development in terms of collecting and implementing feedback. Direct feedback from users of the document may be collected directly through the application during the execution of an event to immediately note needed changes or improvements. KPI and other operation data may also indicate where there is issue with the delivered benefit, whether in terms of efficacy, variability, or efficiency. In whatever way the improvement requirement is identified, subsequent revisions of the specification may be drafted, peer reviewed and tested ahead of their eventual release to production.

### E. Unambiguous

Differentiated benefit in this dimension stems from enforcing the specification and accompanying work instructions strictly in the manner as defined. Certainly, in the discussion heretofore all of the principles aid in avoiding ambiguity. It is uniquely in the character of the decision support system to compel the given user in a manner according to the situation, data, and rule sets within the maintenance specification in a way that no static document is able.

One significant benefit is in the application of data specifications and how that might subsequently direct tasks. If a parameter is out of tolerance, for example, additional work steps may be required to remedy the situation. A static document may present the decision point, specification limits and work instructions for each situation. An online decision support system, however, may collect the data and then automatically apply the specification and rule set to determine what steps to show automatically. This removes any guesswork or element of human error, and also may restrict the option to the technician in terms of what steps may be taken.

Another shared benefit is in how specifications may be defined once and used in different contexts. It is a common practice to define events with different periodicity with shared or similar tasks definition, e.g., weekly, monthly and quarterly events. In defining the events once to be used across those contexts, each task may be marked as applying to only a subset of those levels. It would then be left to the technician to decipher which of the line items apply to the current context. With a system like FabRecover those lines that are superfluous in the current context are not shown, thereby only provide those items that are strictly significant for the current event.

## IV. RESULTS AND DISCUSSION

A number of semiconductor manufacturers have implemented maintenance improvement efforts with FabRecover into practice and have shared tangible benefits that we directly attributed or related to use of the application as the standard for supporting the improved maintenance activities. In much the same manner as previous, the presentation of these results will be in the content of C4U criteria as a method to illustrate the nature of the benefits.

TABLE I.     EXAMPLE FABRECOVER IMPLEMENTATION BENEFITS

| Clear | Convert static documents to online work instructions | 10% increase in First Time Right (FTR) |
|---|---|---|
| Clear, Correct | Four specs combined into a single online document | 75% reduction in document count |
| Concise, Correct | Work instructions reviewed for applicability and benefit; content reduced where applicable | 40 - 68% reduction in wrench time |

| Unambiguous | Decision points and specifications automated to drive task display | 74% less variability |
|---|---|---|
| Correct | Entire PM event eliminated; tasks consolidated | 50% reduction in planned downtime |
| Unambiguous | Feedback calculations automated from user data entry | Eliminated errors in machine constant updates |
| Unambiguous | Troubleshooting guide implemented to identify common, cheaper fixes before costly parts changes | Reduced cost in Diffusion contamination control |

Table I summarizes some representative benefits from the application of Smart Manufacturing concepts and capabilities to the simple practice of specification management from the various FabRecover deployments from which benefits statements were recorded. In all cases the data was collected within the system during maintenance execution, as well as combined with data from other manufacturing systems as needed, to assess and calculate the benefits stated herein. This is an indication of how execution as well as continuous improvement cycles may benefit from such a decision support framework and support of Smart Manufacturing principles.

A related benefit currently outside the scope of management with the application is a decrease in maintenance costs in several dimensions.

- Lower overall execution time and improved FTR both increase the tool availability, leading directly to high factory productivity and/or lower cycle times.

- Decreased variability in event duration provides an ability for more accurate prediction, thereby increasing efficiency in both maintenance staff and production planning.

- Unscheduled events may be reviewed in the same context as scheduled events, leading in some

cases to more effective preventative measures to avoid the form. This conversion to scheduled events further increases predictability and overall tool availability.

## V. CONCLUSION

Smart Manufacturing as a concept, and more increasingly as a practice within the semiconductor industry, drives benefit in a host of dimensions. Cutting edge applications such as Machine Learning, Artificial Intelligence, data mining and other such modern computational methods often take center stage as differentiating capabilities unto themselves. These can provide important benefits to several contexts, and do represent the promise of disruptive innovation, but not all problems require that level of complexity to achieve significant benefit.

The practices of defining and executing planned maintenance activities for semiconductor manufacturing toolsets can benefit significantly from the more foundational elements of Smart Manufacturing solution development. Improved information delivery, data collection and aggregation in terms of Sensing can yield significant gains directly simply in terms of content delivery and management. Larger factory context and information may further drive continuous improvements by facilitating more direct and complete assessment of delivered maintenance benefit. Finally, this level of interoperability holds the promise of more complete modeling and prediction of maintenance execution, such that all events are done at the right time by the right people and in the right way for the right reasons.

## REFERENCES

[1] Thomas, D. (2018), The Costs and Benefits of Advanced Maintenance in Manufacturing, Advanced Manufacturing Series (NIST AMS), National Institute of Standards and Technology, Gaithersburg, MD, [online], https://doi.org/10.6028/NIST.AMS.100-18.

[2] C. H. Waddington, "Operational Research Against the U-Boat," Elek Science, London, 1973.

[3] J. P. Ignizio, "The Waddington Effect, C4U-Compliance, and Subsequent Impact on Force Readiness," Phalax, Sept. 2010, p. 17-21.

[4] J. Behnke, H. Smith and C. Bode, "Be Smart About Smart Manufacturing," Semiconductor Digest Magazine [online], https://www.semiconductor-digest.com/be-smart-about-smart-manufacturing/, accessed June 2021.

# Application of Natural Language Processing in Semiconductor Manufacturing

Daisuke Kobayashi
*AI Application Study Group*
*Sony Semiconductor Manufacturing Corporation*
Kumamoto, Japan
daisuke.f.kobayashi@sony.com

Shunsaku Yasuda
*AI Application Study Group*
*Sony Semiconductor Manufacturing Corporation*
Kumamoto, Japan
shunsaku.yasuda@sony.com

Takashi Iuti
*AI Application Study Group*
*Sony Semiconductor Manufacturing Corporation*
Kumamoto, Japan
takashi.iuchi@sony.com

Shiho Ito
*AI Application Study Group*
*Sony Semiconductor Manufacturing Corporation*
Kumamoto, Japan
shiho.ito@sony.com

*Abstract*— Recently, natural language processing has been making great progress in the AI field and is attracting attention. This paper describes the application of natural language processing in semiconductor manufacturing. No numerical or image data were used in any of the analyses. Using the natural language processing engine developed by SONY, we were able to analyze trends in quality troubles and extract features of manufacturing equipment from text data alone by performing various natural language processing represented by Bag-of-Ngrams and Chi-square test. By this, it can contribute to quality and productivity improvements from a different perspective.

*Keywords— natural language processing, baf-of-ngrams, tf-idf, x-squared test, burst analysis*

## I. INTRODUCTION

Many experts have reported a lot of cases in which big data accumulated daily in semiconductor manufacturing factories has been meaningfully applied to various prediction, control, and anomaly detection[1],[2],[3],[4],[5],[6]. On the other hand, there is also a lot of textual information on quality troubles and equipment maintenances at the manufacturing site. These data may contain hidden trends that cannot be obtained from conventional numerical or image data, and we cannot ignore the use of natural language processing to improve yield and productivity.

In this paper, we report two examples of analysis of quality troubles and equipment maintenances information using an in-house natural language processing engine developed by SONY.

## II. METHODS

Natural language processing is one of the AI technologies for analyzing text, and generally refers to term frequency-based statistical analysis based on series data expressed through morphological analysis and applied analysis with neural language model represented by BERT and GPT. The engine developed by SONY contains many statistical and machine learning elements shown in Fig.1, and its GUI is shown in Fig.2.

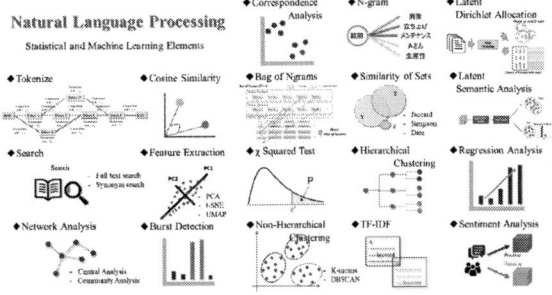

Fig.1 Statistics and machine learning elements used in natural language processing.

Fig.2 The natural language processing engine developed by SONY.

### A. Analysis of quality troubles

To research the trends in occurrence of quality troubles, we first constructed the corpus for analysis by performing preprocessing such as normalization and cleansing on text containing the outline of the quality troubles and incidental information. Next, since many uncommon technical terms are used in the semiconductor domain, we conducted a morphological analysis enhanced to the semiconductor domain using the semiconductor dictionary consisting of several tens of thousands of words that we have constructed independently. Finally, after post-processing such as synonym conversion and noise removal, and after quantification of documents called Bag-of-Words and Bag-of-Ngrams, key phrases weighting called TF-IDF is used to extract features.

Bag-of-words, a numerical representation commonly used in natural language processing, is a matrix representation of text

978-1-6654-7134-3/22 $31.00 © 2022 IEEE

data consisting of documents and words in a corpus, with it has a frequency of occurrence in the value. A generalization of this is Bag-of-Ngrams, where Bag of words is a special case of Bag-of-Ngrams (N=1, token = word). Since it is difficult to analyze Bag-of-words alone with context, we also used Bag-of-Ngrams as shown in Fig.3.

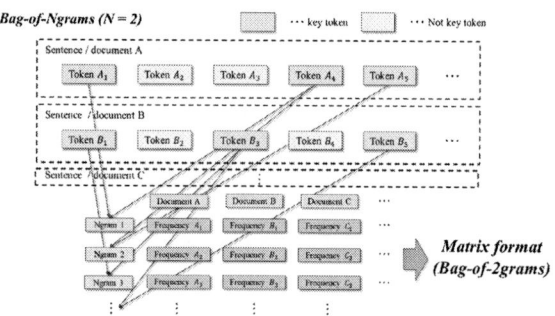

Fig.3 Bag of Ngrams

TF-IDF is represented by term frequency in the role of local weights and inverse document frequency in the role of global weights by the following equation:

$$TFIDF(t_i, d_j) = \frac{freq(t_i, d_j)}{\sum_{t_k \in d_j} freq(t_k, d_j)} log_e \left( \frac{|D|}{|d:t_i \in d| + \alpha} \right) + \alpha \quad (1)$$

where $freq(t_i, d_j)$ denotes the frequency of occurrence of term $t_i$ in document $d_j$, $|D|$ denotes the total number of documents in the corpus, $\alpha$ denotes the smoothing factor. Fig.4 is the result of visualizing the TF-IDF, and this result quantifies that each factory has trends in occurrence of quality troubles. For example, when we check by process, process A is the keyword for trouble documents in three factories, and some kind of abnormality related to surface or appearance is the keyword in two factories.

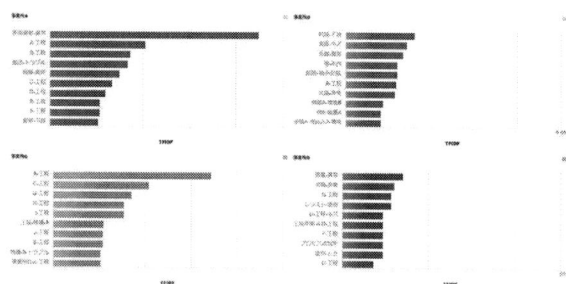

Fig.4 Top 10 key phrases weighted by TF-IDF regarding to quality troubles on the factory.

To further research how these key phrases are related among factories, correspondence analysis was performed on the TF-IDF matrix to visualize items that are strongly related or not characterized, as shown in Fig.5. The results show that while keywords related to many processes, such as A, G, and D, tend to be common to all factories, related keywords such as process N, and E, which are typical for each factory, were also identified.

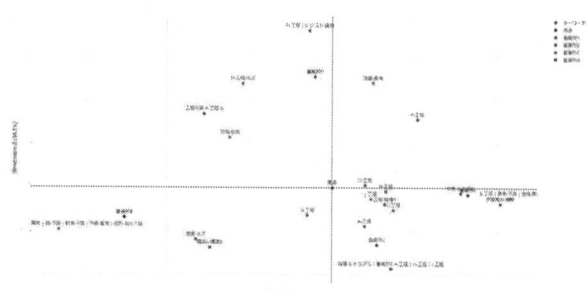

Fig. 5 Relationship of key phrases of quality troubles among the four factories.

### B. Analysis of equipment maintenances

In the same way as in Methods A above, we constructed and quantified the corpus for analysis to research trends in failure status, anomalous correspondence details, etc. between two models that perform the same WET process. To extract key phrases for which statistically significant differences exist, we performed key phrases scoring by the χ-squared test focusing on their occurrence rate and extracted those that showed large significant differences between the two models, the results of which are shown in Fig.6. The χ-squared score is defined as follows:

$$\chi^2_{yates} = \sum_{i=1}^{N} \frac{(|O_i - E_i| - 0.5)^2}{E_i} \quad (2)$$

where $O_i$ denotes the observed frequency of key phrases, $N$ denotes the number of distinct events, $E_i$ denotes an expected frequency of key phrases.

The most noteworthy result in this analysis is not the χ-squared score, but rather to see what key phrases appear at the top of the χ-squared score, and also to compare the number of occurrences of those key phrases using an effect size such as odds. This result confirmed that there were significant differences in key phrases related to contamination testing and chemical flow rates, etc. in the equipment maintenances. For example, the odds ratio for "contamination test" is 8.86, indicating that 8.86 times more contamination tests are performed in model D than in model E.

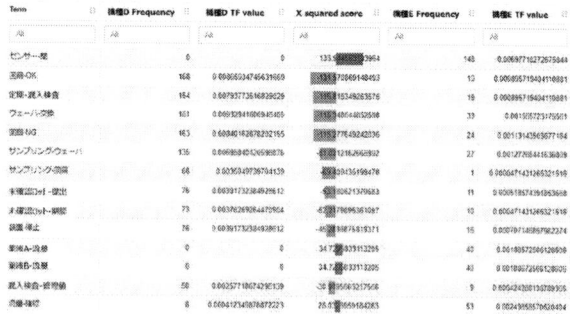

Fig.6 Key phrases with significant differences between two models that perform the same WET process.

On the other hand, if there is some kind of trouble that has been gradually increasing, or if there has been a sharp increase recently in some case, we need to grasp the trend and take countermeasures. To research if there were periods of spikes in specific failures or anomalous correspondence, we performed the burst analysis using an infinite Markov model in these key phrase streams. The burst analysis is one of the unsupervised algorithms used in time series analysis, which classifies the trend of key phrase occurrence into stationary or hierarchical non-stationary states. Fig.7 shows an example of the results of extracting key phrases that increased sharply at a particular time.

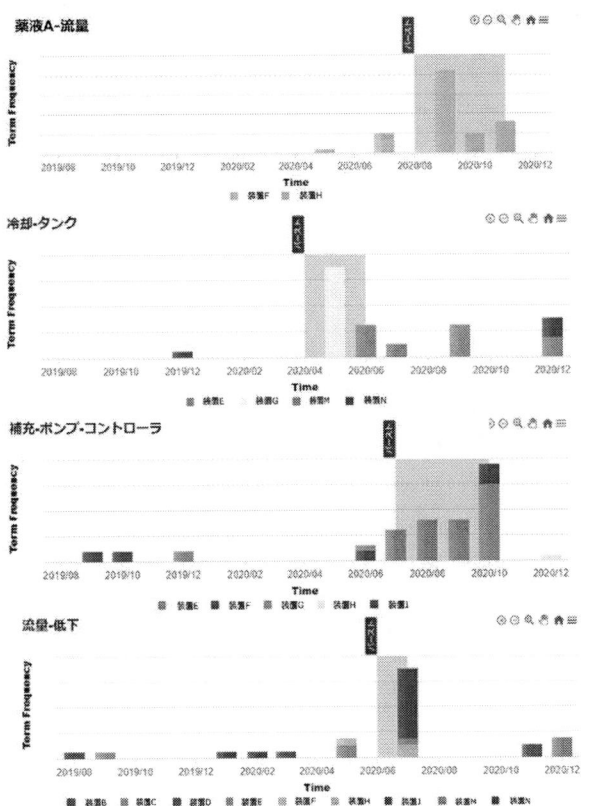

Fig.7 An example of key phrases with periods of steep increase extracted by burst analysis.

## III. RESULT AND DISCUSSION

Fig.4 and Fig.5 show that the tendency for quality troubles to occur differed greatly among factories, and we have linked this knowledge to feedback to the manufacturing factory. In addition, Fig.6 shows that there were large differences on dust abnormality between the two models in the equipment maintenance information of the WET process. In fact, model E, the successor to model D, is a model with enhanced dust control, and this feature could be seen in the text. Fig.7 also shows that the response history for abnormal chemical A flow rate and replenishment pump controllers increased rapidly at a particular time, which is consistent with the findings of the

WET engineers and suggests the possibility that the response history can lead to enhanced countermeasures and prediction of failures.

## IV. CONCLUSION

We analyzed the text of quality trouble information and equipment maintenance information using natural language processing. We extracted very beneficial findings from various perspectives and demonstrated the utility of natural language processing in semiconductor manufacturing. The feedback of these findings to on-site engineers can contribute to quality improvement and productivity improvement.

## ACKNOWLEDGMENT

The authors would like to thank S. Yasuda and many others at Sony Semiconductor Manufacturing Corporation for their support in writing this paper.

## REFERENCES

[1] K. Nomura, T. Okazaki, S. Yasuda, A. Kawashima, H. Tani and K. Masuda: "Virtual Metrology of Dry Etching Process Characteristics using EES and OES", PC-O-018, AEC/APC Symposium Asia 2011

[2] S. Yasuda, T. Okazaki and K. Nomura: "The VM·APC activities in Sony Semiconductor", PC-O-22, AEC/APC Symposium Asia 2013

[3] T. Uemura, T. Yoshimoto, T. Okazaki, K. Nomura, M. Ikeda, S. Yasuda and Y. Tanaka: "Prediction and Stabilization of MOSFET Threshold Voltage by VM-APC using Factory Data", MC-O-30, AEC/APC Symposium Asia 2015

[4] T. Okazaki, K. Okusa, K. Yoshida: "Prediction of the Number of Defects in Image Sensors by VM using Equipment QC Data", PC-O-66, ISSM 2018

[5] M. Ikeda, T. Okazaki, S. Yasuda: "Implementation of VM-APC Automated Execution System for Cu-CMP Process", EPC-O-19, AEC/APC Symposium Asia 2019

[6] R. Kurosawa, K. Okazaki, S. Yasuda, M., Ikeda, H, Okawa, Y. Miyaji, "Anomaly Detection of SEM images by using DeepLearning", TDA-007, AEC/APC Symposium Asia 2021

[7] J. Kleinberg, "Bursty and Hierarchical Structure in Streams", ACM SIGKDD International Conference on Knowledge Discovery and Data Mining 2002

## AUTHOR BIOGRAPHY

**Daisuke Kobayashi** (currently 26 years old) joined Sony Semiconductor Manufacturing Corporation in 2020. He works as a data scientist on AI development task such as natural language processing and image recognition.

**Syunsaku Yasuda** has been engaged in statistical development using semiconductor manufacturing data, and has developed Virtual Metrology technology using equipment data since 2006. Currently, he is promoting the development of AI application technology and the training of AI human resources in the semiconductor manufacturing field.

**Takashi Iuchi** joined Sony Semiconductor Manufacturing Corporation in 2018. He is engaged in system implementation of Virtual Metrology technology and natural language processing technology.

**Shiho Ito** joined Sony Semiconductor Manufacturing Corporation in 2021. She is engaged in text analysis from survey to manufacturing and development of natural language processing.

# Advanced Process Control Model for Trench Shape of Power Devices

Takumi Ito
Advanced Semiconductor Device
Development Center,
Toshiba Electronic Devices & Storage
Corporation
1-1, Iwauchi-machi, Nomi, Ishikawa,
923-1293, Japan
takumi8.ito@glb.toshiba.co.jp

Wang Xueting
Advanced Semiconductor Device
Development Center,
Toshiba Electronic Devices & Storage
Corporation
1-1, Iwauchi-machi, Nomi, Ishikawa,
923-1293, Japan
xueting1.wang@toshiba.co.jp

Yasuhisa Oomuro
Advanced Semiconductor Device
Development Center,
Toshiba Electronic Devices & Storage
Corporation
1-1, Iwauchi-machi, Nomi, Ishikawa,
923-1293, Japan
yasuhisa.oomuro@toshiba.co.jp

Kazutaka Nagashima
Advanced Semiconductor Device
Development Center,
Toshiba Electronic Devices & Storage
Corporation
1-1, Iwauchi-machi, Nomi, Ishikawa,
923-1293, Japan
kazutaka.nagashima@toshiba.co.jp

*Abstract*—In the semiconductor manufacturing, the manufacturing equipment is managed via the quality control (QC). The shape of the pattern is checked whether it meets the specification. If the shape is out of the specification, some recipe parameters are modified so that the shape meets the specification. The calculation method of the recipe parameters depends on the know-how of the individual engineers, which causes difficulties in the QC. We develop the automatic calculation of the optimal recipe parameters with Advanced Process Control (APC) model in order to solve these problems.

*Keywords*—*Semiconductor manufacturing, Advanced process control model*

## I. INTRODUCTION

In the semiconductor manufacturing, the quality control (QC) has been performed periodically to manage the condition of the manufacturing equipment in every few production lots (Fig. 1). In the QC, the typical shape is fabricated on test wafers with the typical recipes. Several points are measured for the processed shape. If the shape is out of the specification, the recipe parameters, which utilized in both the QC and the product lots, are modified so that all QC items are inside of the specification in the next QC. The QC without the modification of the recipe parameters is named "the periodic QC," while the QC with the modification of the recipe parameters is named "the non-periodic QC". In the conventional method, the new recipe parameters are calculated manually from the latest QC results by engineers based on their individual know-hows. It causes individual differences via each engineer and difficulty in the succession of their know-how to other engineers.

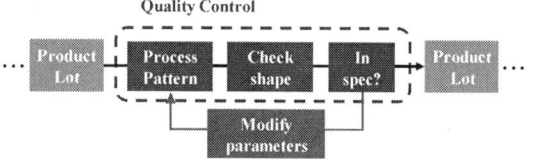

Fig. 1 The schematic image of QC.

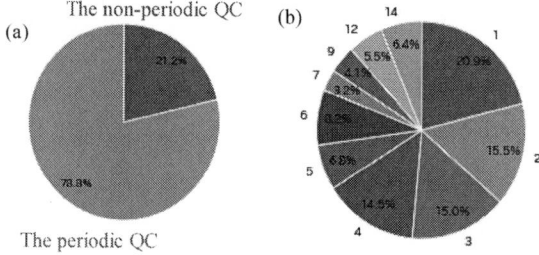

Fig. 2(a) The pie chart with the non-periodic QC (blue) and the periodic QC (orange).
(b) The breakdown of QC data with parameters modification by the number of total continuous QC with parameters modification.

When the parameter modification cannot be completed immediately, the QC with the modification of the recipe parameters are performed several times continuously. Fig. 2 (a) shows the percentage of the non-periodic QC in blue color and the periodic QC in orange color. The periodic QC (orange) occupies 78.8% of the total QC, while the non-periodic QC (blue) occupies 21.2%. Fig. 2 (b) shows the breakdown of the non-periodic QCs with the number of the QCs which are performed continuously with the modification of the recipe parameters. 20.9% is completed with the single modification, while 79.1% needs the multiple modification and QCs so as to make all the QC results inside of the specification. This result means that it is sometimes difficult to find the optimal recipe parameters with single calculation in the conventional method. If the optimal parameters are estimated in the single time, about 50% of the non-periodic QC (about 10% of the total QC) can be reduced.

In this paper, we develop the APC (Advanced Process Control) model to calculate the optimal recipe parameters automatically with the prediction models of the QC items. Normally, the APC model is modeled with the QC item and the recipe parameters [1-3]. The APC model, however, for the optimization of the recipe parameters are difficult due to so many QC items, compared to the modifiable recipe

978-1-6654-7134-3/22 $31.00 © 2022 IEEE

parameters. We utilize the prediction models of the QC items to estimate the optimal recipe parameters for the equipment condition. The input data for the prediction models of the QC items are the recipe parameters and the equipment log data. The prediction models output the predictions of the QC items. With these prediction models and the equipment log data, the optimal recipe parameters can be explored so as to the predictions become close to the target values of the QC items. This method can solve the individual difference between engineers and also solve the problem in the succession of the know-how. In comparison with the conventional method, this method utilizes the equipment log data in addition. It helps finding the optimal parameters for the current equipment condition and can reduce the non-periodic QCs by calculating the optimal recipe parameters in single time.

## II. PREDICTION MODEL

We choose the RIE equipment as a motif which processes trenches of the power devices. In this equipment, the QC is performed periodically in every few product lots whose frequency depends on the stability of the equipment. In the QC, widths at multiple points and a depth of the trench are measured (Fig. 3(a)). Due to the position dependence of the wafer surface, the measurements are performed in the different five points of the wafer surface positions respectively (Fig. 3(b)). On the other hand, there are only five recipe parameters for the modification and it causes the difficulties in the conventional method. They are managed so as to be in the specification. The modification of recipe parameters are performed when some QC items are out of the specification.

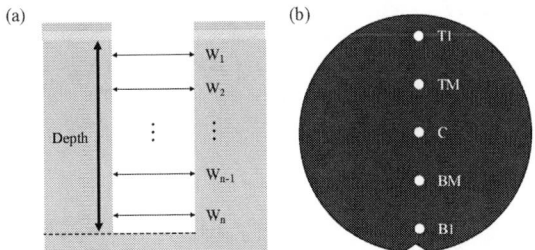

Fig. 3 (a) Schematic image of the trench and the QC items. (b) The schematic image of the measurement point in the wafer surface.

To train the prediction models of the QC items, we use the equipment log, the QC results and the recipe parameters data measured in the past several years. The prediction model is modeled in each QC item and each measurement position respectively. The input data of the prediction model for $W_n$ include the equipment log, the recipe parameters and the prediction values of the $W_1$ to $W_{n-1}$ in the same wafer position. It is because the deeper widths and depth are affected by the shallower ones. The input data for the depth includes the predictions of all widths. For the prediction model, we utilize the stacking model method which is an ensemble modeling technique that involves the combination of data from the predictions of multiple different models. The stacking model is divided into two layers as shown in Fig. 4. In the first layer, different multiple models predict the QC item and they become the inputs for the second layer model. The output of the second layer model is the final prediction of this stacking model. This enables the high accuracy for the target item with the complicated mechanisms such as the depth of the trench.

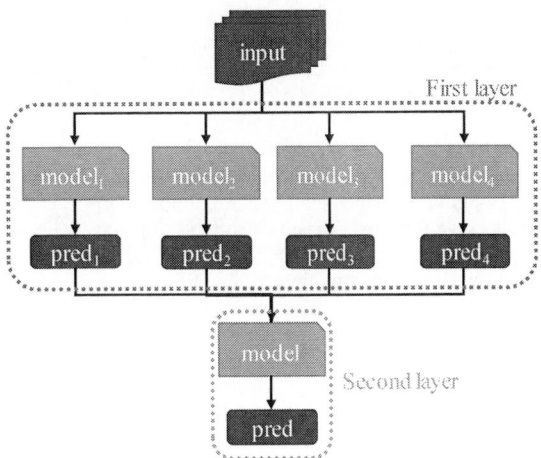

Fig. 4 The schematic image of the stacking model.

Here, we utilize the models of the linear regression, k-NN, LASSO, XGBOOST, LightGBM for the first layer of the stacking model. As the second layer model, we utilize the Ridge regression model. Fig. 5 shows the coefficients of determination ($R^2$) of the all prediction models for all QC items. Normally, the model with $R^2 > 0.6$ can be used for the prediction. Fig. 5 shows that almost all of models have larger $R^2$ than 0.6 and they have enough accuracy for the prediction model.

|  | T1 | TM | C | BM | B1 |
|---|---|---|---|---|---|
| $W_1$ | 0.85 | 0.78 | 0.75 | 0.79 | 0.82 |
| $W_2$ | 0.8 | 0.73 | 0.69 | 0.73 | 0.74 |
| $W_3$ | 0.69 | 0.59 | 0.78 | 0.74 | 0.77 |
| ⋮ | ⋮ | | ⋮ | | |
| $W_{n-2}$ | 0.78 | 0.57 | 0.72 | 0.66 | 0.81 |
| $W_{n-1}$ | 0.81 | 0.66 | 0.73 | 0.77 | 0.83 |
| $W_n$ | 0.76 | 0.67 | 0.75 | 0.77 | 0.82 |
| Depth | 0.78 | 0.72 | 0.83 | 0.83 | 0.74 |

Fig. 5 Coefficients of determination ($R^2$) for the prediction models.

## III. OPTIMIZATION OF THE RECIPE PARAMETERS

Fig. 6 shows the schematic image of the system configuration of the automatic optimization of the recipe parameters with the prediction model. First, the prediction models output the prediction values of all QC items from the latest equipment log and the recipe parameters (initially random). Second, the "Gap score" is calculated to quantize the

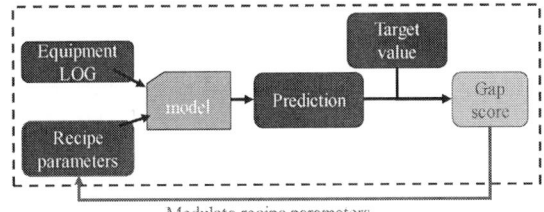

Fig. 6 Schematic image of the system configuration of the automatic optimization.

gap between the target values and the predictions of the QC items. Third, the recipe parameters are modified so as to make Gap score smaller and make the prediction close to the target values. In the optimization process, these three steps are repeated 30 times. This optimization system explores the optimal recipe parameters, which make Gap score the smallest for the latest equipment log. For the optimization process, we utilize Optuna which is the hyper parameter optimization framework which uses the Bayesian Optimization. With this method, we can calculate the optimal recipe parameters combination for current equipment condition.

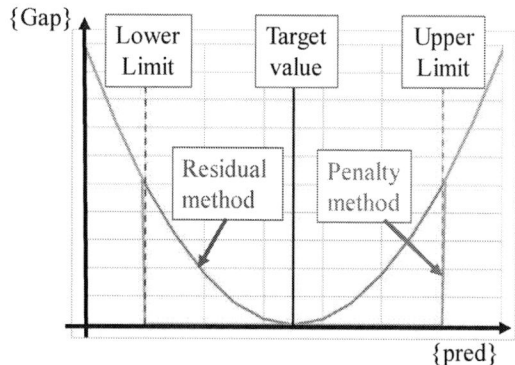

Fig. 7 The schematic image of the two definitions of the Gap score. The horizontal axis shows the prediction {pred} and the vertical axis shows the contribution to Gap score {Gap}.

In this method, the definition for the Gap score is important for the recipe parameters optimization. Here, we define two different definitions of the Gap score and compare the results calculated with these definition. Fig. 7 shows the schematic image of the two different Gap score definitions. One method is named as "Residual method" (colored orange in Fig. 7). In Residual method, the contribution to the Gap score for each QC item ({Gap}) is calculated from the prediction ({pred}) and the target value ({target}) as below for each QC item,

$$\{Gap\} = \left( \frac{\{pred\}-\{target\}}{\{target\}} \right)^2 \quad (1)$$

Here, the residual value between the prediction and the target value is divided by the target value in order to compensate the difference of the order between different QC items. With this method, the recipe parameters are optimized so that the all QC items get closer to the target values equally. The other method is named as "Penalty method" (colored green in Fig. 7). When the prediction is outside of the specification for the QC, the contribution to the Gap score becomes the same value with Residual method. When the prediction is inside of the specification, the contribution becomes 0. In this method, it is the most important that the all prediction values are inside of the specifications and the prediction values are not made close to the target values, which is different from the Residual method.

## IV. RESULTS

To verify the effect of the APC model, we perform the numerical simulation of them with the past QC data. Fig. 8 shows the schematic image of the numerical simulation flow. We focus on the QCs with the modification of the recipe parameters. The left flow surrounded by the blue broken line

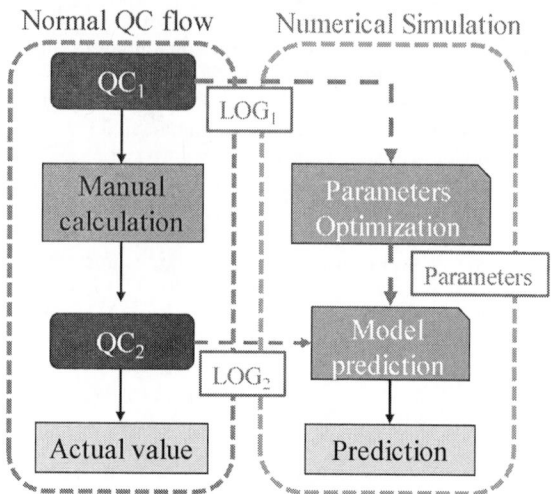

Fig. 8 The schematic image of the flow for the numerical simulation.

shows the normal QC flow with the modification of the recipe parameters. In the $QC_1$, some QC items are outside of the specification and the new recipe parameters are calculated by the engineer. The new recipe parameters are applied on the equipment and the $QC_2$ is performed. In the following, the results in $QC_2$ is called as "Actual value". In the numerical simulation (the right flow in Fig. 8, surrounded by the green broken line), the optimal recipe parameters are calculated from the equipment log of the $QC_1$. Then, the prediction is calculated with the recipe parameters, calculated by the APC model and the equipment log for the $QC_2$. Its output is called as "Prediction" from now on. In the numerical simulation flow, we calculate the optimal recipe parameters and the Prediction with the different two Gap scores. With this numerical simulation, we can compare the actual QC results and the results when the APC model is utilized.

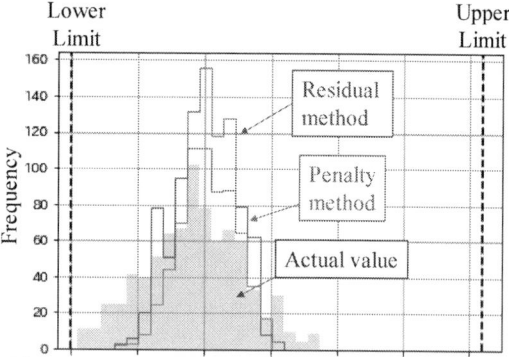

Fig. 9 The distribution of the actual values for a QC item (blue) and the numerical simulation results for the Residual (orange) and the Penalty method (green)

Fig. 9 shows the distribution of the actual value for a QC item (blue) and the prediction calculated with different two sets of the recipe parameters optimized by Residual method (orange) and Penalty methods (green). In comparison between the actual value and the predictions, the predictions distribute narrower than the actual values and are located far from the lower limit of the specification. It means the recipe parameters calculated by the APC model improve the shape of the trench.

978-1-6654-7134-3/22 $31.00 © 2022 IEEE

In comparison between the predictions for the two different definitions of the Gap score, the distribution of the Residual method gets narrower than that of the Penalty method. As shown in Fig. 7, {Gap} inside the specification is 0 in the Penalty method, while that in the Residual method gets smaller so as to be closer to the target value. Thus, the predictions tend to gather closer to the target value in the Residual method. In the other QC items, same trends are observed in the numerical simulations.

In the QC after the modification of the recipe parameters, all QC items have to be inside of the specifications to restart processing the production lots. Fig. 10 shows the percentage of the QC data which the all QC items are inside of the specification after the modification of the recipe parameters. 31% of the actual QC results (blue) fulfills the requirement. It means that almost all of the modifications are performed more than 2 times to restart processing the production lots. Those for the numerical simulation results become twice higher than that for the actual value, with 63% for the Residual method (orange) and 69% for the Penalty method (green). It means

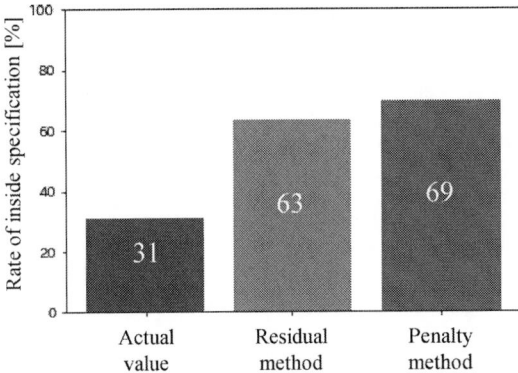

Fig. 10 The percentage of the actual QC and the numerical calculation data which the all QC items are inside of the specification.

that the automatic calculation of the recipe parameters can provide the optimal parameters in single time and reduce the number of the non-periodic QC. In comparison between the Residual method and the Penalty method, the percentage for the Penalty method gets higher than that for the Residual method. It is because that the prediction values in the Residual method are made closer to the target values too much and sometimes the other QC items become outside of the specification or close to the specification. It is not needed to make the all QC items inside of the specifications, because it makes the predictions of the other QC items outside of the specification. On the other hand, in the Penalty method, unnecessary gathering of the predictions to the target values does not occur, which contributes to boost the percentage.

## V. Conclusion

We successfully developed the APC model for the automatic optimization of the recipe parameters in the QC of the equipment for the trench process of the power devices. We demonstrated the optimization of the recipe parameters with APC models. It is verified that these parameters make the predictions of the QC items more suitable than the actual values in the numerical simulation. For the future work, we would like to apply the calculated parameters to the manufacturing equipment actually in order to get the mass data and check the QC items with the APC model and those with the manual modulation.

## Acknowledgment

The authors would like to thank Mr. Shiozawa, Mr. Susuki and Mr. Kunizaki for the cooperation in building the system for the APC model. The authors also would like to thank Mr. Kato and Mr. Miyashita for the discussion about this paper.

## Reference

[1] Xueting Wang et al., AEC/APC Symposium Asia 2021, TDA-011.

[2] Xueting Wang et al., AEC/APC Symposium Asia 2019, TDA-014.

[3] M. Kano, "Data-based Process Modeling" Journal of the Society of Instrument and Control Engineers, vol. 49(2), pp 101-106, 2010.

# Principal Component Analysis based GaN transistor live health Monitoring

Florian Chalvin
Research center
Rohm co. ltd.
Kyoto, Japan
florian.chalvin@dsn.rohm.co.jp

Yoshinori Miyamae
Research center
Rohm co. ltd.
Kyoto, Japan
yoshinori.miyamae@dsn.rohm.co.jp

Yoshiaki Oku
Research center
Rohm co. ltd.
Kyoto, Japan
yoshiaki.oku@dsn.rohm.co.jp

Ken Nakahara
Research center
Rohm co. ltd.
Kyoto, Japan
ken.nakahara@dsn.rohm.co.jp

*Abstract*—**Adoption of next generation semiconductors is still low, partly due to limited knowledge from the reliability point of view. To help solving this problem we introduce a way to track transistor degradation in real time using PCA analysis. By using this method, it is possible to detect when a transistor is no longer operating nominally from easily obtained voltage measurements.**

*Keywords—Device health, unsupervised, live monitoring*

## I. INTRODUCTION

We introduce an application of the principal component analysis (PCA) method to individually tailored GaN transistor health monitoring. GaN transistors are promising great energy savings thanks to increased efficiency and are becoming widely adopted in our environmentally conscious society. However, while we have a deep knowledge of the behavior of Silicon based device with regard to fatigue and failures this knowledge does not exist yet for new generation devices such as GaN transistors. In this context live monitoring of device health is especially interesting. It has been shown that due to device characteristics variance it is necessary to have health monitoring methods that can accommodate each device own characteristics [1]. We show that by using a PCA derived 2D projection of the device operating point it is possible to detect anomalous behavior [2].

To get the data we ran a high frequency switching experiment on 11 EPC8010 GaN transistors. The devices were held at an ambient temperature (Ta) of 100°C and switched continuously at a frequency of 6.78 MHz, a common working frequency for high efficiency GaN-based circuits [3][4]. The transistors were controlled by LT4440-5 (Linear Technology) gate drivers with a 4.7 ohms resistor in series. Drain voltage was set at 30V with a 200 ohms resistor in series. A schematic of the circuit is shown in Fig. 1. Initially the gate voltage was set to 5V and was increased to 6V, then to 7V on half of the devices to induce noticeable damage in an acceptable timeframe. Experimental conditions are summarized in table I. Periodically the devices characteristics (IV curve, threshold voltage, gate leak, drain leak, and on-state resistance (Ron)) were measured on a curve tracer to have an objective assessment of the degradation. Waveforms for the drain-source voltage ($V_{ds}$) and drain current ($I_{dc}$), actually the voltage across the 200 ohms lead resistor R2, were measured on an oscilloscope at the same time and used for the health monitoring as curve tracer measurements would not be available on a live system. Examples of waveforms measured are shown in Fig. 2

## II. METHOD

To carry out health monitoring of the devices we use the measured $V_{ds}$ and $I_{dc}$ data. First, each channel is standardized

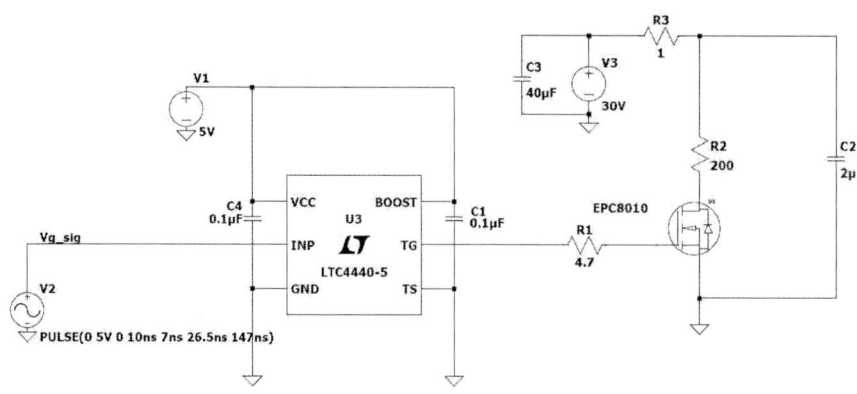

Fig. 1: Circuit schematics

978-1-6654-7134-3/22 $31.00 © 2022 IEEE

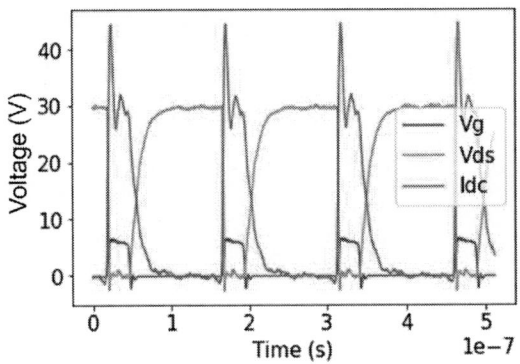

Fig. 2: Waveforms measured

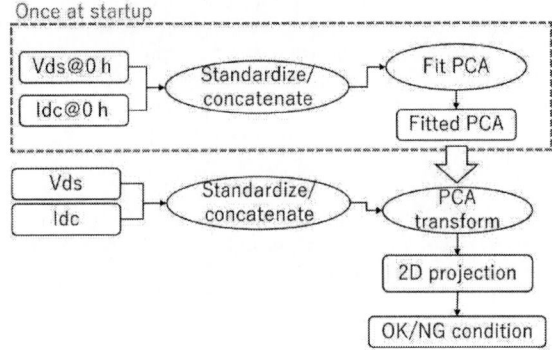

Fig. 3: Method flowchart

using (1). Where x' is the standardized feature and x are the

$$x' = \frac{x - mean(x)}{std(x)} \quad (1)$$

measured values.

After standardization both channels are concatenated an used as input into the PCA algorithm. We fit the PCA on the first measurement carried out for each device (corresponding to values at 0h time). Then we apply the transformation to every subsequent measurement while normalizing the every PCA output with regards to the output of the transformation of the first measurements. For easily interpretable results only 2 dimensions for the output are kept, these 2 values give coordinates to plot the device operating point on a plane. A flowchard of the method is shown in Fig.3. The initial measurements at 0h will always result in a point located at the (1,1) coordinates on the plane. As the device is assumed to be healthy at the beginning of the experiment, the area around this initial point represents a device that is working nominally. A drift away from that point indicates degradation. Once the drift in the points location away from the (1,1) origin reaches a sufficient magnitude we can assume that the device is no longer functioning nominally.

TABLE I. EXPERIMENTAL CONDITIONS

| Operating conditions | Experiment duration | | | | |
| --- | --- | --- | --- | --- | --- |
| | *Initial* | *From 1800h* | | *From 2604h* | |
| | | 1 to 5 | 6 to 11 | 1 to 5 | 6 to 11 |
| Device nb. | 1 to 11 | 1 to 5 | 6 to 11 | 1 to 5 | 6 to 11 |
| Device model | EPC8010 | | | | |
| Switching frequency (duty rate) | 6.78MHz (18%) | | | | |
| Gate voltage (rated 6V) | 5V | 5V | 6V | 5V | 7V |
| Ta | 100° C | | | | |
| Drain current (rated 2.7A) | 0.15A | | | | |
| Drain source voltage ( rated 100V) | 30V | | | | |

## III. RESULTS

This section will discuss the results obtained following the method described previously. The graph in Fig. 4 shows result for a healthy device, with operating points clustered near the (1,1) coordinates. Some stray points are observed due to noise and errors in the measurements due to the experimental setup used. The graph in Fig. 5 shows results for a damaged device. We can observe the operating point drift in a damaged device. Using these graphs, it is possible to set a threshold for deciding whether a device is still usable or not. Each application will need its own threshold depending on the acceptable risk, but we suggest that a device is no longer functional when more than 3 consecutive measurements fall outside of a circle of

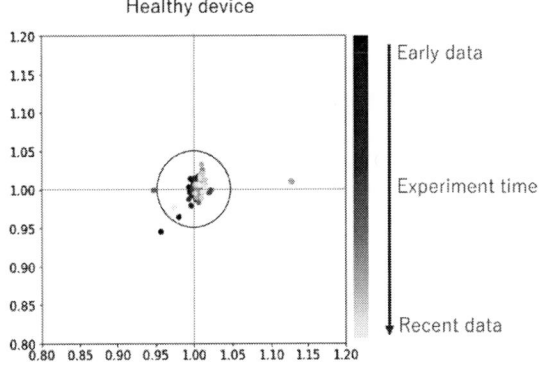

Fig. 4: PCA output for a healthy device

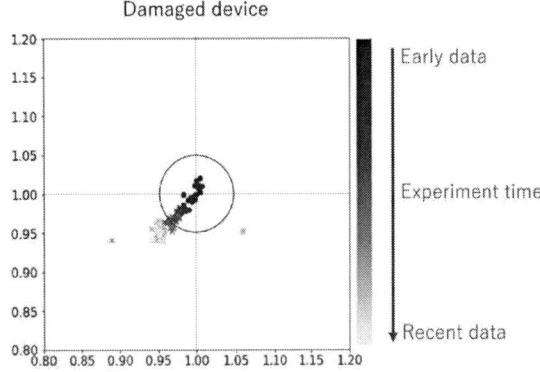

Fig. 5: PCA output for a damaged device

radius 0.05 centered on the original point at coordinates (1,1), this is the point where curve tracer measurements indicated device anomaly in our experiment. As explained in section I these curve tracer measurements were carried out at the same time as the waveform measurement to have an accurate image of the devices state throughout the duration of t he experiment.

## IV. CONCLUSION AND FUTURE WORKS

We have shown a way to assess device health in real time with no interruption of the device operation by using PCA to obtain a 2D image of the device operating point. A graphical threshold was set to determine when a device is no longer working within the specified datasheet values. This method is tailored for each device, allowing accurate health assessment despite device characteristics variability. However more data is needed to test the method on various failure modes and judge the validity of the threshold selected in these various situations. In the future we hope that this method can provide an accelerated way to detect devices prone to early failures among newly manufactured components, reducing the so called "infant mortality" and increasing circuits reliability.

## REFERENCES

[1] D. McMenemy, W. Chen, L. Zhang, K. Pattipati, A. M. Bazzi and S. Joshi, "A Machine Learning Approach for Adaptive Classification of Power MOSFET Failures," 2019 IEEE Transportation Electrification Conference and Expo (ITEC), 2019, pp. 1-8

[2] Lall, Pradeep, and Tony Thomas. "PCA and ICA based prognostic health monitoring of electronic assemblies subjected to simultaneous temperature-vibration loads." International Electronic Packaging Technical Conference and Exhibition. Vol. 58097. American Society of Mechanical Engineers, 2017.

[3] Jiang, Ling, and Daniel Costinett. "A high-efficiency GaN-based single-stage 6.78 MHz transmitter for wireless power transfer applications." IEEE Transactions on Power Electronics 34.8 (2018): 7677-7692.

[4] Gu, Lei, et al. "6.78-MHz wireless power transfer with self-resonant coils at 95% DC–DC efficiency." IEEE Transactions on Power Electronics 36.3 (2020): 2456-2460.

# Dynamic AI Computation Tasks with SECS/GEM in Semiconductor Smart Manufacturing

Hung H Nguyen
*Engineering Department*
*Yield Engineering Systems, Inc.*
Fremont, California, USA
hnguyen@yieldengineering.com

*Abstract—* **Semiconductor manufacturing has data management systems comprising multiple layers, including the cloud layer, the edge layer, and the equipment or device layer, which perform different functions in the system. The equipment layer performs data monitoring and detection of faults—the cloud layer and the edge layer help perform computational tasks. Performance of the computational tasks at the equipment layer is beneficial because they help achieve real-time response to the production and reduce the delays caused by data transfer from the equipment layer to the edge or cloud layer. In semiconductor manufacturing, the host computer located at the edge layer communicates to the equipment through Secs/Gem communication protocol. According to the results from our experiment, it is more efficient and effective to perform data analysis at the equipment level. This paper proposes a new Secs/Gem protocol for performing dynamic AI tasks on the equipment. The protocol allows the host to dynamically assign tasks of analyzing data to the equipment, and the equipment reports the results back to the host.**

*Keywords—edge computing, secs, gem, dynamic analysis, real-time, smart manufacturing, semiconductor*

## I. Introduction

### A. Background

With the rapid development of machine learning and artificial intelligence, advanced data analysis has become essential in smart manufacturing, especially in the semiconductor industry, since it helps in early detection of failures in the process [1]. A massive amount of data is collected during manufacturing processes, and analyzing the data in real-time is critical for improving quality and productivity.

The use of multiple layers systems is the most common way of managing smart manufacturing: the equipment layer, the edge layer, and the cloud layer [2]. In semiconductor manufacturing, the equipment layer includes many types of equipment such as process equipment, metrology equipment, and test equipment. The equipment layer is responsible for collecting data during processing and testing. The data will be transferred to the edge layer for analysis and production control. Some data will be transferred to the cloud layer for long term storage, machine learning and artificial intelligence computing tasks. This conventional method of organizing data and analysis creates latency due to the delay in data transfer between layers of the smart manufacturing system. Another problem with this method is that the network bandwidth gets overloaded with vast amounts of data transferring from equipment in the factory to the edge servers.

Some computing tasks can be performed at the equipment layer to reduce the latency of transferring data from the equipment to the edge. The issue is that computing tasks need to be dynamically decided by the edge layer during material processing. The equipment layer is not able to identify which computing tasks need to be performed at a given time during processing.

### B. Motivation

The issues motivate us to propose a communication protocol in Secs/Gem messages for the host to assign data analysis tasks to the equipment and collect the results for making decisions for the next steps in the process. In semiconductor manufacturing, different types of data analytics need to be performed on process data, including process data monitoring, statistical process control, resources scheduling, and failure prediction [3]. Some of those computing tasks can be performed at the equipment level to achieve real-time response.

Without instruction from the host, the equipment always performs certain pre-defined analyses at some pre-defined stages during material processing. It is hence critical for the host to be able to send requests to the equipment to perform data analyses and report the results. The proposed system will enable dynamic analysis computation tasks in semiconductor manufacturing.

## II. Related Works

Some significant efforts have been made to perform data analysis at the equipment level. In [4], the authors introduced the Equipment level Fault Detection and Classification (FDC) System to perform optimization and anomaly detection of semiconductor equipment. In this design, the equipment transfers some data to the Equipment level FDC System locally instead of the host. The Equipment Level FDC System can perform data analysis in a shorter time since it does not add data transfer latency into the analysis process and can react faster to any issues in material processing.

## III. Proposal

In semiconductor industry, factory automation is usually implemented through Secs/Gem interface. The Secs/Gem interface supports a set of standards which were defined by the Semiconductor Equipment and Materials International organization (SEMI). In smart manufacturing system, the equipment belongs to the equipment layer, and the host, which controls the equipment and collects data though Secs/Gem interface, belongs to the edge layer.

The current Secs/Gem interface does not include a protocol for implementing dynamic data analysis to support smart manufacturing in the semiconductor industry. This paper proposes a protocol for the host in the edge layer to assign data analysis tasks to the equipment and collect the results through Secs/Gem interface.

We propose to add a new stream, which includes a collection of Secs messages, to support dynamic data analysis during processing. The new stream provides an interface for

978-1-6654-7134-3/22 $31.00 © 2022 IEEE

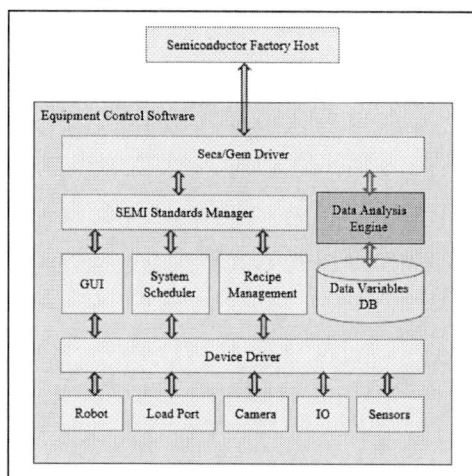

Fig. 1. System Architecture

TABLE I. LATENCY OF TRANSFERRING DATA VARIABLES WITH SECS/GEM

| Number of Data Variables | Assigning Values to Data Variables Time (seconds) | Transferring Data Variables to Host Time (seconds) |
|---|---|---|
| 1000 | 0.33 | 1.48 |
| 2000 | 0.83 | 2.15 |
| 3000 | 1.17 | 3.65 |
| 4000 | 1.84 | 4.92 |
| 5000 | 2.02 | 5.84 |
| 6000 | 2.31 | 6.91 |
| 7000 | 2.87 | 7.38 |
| 8000 | 3.19 | 8.53 |
| 9000 | 3.45 | 9.28 |
| 10000 | 3.66 | 10.26 |

the host to define computing tasks and send the tasks to the equipment. The new stream also defines an interface for the equipment to acknowledge the tasks and report the results to the host. A computing task can be a simple calculation, a predefined statistical process control rule, or a python script with input parameters.

### IV. EXPERIMENTS FOR COLLECTING LATENCY DATA

To carry out experiments, we set up a small network that includes two computers with CPU Intel i7 Processor, 16G RAM, 1TB HD, and Windows 10 Professional. The computers were connected through a gigabit Ethernet switch. We developed a basic Secs/Gem message transfer program for testing purposes. The host computer executed Secs/Gem in active mode and the equipment computer executed Secs/Gem in passive mode. We collected data variables by using data collection method defined in [5, 6]. The experiments were conducted with different numbers of data variables. Table I shows the results we collected in the experiments.

### V. SYSTEM DESIGN

To reduce the latency of data analysis tasks, we design a system that can minimize data transfer from the equipment to the host. Fig. 1 shows the system design that supports dynamic analysis computations using Secs/Gem. In this design, the host can send Secs messages to the equipment to define analysis computing tasks and link the tasks to some specific events. On the equipment, we add a new software module, Data Analysis Engine, that executes the analysis tasks and reports the results to the host. With this design, it is not necessary to send data variables to the host for analysis.

### VI. DESIGN OF THE STREAM 22 IN SECS/GEM PROTOCOL

We propose a new collection of messages, Stream 22, in Secs/Gem interface to handle all dynamic data analyses communication between the host and the equipment. Data analysis tasks include a method to perform data analysis, a data variable list used for the task, and a data variable list for storing the results. Like dynamic report configuration in Secs/Gem protocol, the task is linked to an event. When the event is triggered, the Data Analysis Engine performs the task and sends the results to the host. Data analysis tasks can be classified as predefined analyses or custom analyses.

#### A. Secs/Gem Protocol for Predefined Analyses

The host can send a request for predefined analysis task to the equipment using keywords. The initial keywords are "min", "max", "average", "stdev", and "sum". More computing keywords can be added into the protocol based on agreements between the chip manufacturers and the equipment providers to support certain analysis tasks. For predefined analyses, we define the following secs messages to setup analysis tasks on the equipment: S22F1, S22F2, S22F3, S22F4, S22F5, and S22F6.

*1) S22F1, Define "predefined" analysis task (H -> E):* The purpose of this message is for the host to define analysis tasks using keyword for the equipment to perform. A list of zero-length following <DATAID> deletes all predefined analysis task definitions and associated links. A list of zero-length following <TASKID> deletes the analysis task TASKID. All CEID links to this TASKID are also deleted.

Structure:

L, 2
    1.   <DATAID>
    2.   L, m
        1.   L, 2
            1.   <TASKID>
            2.   L, 3
                1.   <KEYWORD>
                2.   L, n #VID list for results
                    1.   <VID>
                    …
                3.   L, p #VID list for analysis task
                    1.   <VID>
                    …
        …

*2) S22F2, Define analysis task acknowledge (H <- E):* Acknowledge or return error from S22F1.

Structure:
<DATACK>

DATACK is Define Analysis Task Acknowledge Code. Valid values of DATACK:

- 0: Accept
- 1: Denied. Invalid format.
- 2: Denied. At least one TASKID already defined.

978-1-6654-7134-3/22 $31.00 © 2022 IEEE

- 3: Denied. At least one VID does not exist.
- 4: Denied. At least one KEYWORD is unknown.
- 5: Denied. Other error.

*3) S22F3, Link event analysis task (H -> E)):* The purpose of this message is for the host to link analysis tasks to a collection event (CEID). A list of zero-length following <CEID> deletes all links to that CEID.
Structure:
L, 2
1. <DATAID>
2. L, m
   1. L, 2
      1. <CEID>
      2. L, n
         1. <TASKID>
         …

   …

*4) S22F4, Link event to task acknowledge (H <- E):* Acknowledge or return error from S22F3.
Structure:
<LATACK>

LATACK is Link Analysis Task Acknowledge Code. Valid values of LATACK:

- 0: Accept
- 1: Denied. Invalid format.
- 2: Denied. At least one CEID link already defined.
- 3: Denied. At least one CEID does not exist.
- 4: Denied. At least one TASKID does not exist.
- 5: Denied. Other error.

*5) S22F5, Task Result Notify (E -> H):* Equipment sends task result to the host.
Structure:
L, 2
1. <TIMESTAMP>
2. L, m
   1. L, 3
      1. <TASKID>
      2. L, n
         1. <RESULT>
         …
      3. <ERRORTEXT>
   …

*6) S22F6, Task Result Acknowledge (H -> E):* The host acknowledges the task result from S22F5.
Structure:
<TRACK>

TRACK is Task Result Acknowledge Code. Valid values of TRACK:

- 0: Accept
- 1: Denied. Invalid format.
- 2: Denied. At least one TASKID does not exist.

- 3: Denied. The number of results does not match the definition of the task TASKID.
- 4: Denied. Other error.

After defining an analysis task and linking the task to a specific event, the host needs to enable the event with S2F37.

### B. Secs/Gem Protocol for Custom Analyses

For complex analysis algorithms, we design a protocol for the host to send the algorithms to the equipment in Python language format and collect the results. In the new stream 22, we define the following secs messages to setup custom analysis tasks on the equipment: S22F7 and S22F8.

*1) S22F7, Define an analysis script task (H -> E):* The purpose of this message is for the host to define analysis script tasks that use python script. A list of zero-length following <DATAID> deletes all analysis script task definitions and associated links. A list of zero-length following <TASKID> deletes the analysis task TASKID. All CEID links to this TASKID are also deleted.
Structure:
L, 2
1. <DATAID>
2. L, m
   1. L, 2
      1. <TASKID>
      2. L, 3
         1. <SCRIPTTEXT>
         2. L, n #VID list for results
            1. <VID>
            …
         3. L, p #VID list for script parameters
            1. <VID>
            …

   …

*2) S22F8, Define analysis script task acknowledge (H <- E):* Acknowledge or return error from S22F7.
Structure:
<DASTACK>

DASTACK is Define Analysis Script Task Acknowledge Code. Valid values of DASTACK:

- 0: Accept
- 1: Denied. Invalid format.
- 2: Denied. At least one TASKID already defined.
- 3: Denied. At least one VID does not exist.
- 4: Denied. Invalid Python script.
- 5: Denied. Other error.

### C. Ad-hoc Analysis

*1) S22F9, Request to perform an analysis task (H -> E):* The purpose of this message is to allow the host to trigger the equipment to perform analysis tasks without waiting for the corresponding events. When an corresponding event is triggered, the tasks will be performed again.

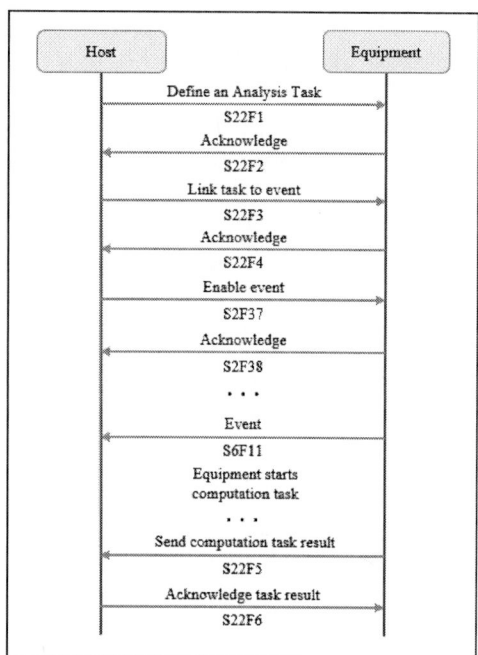

Fig. 2. Communication Diagram

Structure:

L, 2

    1.   \<DATAID>

    2.   \<L, m

        1.  \<TASKID>

        …

*2) S22F10, Acknowledge the request to perform an anlysis task (H <- E):* Acknowledge or return error from S22F9.

Structure:

\<REQTASKACK>

REQTASKACK is Request Task Acknowledge Code. Valid values of REQTASKACK:

- 0: Accept

- 1: Denied. Invalid format.

- 2: Denied. At least one TASKID does not exist

- 3: Denied. Other error.

*D. Usage Scenario Example*

Fig. 2 shows a communication diagram of a scenario for the host to set up an analysis task.

*1) Define an analysis task:* The host sends S22F1 to define an analysis task.

*2) Acknowledge:* The equipment sends S22F2 to acknowledge the message S22F1 from the host.

*3) Link the analysis task to an event:* The host sends S22F3 to link the analysis task to an event

*4) Acknowledge:* The equipment sends S22F4 to acknowledge the message S22F3 from the host.

*5) Enable the event:* The host sends S2F37 to enable the event.

*6) Acknowledge:* The equipment sends S2F38 to acknowledge the message S2F37 from the host.

*7) Perform data analysis:* When the event is triggered on the equipment, the Data Analysis Engine on the equipment performs the analysis task.

*8) Send results to the host:* When the Data Analysis Engine finishes the data analysis task, it sends S22F5 to report the results to the host.

*9) Acknowledge:* The host sends S22F6 to acknowledge the message S22F5 from the equipment.

## VII. CONCLUSIONS

In this paper, we conduct experiments to measure the latency when transferring data variables from the equipment to the host by using Secs/Gem messages. The experiments show that latency can significantly affect productivity in semiconductor manufacturing. We also propose a new system design with Data Analysis Engine and a new Stream 22 in Secs/Gem interface for the host to setup the equipment to perform real-time data analyses.

## ACKNOWLEDGMENT

This study was conducted with the support of Yield Engineering Systems, Inc., Fremont, California, USA.

## REFERENCES

[1] M. Khakifirooz, M. Fathi, and K. Wu, "Development of smart semiconductor manufacturing: Operations research and data science perspectives," IEEE Access, vol. 7, pp. 108 419–108 430, 2019.

[2] X. Li, J. Wan, H.-N. Dai, M. Imran, M. Xia, and A. Celesti, "A hybrid computing solution and resource scheduling strategy for edge computing in smart manufacturing," IEEE Transactions on Industrial Informatics, vol. 15, no. 7, pp. 4225–4234, 2019.

[3] M. Ghahramani, Y. Qiao, M. C. Zhou, A. O'Hagan and J. Sweeney, "AI-based modeling and data-driven evaluation for smart manufacturing processes," in IEEE/CAA Journal of Automatica Sinica, vol. 7, no. 4, pp. 1026-1037, July 2020, doi: 10.1109/JAS.2020.1003114.

[4] N. Kim, H. Choi, J. Chun and J. Jeong, "Introduction of equipment level FDC system for semiconductor wet-cleaning equipment optimization and real-time fault detection," 2022 33rd Annual SEMI Advanced Semiconductor Manufacturing Conference (ASMC), 2022, pp. 1-4, doi: 10.1109/ASMC54647.2022.9792476.J. Clerk Maxwell, A Treatise on Electricity and Magnetism, 3rd ed., vol. 2. Oxford: Clarendon, 1892, pp.68–73.

[5] Semiconductor Equipment and Materials International. SEMI E30-0418, Specification for the Generic Model for Communications and Control of Manufacturing Equipment (GEM)

[6] Semiconductor Equipment and Materials International. SEMI E5-0219, Specification for SEMI Equipment Communications Standard 2 Message Content (SECS-II)

## AUTHOR BIOGRAPHY

**Hung H Nguyen** has been with Yield Engineering Systems, Inc. (YES) since 2020. He is currently Sr. Director, Head of Software Engineering and is leading software development for semiconductor equipment. Prior to joining YES, he held various engineering and management positions at Applied Materials, SanDisk, and Western Digital.

# Application of Big Data Science in High Reliability Automotive Wafer Yield Management System and Failure Analysis

Chia-Cheng Kuo
Equipment Engineering Department
Taiwan Semiconductor Co., Ltd.
Yilan , Taiwan (R.O.C.)
0000-0001-8758-0885
oke.kuo@ts.com.tw

Po-Chih Chen
Product Engineering Department
Taiwan Semiconductor Co., Ltd.
Yilan , Taiwan (R.O.C.)
0000-0002-0788-5919
po-chih.chen@ts.com.tw

Chang-Tsun Tseng
Product Engineering Department
Taiwan Semiconductor Co., Ltd.
Yilan , Taiwan (R.O.C.)
0000-0002-8835-7340
zen.zeng@ts.com.tw

*The system automatically produces yield reports, shipping reports, real-time yield monitoring and statistics, abnormal cause statistics, Yield Chart, etc., to assist manufacturing, process, integration and other personnel to quickly Obtain the finished product/work in process yield report, and grasp the product yield information in real time, and find possible yield problems in real time. With the yield analysis tool provided by this system, we can quickly find the possible abnormal reasons to reduce the abnormal yield rate and the impact of production lines and shipments.*

*Keywords—Big Data, Intelligent System, High Reliability, Automotive, Yield Management*

## I. INTRODUCTION

Big data science has been applied to various industries in recent years, through the powerful computing power of computer processors and database's storage and calculation capabilities, after the integrated application of statistical theory combined with algorithms and high-speed computing, many problems that were originally difficult to explain and predict. Through the development and application of big data science, we can get the answer about high reliability automotive wafer yield management system and failure analysis.[1]

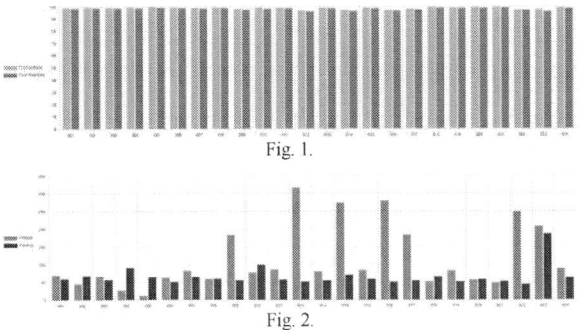

Fig. 1.

Fig. 2.

## II. WHAT WE FACED

We applied big data science to establish yield management system and failure analysis, and got quite good results, from the collection and calculation of yield data to the generation of yield report, all of which are very fast. And the data is checked through the calculation of computer algorithms. It is more rigorous and improves the reliability of yield data.[2]

In the part of probing test, we have obtained a lot of electrical testing data, such as BIN, CONTA, CONTC, POLAR, VF1, VF2, DVF, TRR, VZ1, VZ2, DVZ1, IR1, and

X/Y-axis, etc. Due to the reliability requirements of automotive products, we get more than seven million probing data per "LOT". Coupled with AOI (Automated Optical Inspection) data, the probing test and defect data obtained by each "LOT" may exceed ten million. Therefore, in order to meet the requirements of reliability and failure analysis, it has become an important task for us to find solutions to difficult problems through big data science.[3]

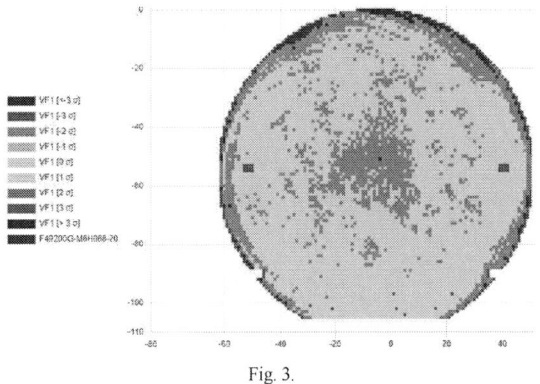

Fig. 3.

## III. THE SOLUTION AND METHODS

First, we collect data from physical equipment, including probing test machines, Automated Optical Inspection machines, etc. We use methods such as EAP (Equipment Automation Program)/SECS/GEM and DCS (Data Collection System) to obtain probe testing and AOI information. Next, the obtained data is properly integrated and stored in a database, and then the data is cleaned, transformed, and explored, and based on this. Analytical models can be built and selected.[4]

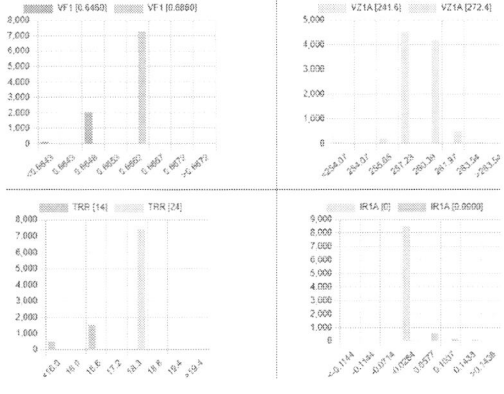

Fig. 4.

978-1-6654-7134-3/22 $31.00 © 2022 IEEE

Finally, we can apply the test qualification preparation to the yield calculation, and quickly and correctly generate the yield report and analysis chart (Fig.1/Fig.2). At the same time, we also provide the probing measurement results, according to the standard deviation, the probing test distribution wafer map and statistical table (Fig.3/Fig.4) were created. In the part of failure analysis, we created the failure distribution wafer map and statistical table (Fig.5/Fig.6/Fig.7), after summarizing all possible statistics; we created a visually integrated analysis plot (Fig.8).

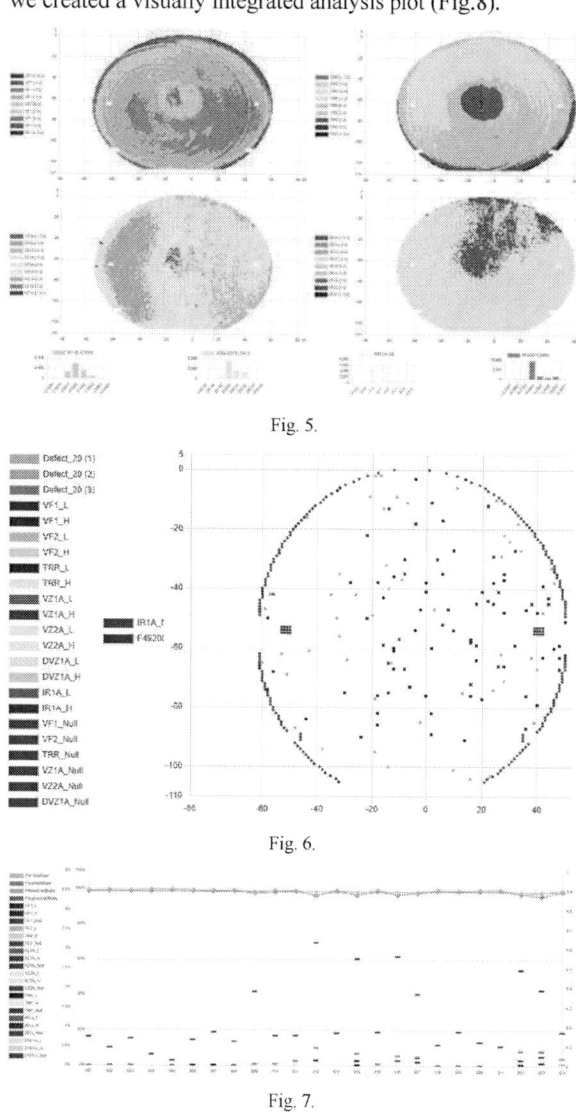

Fig. 5.

Fig. 6.

Fig. 7.

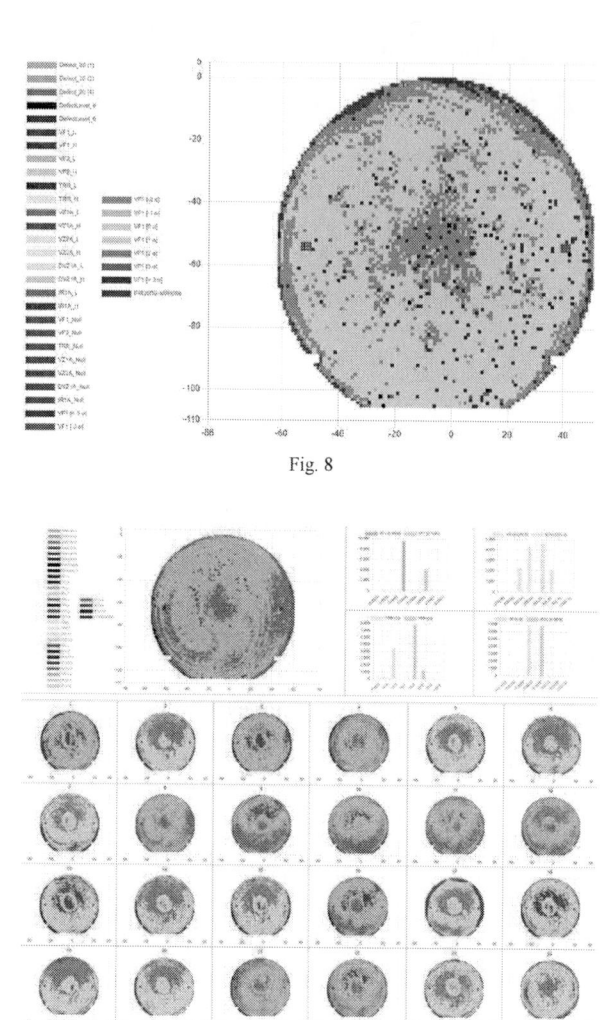

Fig. 8

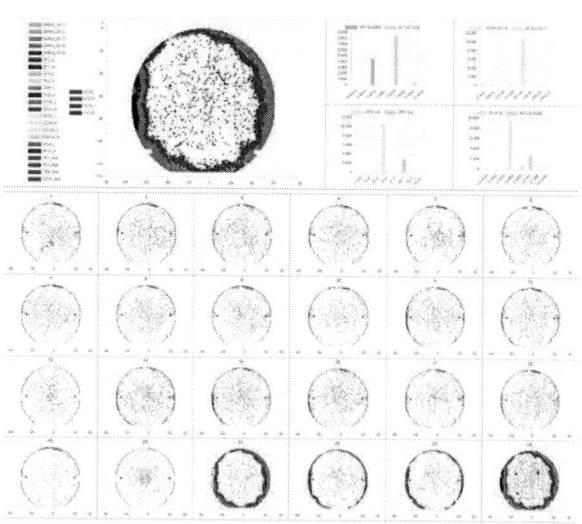

Fig. 9

Fig. 10

## IV. RESULTS AND CONCLUSION

The raw data of the measuring machine is imported and integrated, and then the system automatically produces yield reports, shipping reports, real-time yield monitoring and statistics, abnormal cause statistics, Yield Chart, etc., to assist manufacturing, process, integration and other personnel to quickly Obtain the finished product/work in process yield report, and grasp the product yield information in real time, and find possible yield problems in real time. With the yield analysis tool provided by this system, you can quickly find the possible abnormal reasons to reduce the abnormal yield rate. The impact of production lines and shipments.

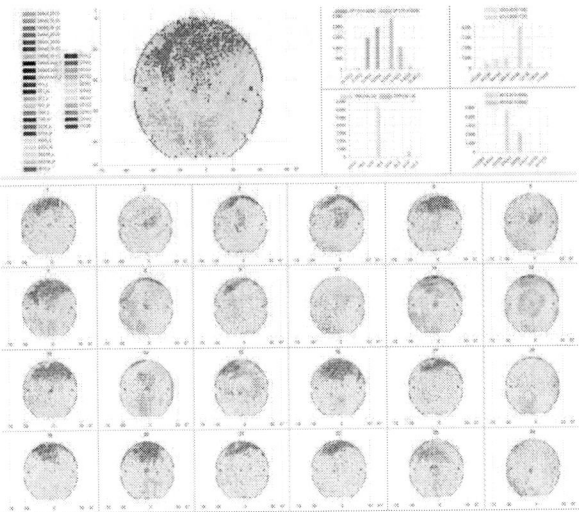

Fig. 11

The difference in labor/system output time, the color before PRB/AOI improvement can only distinguish green (good) and red (bad). After the improvement, Bad Die distinguishes High and Low according to Spec, and the electrical data of the entire Wafer All Die is also divided according to The standard deviation of each measurement value is used to distinguish the color gradient, and the comparison screen of the Bad Die and All Die STD of different Wafers with the Lot is integrated, so as to facilitate the relevant engineers to analyze the causes of the Bad Die(Fig.9/Fig.10).

And integrate the Defect of AOI with the actual photos taken. The screen is to facilitate the relevant engineering personnel to analyze the cause of the Defect. The functions include: all the Defects on a single wafer, all the Defects in the same Lot, and the same coordinates. In order to judge the cause of Defect, and then integrate Bad Die and Defect's E-Map to further analyze the relationship between Defect Die and Bad Die(Fig.11).

Through big data science and high-speed computing, we can quickly obtain the results of yield analysis, display a large amount of measurement data graphically and quantitatively, and manufacture high reliability automotive wafer products.

## REFERENCES

[1] Jiawei Han、Micheline Kamber、Jian Pei（2011），《Data Mining Concepts and Techniques》，Third Edition，Burlington：Elsevier Science, pp. 6.

[2] Anany Levitin（2011），《Introduction to the Design and Analysis of Algorithms》，Third Edition，Boston：Pearson, pp. 3.

[3] Stuart Russell、Peter Norvig（2010），《Artificial intelligence：a modern approach》，Third Edition，Upper Saddle River, N.J：Prentice Hall, pp. 2.

[4] Kenneth H. Rosen（2012），《Discrete Mathematics and Its Applications》，Seventh Edition，Boston：McGraw-Hill, pp. 191.

## AUTHOR BIOGRAPHY

**Chia-Cheng Kuo** Graduated from the Graduate Institute of National Development, National Taiwan University with a master's thesis title: The Application and Development of Big Data Analysis in Intelligent Transportation Systems: A Case Study of Taiwan's Freeway No.5 and Suhua Highway. He joined the Automation Engineering Department of Powerchip Technology Corporation in 2003, And obtained the invention patent in the United States and Taiwan: AUTOMATED MATERIAL HANDLING SYSTEM AND METHOD (US 8219242B2/TW I373090) in 2012.He joined Intelligent System Department of CECI Engineering Consultants, INC. in 2019 (Taiwan's Suhua Highway Improvement Project - Traffic Control Center), Currently serving Taiwan Semiconductor Co., Ltd. to join in 2021.

**Po-Chih Chen** Graduated from the Institute of Lighting and Energy Photonics of the National Yang Ming Chiao Tung University and joined Taiwan Semiconductor as an engineer in 2020. Currently, I am mainly responsible for defect analysis and electrical measurement, good at material analysis and component fabrication.

**Chang-Tsun Tseng** Graduated from the Institute of Institute of Electro-Optical Engineering, National Tsing hua University. Master's thesis title: Study of GaN UV PIN Photodiodes and HEMTs. Joined PHOENIX SILICON INTERNATIONAL CORPORATION in 2020, as a yield improvement project, mainly responsible for defect analysis and yield improvement. in 2021, Currently working at TAIWAN SEMICONDUCTOR, mainly responsible for Diode process integration and product development.

# Equipment Sensor Data Cleansing Algorithm Design for ML-Based Anomaly Detection*

Yun-Che Hsieh
*National Taiwan University*
1, Section 4, Roosevelt Rd., Taipei,
Taiwan, 10617
r07921104@ntu.edu.tw

Chieh-Yu Chen
*National Taiwan University*
1, Section 4, Roosevelt Rd., Taipei,
Taiwan, 10617
f07921016@ntu.edu.tw

Da-Yin Liao
*Straight & Up Intelligent Innovations
Group Co.*
San Jose, CA 95113, USA
eliao@miicg.com

Peter B. Luh
*National Taiwan University*
1, Section 4, Roosevelt Rd., Taipei,
Taiwan, 10617
peter.luh@uconn.edu

Shi-Chung Chang[+]
*National Taiwan University*
1, Section 4, Roosevelt Rd., Taipei,
Taiwan, 10617
scchangee@ntu.edu.tw

*Abstract*—Anomaly detection (AD) by exploiting machine learning (ML) of equipment sensory data can make significant contributions to yield improvements. Data cleansing is critical to provide ML-based AD with fixed-length input without distortion of data characteristics. We present a novel data cleansing design. Design innovations are: process step and mode-based input data length determination, importance indicator of sample data based on relative difference, and data cleansing priority by exploiting importance indicator and entropy. Experiment results demonstrate our cleansing design is superior to two frequently used methods in preserving data characteristics for effective AD by using an unsupervised ML approach.

*Keywords*—data cleaning, data characteristics, indicator of sampling point importance, data imputation, data discard, anomaly detection

## I. INTRODUCTION

In semiconductor fabrication, early equipment anomaly detection (AD) can not only reduce the capacity loss of a fab, but also improve final product yield. AD usually identifies the conditions of equipment by utilizing equipment sensory data (ESD). With the increases in numbers and types of sensors and sampling rates, ESD is generated in high volume and rate beyond what even veteran equipment engineers could effectively handle. Machine learning (ML) has therefore been widely adopted to extract normality features of ESD items of tools and assist engineers with or automate real time AD [1, 2].

ESD of a semiconductor tool under normal conditions shows apparent periodicity in repetitively processing wafers of a recipe. A tool may collect hundreds of sensor data items (SVIDs) from various sensors at a time [3]. Fig. 1 shows a data cycle of one SVID collected from a tool processing a specific recipe, which consists of a few process steps. Different process steps result in segments in a SVID data cycle. Fig. 2 depicts two cyclic SVID signals from concatenating the ESD of 5 wafer processing.

Among the ML methods for normality extraction and AD from ESD, unsupervised learning (UL) techniques mine ESD without requiring much prior knowledge about the normality features. Fig. 3 presents a UL-based framework of detecting in-line equipment anomaly, named Spectral and Time Auto-

encoder Learning for Anomaly Detection (STALAD) [1]. STALAD first adopts stacked auto-encoders (SAE) to learn statistical features of ESD normality. It then performs hypothesis testing of individual SVIDs wafer by wafer for real-time anomaly detection.

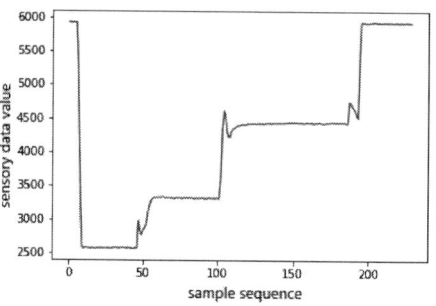

Fig. 1. ESD of an SVID cycle

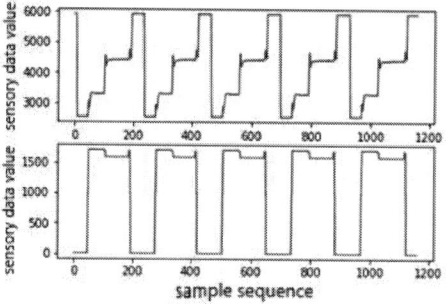

Fig. 2. Concatenated cycles of two SVIDs

STALAD [1] requires a data cleansing method to make all input training data vectors have the same number of data samples [4], as Fig. 4 depicts. Despite the apparent periodicity of ESD, the number of data samples of each cycle may not be exactly the same because of data acquisition and communication losses or process variations. Data cleansing for same size input vectors to STALAD may distort the features of ESD, which in turn may affect the normality features extracted by STALAD and hence the anomaly detection performance.

We present in this paper a novel design of data cleansing algorithm for ML-based AD. Our design consists of three parts and makes the least changes of the original ESD while tackling the following three issues:

i1) How to identify recipe cycle and step data lengths?

---

\* This work was supported in part by the Ministry of Science and Technology, Taiwan, R.O.C., under grants MOST 110-2221-E-002-144 and MOST 111-2221-E-002-184.

[+]Correspondence author

978-1-6654-7134-3/22 $31.00 © 2022 IEEE

i2) What are the least important data samples for characterizing a recipe cycle?

i3) How to delete or impute among the least important data samples to achieve the same size input data requirements of ML-based AD?

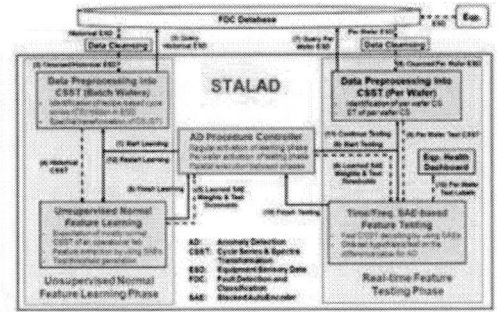

Fig. 3. STALAD framework

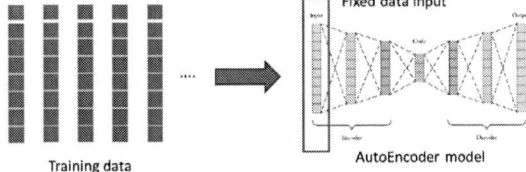

Fig. 4. A stacked auto-encoder model

The cleansing is among sample data cycles of the same SVID from processing the same recipe, namely, by individual SVIDs of individual recipes. To make the presentation concise in the following discussions, a cycle refers, in short, to ESD samples obtained from processing a wafer for a specific recipe while and a step refers, in short, ESD samples collected from processing a process step.

## II. DATA LENGTH AND IMPORTANCE DETERMINATION

This section presents how our cleansing method design addresses issues i1) and i2).

M1) *Individual step data length determination based on the mode of sample point numbers of the same step among all cycles:*

There are two advantages of defining data length by individual steps in a cycle rather than by the cycle. 1) This ensures that after cleansing, there is data of each process step in a cycle. If we only consider data length by the cycle, a cycle of longer data length may suffer the jeopardy from discarding data of some steps entirely in order to have the same length with the shorter data cycles. 2) ESD of a process step contributes to one same segment among the ESD sample sequences of different cycles. Extracting and comparing step data features among various cycles is a foundation of ML-based anomaly detection.

Our design for handling i1) first identifies cycles in a SVID and all process steps in a SVID cycle. A step has a number of data samples, which may vary among steps of the same kind in cycles. Definition of a step data length adopts the statistic notion of mode, which is the most frequently occurred sample numbers among the same kind of steps over the cycles as illustrated in Fig. 5. Adjusting step length to the mode leads to minimal number of changes to the same kind of steps over individual cycles. The fixed length of steps in each cycle adds up to make all cycles in a fixed length.

M2) *Sample data importance indication by using difference ratio between two adjacent data points in a sample sequence*

Analyses of real ESD show that data cycles of one recipe have quite similar time series patterns, which are characterized by value changes and inter-change durations [5]. Motivated by such observations, we define the importance indicator for each sample data as the relative difference, i.e., ratio, between the sample and the next sample in sequence:

$$d_{s,w,j,k} = \left| \frac{X_{s,w,j,k+1} - X_{s,w,j,k}}{X_{s,w,j,k}} \right| \qquad (1)$$

where $X_{s,w,j,k}$ and $X_{s,w,j,k+1}$ are the k-th and the (k + 1)-th sampling points of step j, cycle w and SVID s. A high ratio implies a relatively significant change and thus high importance. Fig. 6 shows a sample sequence of a cycle, where the x-axis is the sample sequence index and y-axis is the sample value. Fig. 7 shows the importance indicator of samples in Fig. 6. The x-axis is the sample sequence index, and y-axis is the importance indicator of SVID s. The larger value change between a sample point and its subsequent point is, the higher importance indicator the sample point has.

## III. UNIMPORTANT DATA DELETION AND IMPUTATION

M3) *Data deletion and imputation by exploiting importance indicator for sample grouping and entropy of sample group for priority of cleansing*:

To retain data patterns as much as possible after cleansing, we segment the time series samples of a SVID cycle into *groups*. A boundary point between two adjacent groups is a point with an importance indicator value (IIV) three times larger than the standard deviation of IIVs of the same recipe cycles. The red dots in Fig. 8 are boundary points while the consecutive black points are in the interior of a group.

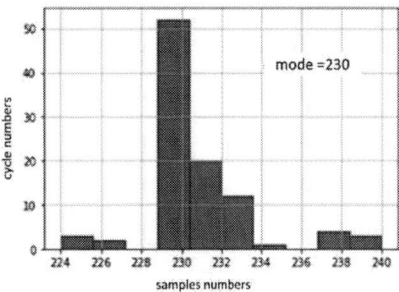

Fig. 5. Number of Cycles vs number of samples

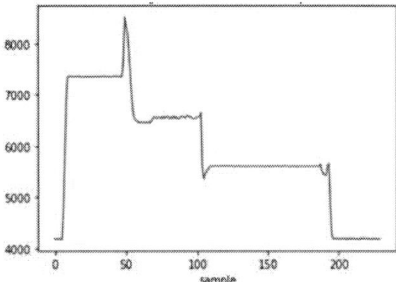

Fig. 6. Sample sequence example

We then calculate the entropy of the distribution of sampling points for each group. Entropy is a measure of uncertainty, and it is related to the amount of information. The larger the entropy is, the more uncertain the sampling values are and the more information they contain, and vice versa.

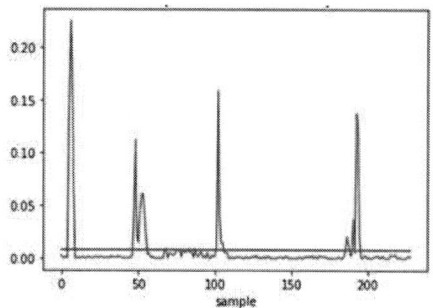

Fig. 7. Difference ratio between two adjacent data

Fig. 8. Boundary points

To *delete* a data sample, we select the *least entropy* group and delete the *least important* sampling points in the group. To *impute* data samples, we define *intervals* in a group as the sampling points in the group evenly divided by the number of sampling points to be filled. We use intervals to determine the filling position in the group *evenly*, and set filling value the *same value as the point before* the filling position. Such imputation does not affect the characteristics defined by M2.

## IV. CLEANSING PERFORMANCE EVALUATION

We integrate our data cleansing method into STALAD anomaly detection framework [1] to evaluate the performance of data cleansing algorithms.

### Performance indices

The performance indices of data cleaning are *sensitivity, false alarm rate* and *anomaly detection time*. Sensitivity denotes the proportion of correctly detected cycle among all the testing data cycles. False alarm denotes the proportion of detected cycle among all the normal testing data cycle. Anomaly detection time is the first cycle of anomaly detection among all the sequential test data cycle

### Dataset description

Evaluating detection sensitivity and time by STALAD requires ESD models to generate ESD data of drift, shift, and spike [6]. There are three types of sample sequences generated: nominal, noise, and anomaly sequences. Superposing a nominal sequence and a noise sequence will form a normal sequence. Superposing a normal sequence with an anomaly sequence forms an abnormal sequence.

### Nominal sequence

A nominal sequence describes the desired target values of ESD. Let $k$ be the time index, and $M_k$ be the real ESD data value at time $k$. The nominal sequence can be represented as $\underline{M} \equiv \{M_k, k = 1, \dots, K\}$, where $K$ is the total number of ESD samples in the sequence.

### Normal sequence

For a normal equipment, it is usually admissible for ESD values to vary slightly from the nominal behavior. A Gaussian white noise is used to describe this variation. Let $\underline{N}$ be a normal sequence and $\sigma$ be the noise level. A normal sequence is the superposition of a nominal sequence with a Gaussian white noise:

$$N_k = M_k + \nu_k, k = 1, \dots, K \qquad (2)$$

$$\nu_k \sim \mathcal{N}(0, \sigma^2) \text{ i.i.d. over } k$$

### Anomaly sequence

By referring to [6], drift, shift, and spike anomaly are modeled in mathematical forms in Table 1 as $\Delta D_k$, $\Delta S_k$, and $\Delta P_k$, respectively. $u(\cdot)$ denotes a unit step function and $\delta(\cdot)$ denotes a unit impulse function. $\alpha$ and $\beta$ are parameters modeling the scale and the occurrence time of the anomaly respectively.

Let $\underline{A}$ be an abnormal sequence, which is a superposition of a normal sequence and an anomaly sequence as follows:

$$A_k \equiv N_k + \Delta A_k, k = 1, \dots, K \qquad (3)$$

$$\{\Delta A_k\} \in \{\{\Delta D_k\}, \{\Delta S_k\}, \{\Delta P_k\}\}$$

TABLE I.    ANOMALY MODEL DEFINITIONS

| Anomaly | Definition |
| --- | --- |
| Drift | $\Delta D_k = \alpha k$ |
| Shift | $\Delta S_k = \alpha u(k - \beta)$ |
| Spike | $\Delta P_k = \alpha \delta(k - \beta)$ |

In the following experiment, normal and abnormal sequences are both generated as ground truth labeled data.

### Performance evaluation

In the experiment, the independent variables are the data cleansing methods. The benchmark cleansing methods for our design (CM3) include trajectory alignment (CM1) [7] and heuristic (CM2) methods. CM1 equalizes data length based on the least number of sampling points in all input data vectors. CM1 removes over-length data samples. CM2 equalizes data length by removing data samples from unimportant ESD suggested by equipment engineers.

In our evaluations, we use the three prominent types of anomaly sequences generated by simulation as input to STALAD for anomaly detection. Analyses of data cleansing effects on detecting different types of anomalies show that anomaly detection is sensitive to the variation of sampling points caused by data cleansing.

In comparison with data cleansing by CM1, we observe in Fig. 9 that data cleansing by CM3 significantly helped increase the sensitivity of STALAD detection of drift anomaly from 0.2 to 0.98. CM3 also helped speed up STALAD in drift detection from the 58th cycle to the 19th as depicted in Fig. 10 and 11 without changing the false alarm rate. Such results are rooted in that our proposed CM3 can characterize abnormal data by abnormal values, which are retained through indicators of sample importance.

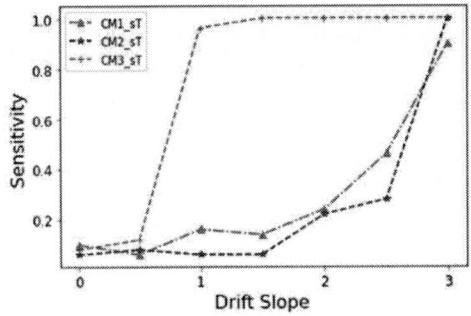

Fig. 9. Sensitivity for drift detection with three different cleansing method

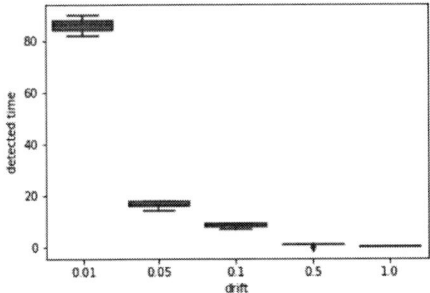

Fig. 10. Detection time for drift detection with our cleansing method

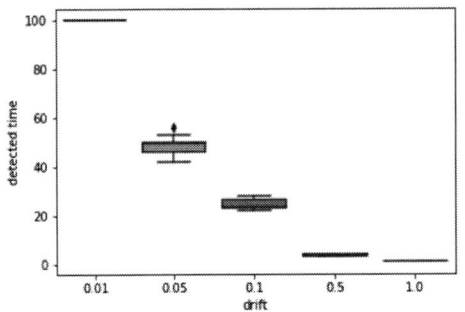

Fig. 11. Detection time of drift by trajectory alignment

Furthermore, the step-based data cleansing of CM3 aligns the process steps, improves the similarity between normal cycle data, lowers the threshold of anomaly detection, and makes AD easier. After data deletion, the cleansed data is obtained by sequentially merging the sample sequence of each step. Fig. 12 shows the original sample sequence, and the red points are deleted points. Fig. 13 compares the original and the cleansed sample sequence. Such results support that data deletion of CM3 by importance indicators well retain data patterns for anomaly detection.

## V. Conclusions

In this paper, we presented a novel data cleansing design for unsupervised ML-based AD by exploiting ESD. Design innovations are: process step and mode-based input data length determination, importance indicator of sample data based on relative difference, and data cleansing priority by exploiting importance indicator and entropy. Experiment results demonstrated that our cleansing design achieved superior performance to two frequently used methods in preserving data characteristics for effective AD.

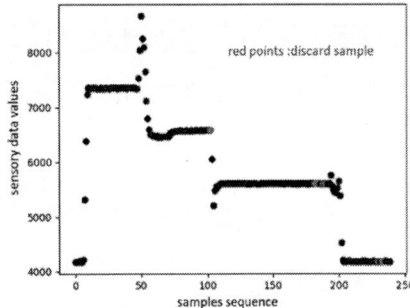

Fig. 12. Sample sequence with discarded samples

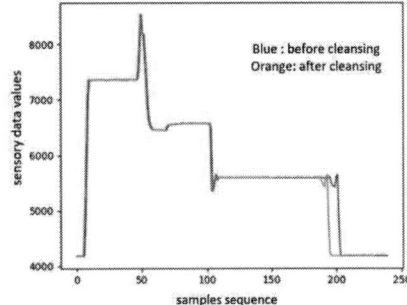

Fig. 13 Data sequence after data cleansing

## References

[1] C. Chen, S. Chang and D. Liao, "Equipment Anomaly Detection for Semiconductor Manufacturing by Exploiting Unsupervised Learning from Sensory Data," *Sensors*, vol.20 ,2020.

[2] Chen-Fu Chien, and Shih-Chung Chuang, "A Framework for Root Cause Detection of Sub-Batch Processing System for Semiconductor Manufacturing Big Data Analytics Sub-Batch Processing System for Semiconductor Manufacturing Big Data Analytics", *IEEE Transactions on Semiconductor Manufacturing*, vol.27, November, 2014.

[3] Hamideh Rostami, Jakey Blue, Claude Yugma, "Equipment condition diagnosis and fault fingerprint extraction in semiconductor manufacturing," in *Proc. of 15th IEEE International Conference on Machine Learning and Applications*, pp.534-539, 2016.

[4] Ki Bum Lee, Sejune Cheon, and Chang Ouk Kim, "A Convolutional Neural Network for Fault Classification and Diagnosis in Semiconductor Manufacturing," *IEEE Transactions on Semiconductor Manufacturing*, vol.30, May, 2017.

[5] Argon Chen and Jakey Blue, "Recipe-Independent Indicator for Tool Health Diagnosis and Predictive Maintenance," *IEEE Transactions on Semiconductor Manufacturing*, vol. 22, November, 2009.

[6] Chieh-Yu Chen, Shi-Chung Chang, Yun-Che Hsieh, "Effective Detection of Prominent Anomalies from Semiconductor Equipment Sensory Data by Using Unsupervised Learning," in *Proc. of The 18th International Conference on Automation Technology (Automation 2021)*, November, 2021.

[7] Q. Peter He, J. Wang, "Statistics pattern analysis: A new process monitoring framework and its application to semiconductor batch processes," in Proc. of *AIChE Journal*, vol. 57, No.1, pp.107-121, January, 2011.

# Secure and Reliable Power Monitoring for Low Consumption Factory Equipment via Programmable IoT Devices

Sergio Garnica and Robert Wieland

Fraunhofer Research Institution for Microsystems and Solid State Technologies EMFT

Hansastr. 27d 80686, Munich Germany

sergio.garnica@emft.fraunhofer.de

*Abstract*—**This paper reports on the implementation of low cost Internet-of-Things enabled power sockets, deployed on the Fraunhofer institute for Microsystems and Solid State Technologies clean room on one of the inspection microscopes used on the CMOS compatible line. The devices were flashed with open source software to ensure local, secure and reliable control without the necesity of an external cloud provider. The architecture of the physical deployment is shown and experimental data is analysed in order to obtain insight into the usage and statistics behind a previously unknown station in the clean room.**

*Index Terms*—**Factory IoT; Big Data**

## I. INTRODUCTION

Power monitoring of low power production machines has been an underutilized tool in semiconductor clean rooms for some time, mainly because of the lack of certified equipment, cost and secure, reliable communication of such devices in the past.

With the popularization of the low cost ESP8266 WiFi chip, which not only provides a WIFI (802.11) capable device but also the SDK to program the micro controller, users and manufacturers have the ability to write their own software stack [1], or to use one of the many libraries available from open source developers[1]. There have been also cases where these chips are used to monitor several environmental parameters [2].

Clean room equipment is generally not easy nor cheap to upgrade, machines such as microscopes, where there is no software or where the original software has no communication capabilities since for its operation it was not required. These machines, which have been kept working and are used for everyday production, represent an opportunity to gain some more insight into the processing happening within the clean room.

This paper will report on the power sensing monitoring performed on an inspection microscope at the clean room on the Fraunhofer institute for Microsystems and Solid State Technologies in Munich, Germany.

[1]https://github.com/esp8266/Arduino

## II. EXPERIMENTAL

### A. Devices and firmware

A 200mm wafer microscope from the company Reichert Jung, Fig. 1, is used to test the power monitoring capabilities of the smart sockets. The connection is made directly before the wall socket, or power extension in this case, as it is possible to see on Fig. 1 lower inset.

Consumer grade and certified standard European power sockets, Fig. 1 upper inset, from Tragant GmbH[2] and Allterco Robotics[3] were used, both with a maximum output power of 2500W and with the ability to perform Ove-the-Air updates, so that there is no need to disassemble the devices before uploading or flashing new firmware.

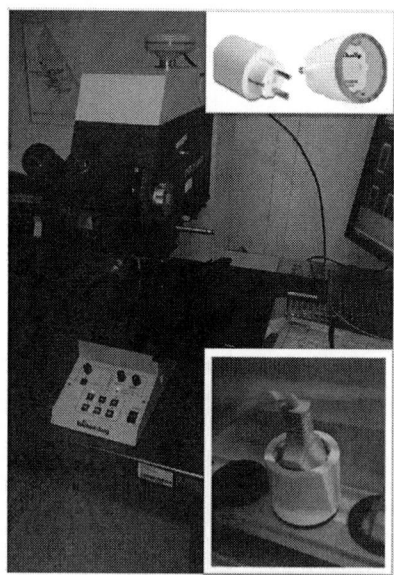

Fig. 1. Microscope used to test power monitoring capabilities of smart sockets. Upper inset: Smart sockets used. Lower inset: Smart socket connection to power extension cord.

[2]https://www.delock.com/produkt/11827/merkmale.html

[3]https://shelly.cloud/products/shelly-plug-s-smart-home-automation-device

The software platform ESPHome[4] was used to create the configuration files for the power sockets, this platform uses PlatformIO[5] to create the firmware that will be used by the devices. These projects are both open source and therefore it is possible to check exactly what kind of software will be running on the sockets when it gets installed.

ESPHome has a high degree of flexibility related to the sensors or actuators that can be employed, for the power sockets in this report the HLW8012[6] sensor can be configured and the measurement of current and voltage can be started, the power and energy used is also calculated.

### B. Architecture

Clean room deployment has to take into account several factors; the power socket should be able to communicate with the Manufacturing Execution System (MES), in order to update the status of the monitored device real-time. It is also necessary that the smart socket communicates to the internal database, based on InfluxDB, where the actual measured power is stored for further analysis. This database can be used in the aggregation of other sensors and devices, creating a so called Big-Data environment.

These cases were covered with different solutions, for one part the smart sockets connected to a WIFI network that was created only for these devices and has no real internet access, it also has an internal VLAN which isolates the sockets from all other devices on the institute.

As an intermediary on the isolated VLAN, several raspberry pi 3B+[7] were also connected to the network and given access to the internal institute network and the MES network via firewall rules. Figure 2 shows a diagram of the communication architecture.

As the programmed firmware offers the possibility to call specific HTTP/s requests directly, on the event of a new measurement, it is beneficial as the Big-Data database can be updated directly from the smart socket, without the need of an intermediary device, as the raspberry pi does for the MES.

### C. Reliability

It is important to address the reliability issues related to the devices, as it is of paramount importance that the smart socket does not shut off the internal relay and therefore leaves the microscope with no electricity.

The configuration of the devices is programmed so that the relay is switched on with a high priority during the boot process of the micro controller. It generally will take less than 2 seconds for the relay to be activated after the smart socket is connected to power.

The socket has a physical button that can be pressed to control the internal relay mechanism in case it is desired to cut the current flow to the device being monitored by the smart socket. In the case being reported here it is not desired to

[4]https://esphome.io
[5]https://platformio.org
[6]https://esphome.io/components/sensor/hlw8012.html
[7]https://www.raspberrypi.com/products/raspberry-pi-3-model-b-plus

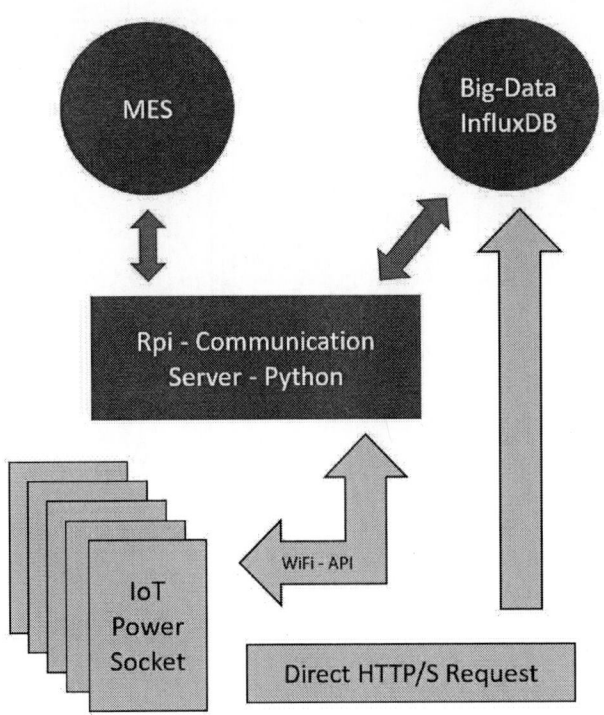

Fig. 2. Clean room communication architecture for the monitoring power sockets, the devices can connect directly to the internal database to record the measured power and a raspberry pi connects to the sockets via python API and updates the corresponding resources on the MES.

ever cut the current to the microscope as it is a machine that has constant usage. The programmed behavior on the smart socket de-activates the physical button on the socket as this feature is not necessary, and so, there is no risk that anyone by accident leaves the microscope without power when pressing the sockets relay button.

An important aspect of IoT devices on an enterprise environment is the ability to change base stations or so called roaming (802.11k,v) which allows the devices to adjust their connection to the Access Point which is currently more beneficial for the communication. This case might happen as new APs get provisioned or new antennas are installed or moved from their previous positions.

The smart socket has also been programmed that in case of not being able to connect to the main WiFi network, it should fallback into a so called Hotspot-mode, which opens up a WiFi network on its own and it is possible to connect to the socket directly. In this mode it is also possible to direct the device to a binary file from which it can perform an OTA update of the firmware, if required.

There has not been any reports, so far, by operators or other clean room personnel about the instability and/or in-operability of machines due to the smart monitoring devices. The devices have been installed for slightly over a year.

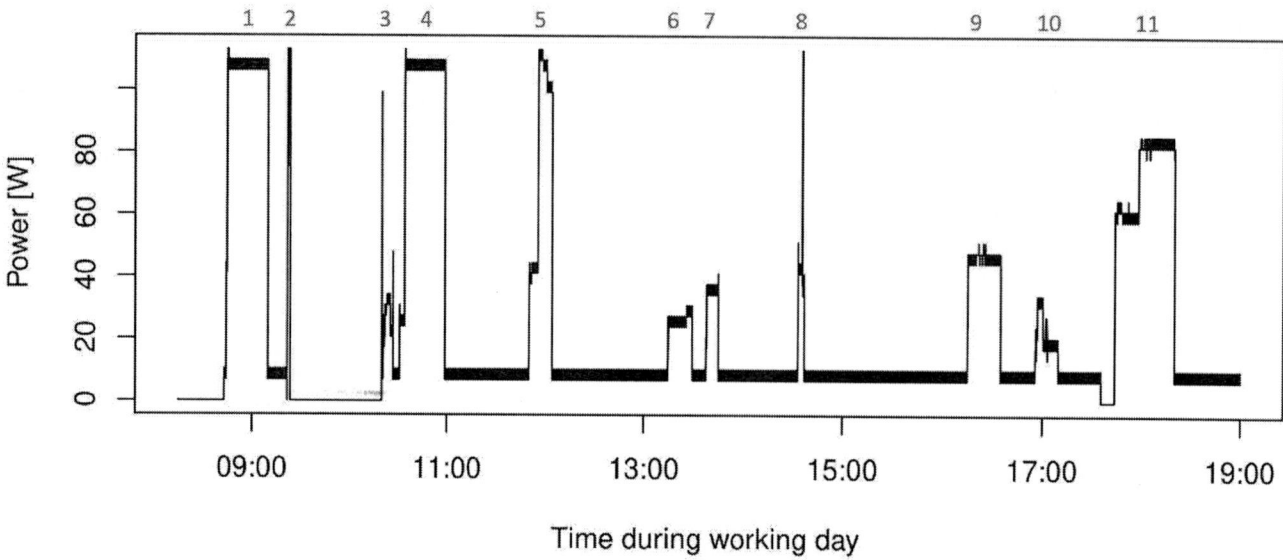

Fig. 3. Visual representation of the power monitoring acquired by the smart power sockets during a working day on the figure 1 microscope, the upper row counts the amount of inspections that happened that day.

## D. Security

One of the main concerns related to Internet-of-Things devices is their security, as many of these devices cannot be properly updated and therefore patched for vulnerabilities that might be discovered, there have been several studies [3]–[5] dealing with security issues on these micro controllers.

In this case, the smart sockets are based on open source software that is constantly maintained, and the updates can be performed via OTA, so that in case of a critical security flaw the devices might be able to get updates and continue their normal operation.

Encryption from the smart socket to the WiFi network is given by the known WPA standard, in enterprise scenarios by WPA-EAP. Connection to the python API on the the device further requires and encryption key which is a 32-byte base64 encoded string, this security measured was programmed by ESPHome based on the Noise Protocol[8]. This ensures that even if someone would gain access to the network, the messages would still be protected.

Physical security is given by the environment of the clean room itself, as it is isolated from external people and only authorized people can access it. If there is a case where someone with access to the clean room would access the device, it would have to physically damage the enclosure, so that it would be noticeable after the fact.

The raspberry pi was configured to test if the smart sockets are still available on the network, in case that the device is not found for a certain time, it will trigger an alert that will let users know that one/some of the smart sockets are not responding so that the appropriate measures can be taken.

[8]https://noiseprotocol.org

## III. Results

Figure 3 shows the results produced by the smart socket connected to the microscope depicted in figure 1. There are several interesting facts that can be extracted from this graph.

Starting from the left hand side of the image, the monitoring states a consumption of zero Watts, which makes sense since the device has not been activated yet. At around 8:30 the microscope is switched on and the inspections for that day start, with 11 inspections in total.

The microscope has a standby mode which let the users turn off the lamp, but keeps the rest of the system on, in this mode the microscope consumes around 10 Watts of power, it is possible to see this behavior between inspection 5-6 or 8-9. As it is also possible to see that there are some operators which prefer to switch the microscope totally off after the inspection is finished, as it is the case after inspection number 2.

The power consumed by the microscope when set to the maximum brightness is around 110 Watts, which inspections 1,2,4,5 and 8 did. It is possible to inspect wafers with a much lower brightness, but since many inspections require also dark field optical search, it is easier for the operators to use a larger brightness in those cases. The minimum brightness at which it is possible to perform an inspection is around 17 Watts, which can be seen on inspection number 10.

As the smart socket is also used to update the status of the resource on the MES system, from Standby to Productive and back, the threshold set for this event to happen is 15 Watts, that is, when the measured consumption of the socket is higher than the threshold, the raspberry pi will set the resource on the MES system as productive, and when the measured value is lower again, the raspberry pi will set the resource to standby. The microscope was on a productive state during this day for around 2 hours and 59 minutes.

From an information point of view, the data collected can provide insight in preventive maintenance, as the microscope was used for around 3 hours, this time can be added on the MES so that it is possible to know how long until the next lamp change is due for the microscope, and if the maintenance time is coming up, then it is also possible to order the lamps before hand so that downtime is kept to a minimum.

Much more sophisticated approaches can be also tested as the InfluxDB has the actual information for the consumed power, so it would be possible to create so-called power-profiles, which are based on the consumption figures. As an example, and as it was commented already, the consumption relates to the brightness of used on the microscope, then it is possible to already know which inspection required dark field and which did not, this provides information on which type of wafer is being processed. Another example would be the length of the inspections, as the Fraunhofer EMFT is a research center which also works with the industry, there are some inspections that require 100% checks, so that the 25 wafer box has to be check, and those inspections require a longer time, but research wafers which only saw little processing might not need so long, as for example inspection number 8 shows.

## IV. CONCLUSION

The introduction of IoT power monitoring smart sockets, for which the firmware has been custom made with tested open source software and easy to configure files, provides a clear advantage for machines that are otherwise not able to provide any reporting information.

Giving not only insight into the processes and providing decision material for maintenance or early failure detection systems, but also creating information which can be used in Big-Data environments to help future decision making.

## V. ACKNOWLEDGMENT

The authors would like to thank all personnel from the Fraunhofer EMFT clean room.

## VI. REFERENCES

[1] Marco Schwartz. *Internet of Things with ESP8266*. Packt Publishing Ltd, 2016.

[2] Ravi Kishore Kodali and Kopulwar Shishir Mahesh. "Low cost ambient monitoring using ESP8266". In: *2016 2nd International Conference on Contemporary Computing and Informatics (IC3I)*. IEEE. 2016, pp. 779–782.

[3] Ivan Vaccari, Maurizio Aiello, Federico Pastorino, et al. "Protecting the ESP8266 module from replay attacks". In: *2020 International Conference on Communications, Computing, Cybersecurity, and Informatics (CCCI)*. IEEE. 2020, pp. 1–6.

[4] Daniel Patricko Hutabarat, Santoso Budijono, and Robby Saleh. "Development of home security system using ESP8266 and android smartphone as the monitoring tool". In: *IOP Conference Series: Earth and Environmental Science*. Vol. 195. 1. IOP Publishing. 2018, p. 012065.

[5] Oleksii Barybin, Elina Zaitseva, and Volodymyr Brazhnyi. "Testing the security ESP32 internet of things devices". In: *2019 IEEE International Scientific-Practical Conference Problems of Infocommunications, Science and Technology (PIC S&T)*. IEEE. 2019, pp. 143–146.

# Systematic search for stabilizing dopants in ZrO$_2$ and HfO$_2$ using first-principles calculations

Yosuke Harashima
*Graduate School of Science and Technology*
*Nara Institute of Science and Technology*
Ikoma, Nara 630-0192, Japan
ORCID: 0000-0003-0705-1583

Hiroaki Koga
*Department of Computational Science and Technology*
*Research Organization for Information Science and Technology*
Minato-ku, Tokyo 105-0013, Japan
ORCID: 0000-0003-4513-956X

Zeyuan Ni
*S-Technology Development Center*
*Tokyo Electron Technology Solutions Ltd.*
Nirasaki, Yamanashi 407-0192, Japan
ORCID: 0000-0002-9815-8652

Takehiro Yonehara
*Department of Computational Science and Technology*
*Research Organization for Information Science and Technology*
Minato-ku, Tokyo 105-0013, Japan
ORCID: 0000-0003-4189-9414

Michio Katouda
*Department of Computational Science and Technology*
*Research Organization for Information Science and Technology*
Minato-ku, Tokyo 105-0013, Japan
ORCID: 0000-0001-7980-5386

Akira Notake
*SDC AI Development Department*
*Tokyo Electron Ltd.*
Sapporo, Hokkaido 060-0003, Japan
ORCID: 0000-0001-5761-882X

Hidefumi Matsui
*S-Technology Development Center*
*Tokyo Electron Technology Solutions Ltd.*
Nirasaki, Yamanashi 407-0192, Japan
ORCID: 0000-0003-1140-4361

Tsuyoshi Moriya
*Advanced Data Planning Department*
*Tokyo Electron Ltd.*
Tokyo 107-6325, Japan
ORCID: 0000-0001-8049-2276

Mrinal Kanti Si
*Center for Computational Sciences*
*University of Tsukuba*
Tsukuba, Ibaraki 305-8577, Japan

Ryu Hasunuma
*Faculty of Pure and Applied Science*
*University of Tsukuba*
Tsukuba, Ibaraki 305-8573, Japan

Akira Uedono
*Faculty of Pure and Applied Science*
*University of Tsukuba*
Tsukuba, Ibaraki 305-8573, Japan
ORCID: 0000-0001-6224-4869

Yasuteru Shigeta
*Center for Computational Sciences*
*University of Tsukuba*
Tsukuba, Ibaraki 305-8577, Japan
ORCID: 0000-0002-3219-6007

*Abstract*—**In this study, we performed systematic search for dopants to stabilize the tetragonal structure of HfO$_2$ and ZrO$_2$ by using first-principles calculations. Whole impurity configurations within the supercell, more than 12,000 systems, are examined and the most stable configurations are assumed to be realized. We reveal the contributions of the dopants to the structural stability of tetragonal phase, and that Si or Ge significantly stabilize the tetragonal phase.**

*Keywords*—*first-principles calculations, high-throughput computing, high-k.*

## I. Introduction

High-k compounds play an important role in semiconductor devices. In metal-oxide-semiconductor field effect transistors (MOSFETs), the high-k compounds are used as insulating films for the gate electrodes. The insulating film for the gate electrode must have sufficient thickness to prevent current leakage, and the leakage degrades device performance. On the other hand, a thicker film is less susceptible to gate voltage. Semiconductor devices are required to have both high susceptivity and current leakage suppression. The dielectric constant must be sufficiently high to allow the gate to operate on even thicker films.

ZrO$_2$ and HfO$_2$ are promising candidates for the high-k dielectrics [1]. The dielectric constants of these compounds are higher than those of SiO$_2$, which is currently used in most

devices. ZrO$_2$ and HfO$_2$ are known to exhibit crystal polymorphism. The monoclinic structures are stable at room temperature and have lower dielectric constants than the

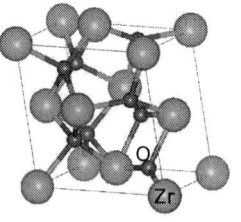 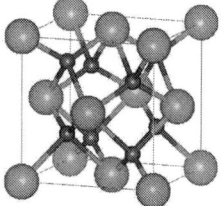

Fig. 1. Crystal structures of monoclinic and tetragonal phases.

tetragonal structures [2]. Their crystal structures are shown in Fig. 1. Stabilization of the tetragonal structure is expected to improve device performance.

One of the strategies of stabilization for a paticular phase is impurity doping. In previous studies [3, 4], Si, C, Ge, Sn, Ti, and Ce have been investigated by means of first-principles calculations. We expand the searching space and systematically perform first-principles calculations for the effects of dopants in ZrO$_2$ and HfO$_2$. We analyze the impurity doping effects on the structural stability of tetragonal phase of ZrO$_2$ and HfO$_2$ considering the configuration of impurities.

## II. METHOD

In this study, we used the projector augmented-wave method for the first-principles calculations (VASP) [5]. The cutoff energy is 520 eV and the k-mesh is $2 \times 2 \times 2$. The exchange-correlation energy functional is approximated by Perdue-Burke-Ernzerhof type generalized gradient approximation (GGA) [6]. The GGA+U method with $U = 4$ eV for Zr, La, Ce, Hf, and 5 eV for Sc, Ti, Y, is used. The effect of U on the electronic structure of $ZrO_2$ was examined in detail in Ref. [7].

Table 1. List of the number of configurations for the doped impurities and corresponding doping concentrations. Examples of compositions are shown. 'Vc' denotes a vacancy.

| Group | Composition | mol% | # of configurations | |
|---|---|---|---|---|
| | | | tetra | monoc. |
| $II^{2+}$ | $CaO:Vc:31ZrO_2$ | 3.125 | 12 | 64 |
| $III^{3+}$ | $Y_2O_3:Vc:30ZrO_2$ | 6.25 | 148 | 992 |
| $IV^{4+}$ | $2SiO_2:30ZrO_2$ | 6.25 | 9 | 23 |
| | $3SiO_2:29ZrO_2$ | 9.375 | 32 | 155 |
| $III^{3+}-V^{5+}$ | $AlPO_4:30ZrO_2$ | 6.25 | 9 | 31 |
| $III^{3+}-VII^-$ | $AlOF:31ZrO_2$ | 3.125 | 12 | 64 |
| $VI^{2-}$ | $ZrS_2:31ZrO_2$ | 3.125 | 22 | 78 |

Table 2. List of doped impurities. Blue letters indicates that doping to $ZrO_2$ only has been calculated.

| Group | Composition | Element |
|---|---|---|
| $II^{2+}$ | $CaO:Vc:31ZrO_2$ | Mg Ca Sr Ba Zn |
| $III^{3+}$ | $Y_2O_3:Vc:30ZrO_2$ | Y La Al |
| $IV^{4+}$ | $2SiO_2:30ZrO_2$ | C Si Ge Sn Ti Zr Hf Ce |
| | $3SiO_2:29ZrO_2$ | Si Ge Sn Ti Zr Hf Ce |
| $III^{3+}-V^{5+}$ | $AlPO_4:30ZrO_2$ | III: Al Ga In Sc Y La<br>V: P As Sb Bi |
| $III^{3+}-VII^-$ | $AlOF:31ZrO_2$ | III: Al Ga<br>VII: F Cl Br I |
| $VI^{2-}$ | $ZrS_2:31ZrO_2$ | S Se |

Supercells consisting of $2 \times 2 \times 2$ conventional cells are used. The lattice constants and inner coordinates are optimized numerically. Effects of doping of groups II, III, IV, V, VI, and VII elements, and their combinations are systematically analyzed. Substitution of both cationic and anionic sites is included. When the substitution results in an excess oxidation number, atoms are removed to adjust it. The impurity configurations are generated by using the program code, supercell [8]. Tables 1 and 2 list the dopants, concentrations, and the number of impurity configurations. The total energies of all possible impurity configurations in the supercell are calculated. The following results are for the most stable configurations in each case.

## III. RESULTS AND DISCUSSION

Energy differences between tetragonal and monoclinic are $\Delta E_{t-m} = E_t - E_m$, where $E_t$ and $E_m$ denote the total energy of the tetragonal and monoclinic structures, respectively. We

performed more than 12,000 calculations and estimated the total energy for each of the substitutions. The calculated

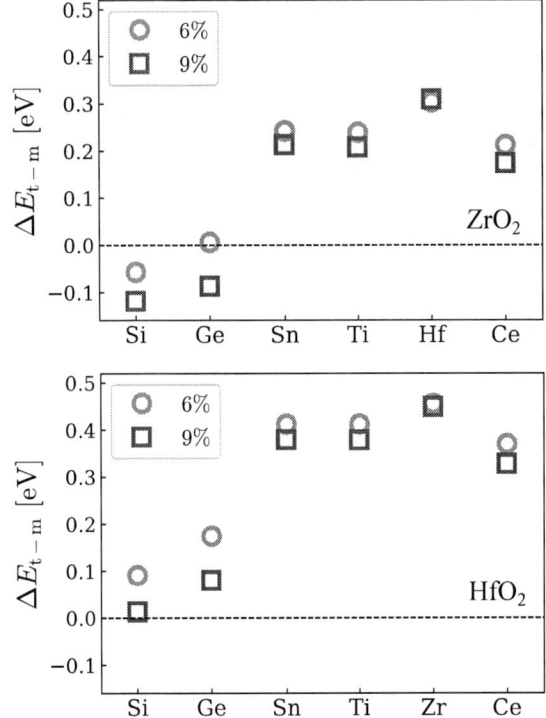

Fig. 2. Energy difference for group IV elements doping. The energy is per conventional unit cell, which contains 12 atomic sites in total.

values for group IV elements are shown in Fig. 2. 6% and 9% Si and 9% Ge doped in $ZrO_2$ stabilize the tetragonal structure. As the energy difference of pristine $ZrO_2$ and $HfO_2$ are respectively 0.293 and 0.466 eV per conventional unit cell, other elements shown in the figure tend to lower the energy differences as well. The energy differences decrease with increasing concentrations, and the changes are nonlinearly correlated with concentrations. The stabilizing trends for Si and Ge are consistent with previous studies [3, 4], while the nonlinearity has not been seen. This is explained by the difference in the treatment of impurity configurations.

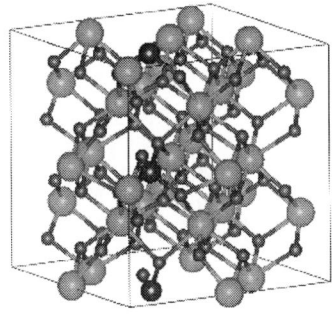

Fig. 3. The most stable configuration for the tetragonal $SiO_2:30ZrO_2$. Si atoms (blue spheres) align linearly.

We found the most stable configuration of Si is that Si atoms align linearly, but are not homogeneously distributed. This inhomogeneous structure is the origin of the nonlinear trend against the impurity concentration, and affects the structural stability. The coordination number of Zr or Hf site changes with the structural transformation (see Fig. 1). This indicates that the interatomic distance is related to structural stability, and unit cell volume is expected to be an important feature in describing the stability.

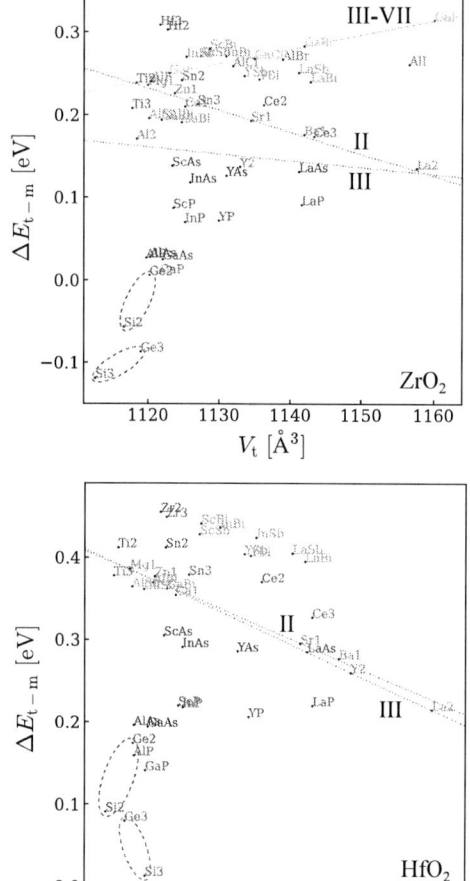

Fig. 4. Energy difference against volume of tetragonal unit cell $V_t$.

Figure 4 shows the energy differences against the unit cell volume. Groups II and III fit lines are shown. It can be seen that these impurities tend to decrease with respect to the cell volume. We also estimated the trend for group III-VII case of ZrO$_2$. In contrast to group II and group III cases, there is an increasing trend. This implies that volume is an important feature in constructing a prediction model of the structural stability, and indicators for elemental groups are required as well. The relationship between volume or atomic radius and

the stabilizing effect of impurity doping has been also discussed in Refs. [9, 10] for permanent magnet compounds, and it is argued that the valency of impurity affects the distance dependence of stability. An index representing the elemental group is likely to be important in constructing a prediction model of structural stability.

## IV. CONCLUSION

The systematic search for stabilizing dopants in HfO$_2$ and ZrO$_2$ is examined. All the impurity configurations, more than 12,000 systems, are taken into account. The most stable configurations are chosen for each dopant and concentration. The energy differences between the tetragonal and monoclinic phases are estimated from first-principles calculations. 6% and 9% Si and 9% Ge doped in ZrO$_2$ stabilize the tetragonal phases. We discuss the nonlinear dependence of the stability on the impurity concentration.

In this study, we focus on the differences between the total energies of tetragonal and monoclinic structures, however, finite temperature effects are neglected. The lattice vibrational free energy and impurity configurational entropy will be discussed in the future work.

## REFERENCES

[1] G. D. Wilk, R. M. Wallace, J. M. Anthony, "High-*k* gate dielectrics: Current status and materials properties considerations," J. Appl. Phys., vol. **89**, pp. 5243, May 2001.

[2] K. Tomida, K. Kita, A. Toriumi, "Dielectric constant enhancement due to Si incorporation into HfO$_2$," Appl. Phys. Lett., vol. **89**, pp. 142902, October 2006.

[3] D. Fischer, A. Kersch, "The effect of dopants on the dielectric constant of HfO$_2$ and ZrO$_2$ from first principles," Appl. Phys. Lett., vol. **92**, pp. 012908, January 2008.

[4] D. Fischer, A. Kersch, "Stabilization of the high-*k* tetragonal phase in HfO$_2$: The influence of dopants and temperature from *ab initio* simulations," J. Appl. Phys., vol. **104**, pp. 084104, October 2008.

[5] G. Kresse, J. Furthmuller, "Efficient iterative schemes for *ab initio* total-energy calculations using a plane-wave basis set," Phys. Rev. B, vol. **54**, pp. 11169, October 1996.

[6] J. P. Perdew, K. Burke, M. Ernzerhof, "Generalized Gradient Approximation Made Simple," Phys. Rev. Lett., vol. 77, pp. 3865, October 1996.

[7] H. Koga, A. Hayashi, Y. Ato, K. Tada, S. Hosokawa, T. Tanaka, M. Okumura, "Facile NO-CO elimination over zirconia-coated Cu(1 1 0) surfaces: Further evidence from DFT+U calculations," Appl. Surf. Sci., vol. **508**, pp. 145252, April 2020, and references therein.

[8] K. Okhotnikov, T. Charpentier, S. Cadars, "Supercell program: a combinatorial structure-generation approach for the local-level modeling of atomic substitutions and partial occupancies in crystals," J. Cheminformatics, vol. **8**, pp. 17, March 2016.

[9] Y. Harashima, T. Fukazawa, H. Kino, T. Miyake, "Effect of *R*-site substitution and the pressure on stability of *R*Fe$_{12}$: A first-principles study," J. Appl. Phys., vol. **124**, pp. 163902, October 2018.

[10] Y. Harashima, T. Fukazawa, T. Miyake, "Cerium as a possible stabilizer of ThMn$_{12}$-type iron-based compounds: A first-principles study," Scripta Mater., vol. **179**, pp. 12–15, January 2020.

# A Novel Approach to Dynamic Line Balance Control and Scheduling with a Digital Twin Production

Hirofumi Tsuchiyama
Inteligent Manufacturing Systems
INFICON
Kawasaki, Japan
hirofumi.tsuchiyama@inficon.com

Holland Smith
Inteligent Manufacturing Systems
INFICON
East Syracuse, NY, USA
holland.smith@inficon.com

*Abstract*— **We have created a line balancing algorithm that uses queueing theory to calculate ideal WIP (Wafer In Process) targets by product and step taking into account the current factory bottlenecks and status, which realize higher equipment utilization, more outs, better WIP bubble/bottleneck management, and reduction of opportunity loss. In this paper, the system architecture and deployment result are described.**
**Keywords—Smart Manufacturing, Line Balance, Factory Scheduling**

## I. INTRODUCTION

With the evolution of Smart Manufacturing and Industry 4.0, the Factory scheduling system integrated Digital twin system is applied widely in world wide. Smart scheduling system can be adjustable to many kinds of factory requirement by GUI and can schedule accurately based on the real time state of equipment and WIP so that the process start/end and lot transfer can be can be by scheduling system. This result 10-15% of factory productivity improvement. [Fig.1]

For further improvement of the factory productivity, Line balance control has been one of the challenges as a fab wide production control. Line balance is broken by bottle neck process, long time Equipment down, wafer scrap, and etc , which creates WIP bubble or WIP shortage. Generating WIP bubble causes WIP shortage where the equipment/process step does not have efficient WIP to run equipment continuously. This results equipment's opportunity loss or SBY(Standby) with no WIP. Decision for Line balance control was normally made by human like a production staff and it is uncertain and totally depends on individual skill in many cases. [Fig.2]

Therefore, we have created an autonomous line balancing algorithm that uses queueing theory to calculate ideal WIP (Wafer In Process) targets by product and step taking into account the current factory bottlenecks and WIP profile across the line.

## II. APPROACH

Fundamental approach to decide line balance is to determine the target WIP for each process step. The following two procedure is to decide the target WIP of the each process step with considering downstream step's WIP level.

Firstly, we need to know the target WIP of each process to consider control Line Balance. Little's law defines the target number of the WIP in process in queueing theory as [1].

Little's Law:
$$(\text{Target WIP})_i = (\text{Through put})_i \times (\text{Cycle time})_i \quad [1]$$
Where;
$(\text{WIP})_i$: WIP in process step i
$(\text{Through put})_i$: Through put in process step i
$(\text{Cycle time})_i$: Cycle time in process step i

The WIP level can be evaluated by Delta WIP.
$$(\text{Delta WIP})_i = (\text{WIP})_i - (\text{Target WIP})_i \quad [2]$$

Hereby, we know if the WIP level is high or low or adequate for this process step.

Secondly, if the next future steps are empty or WIP shortage, we need to consider to push the WIP so that the equipment in these steps have enough WIP to run. To know the target WIP with considering the future step WIP level, the logic is configured to look ahead a certain number of days (default =3) of average cycle time. The WIP delta for each future step can be weighted based on how far a way it is. The sum (or weighted average) of the future WIP delta is considered the Future WIP delta for the given step. The sum of WIP Delta and future step WIP is the efficient WIP Delta.

$$(\text{WIP Delta})\text{eff}_i = (\text{Delta WIP})_i + (\text{future WIP delta})_i \quad [3]$$

Lastly, $(\text{WIP Delta})\text{eff}_i$ is normalized to determine the priority to be processed for this product/process step. This is called "Line Balance Score"

$$(\text{Normalized WIP Delta})\text{eff}_i = (\text{WIP Delta})\text{eff}_i / \text{WIP average} \quad [4]$$

In actual practice in high mix production line, autonomous control of line balance is composed by following 6 steps to compute the control algorithm.

Step1: Set up Common step
High mix factory has hierarchy of the products. Route group ties the same category of the product or technology. Route Group is composed by multiple Route Family which ties the similar process flow. Route family is composed by Route which is a specific process flow. Route has each product for different mask set. It is important to tie the Routes together appropriately and create Commons Step flow to apply the little law's control algorithm effectively.

[Fig 3] shows the hierarchy of the product and example of the process step. For example, Ash1 is common step across routes. Ash2 is the Common step in the same route family. ClnA is same processes within and across the routes. Target WIP calculation depends on how to tie these elemental process family and create Common step flow across the product.
Common step flow can be tied within specific one route. But in this case, WIP amount will not be enough to control the line balance for thousands of process steps. Also, Common step can be defined within or across the whole Route group. In this case, we can get high level of WIP but tying unsimilar product/technology together will result to create wrong common step flow and destruct the control. That way, we recommend to tie the processes within the process family.

Step2: Calculate the target WIP
The target WIP is calculated in proportion to each step's process time within it's route section according to Little's law. [Fig 4]

Step3: Calculate the WIP Deltas
The WIP Deltas is calculated by taking Delta value between the Current WIP and Target WIP based on the equation [2]

Step4: Calculate the Weighted Future WIP Deltas
Logic is configured to look ahead a certain number of days (default = 3) of average cycle time or number of steps (10).

The WIP delta for each future step can be weighted based on how far away it is, or each future step can be treated equally. [Fig.5]

Step5: Calculate the efficient Delta WIP by equation [3]
Efficient WIP Delta is simply subtracted the value of Weighted Future WIP Delta from WIP Delta. This value is the revised WIP Delta value which is considered the future step's WIP level. For example, If the future step WIP is less than target, the value is increased, which means this process step need to push the WIP to next step.

Step 6: Calculate of the Line Balance Score by equation [4]
The line balance score is normalized from -10 to 10 depends on the efficient Delta WIP. Giving bigger value of Line balance Score means to encourage to increase more move at this step.

## III. RESULT

We feed the results of this algorithm into a factory scheduler in our Digital Twin production system to drive performance. The result shows that the line linearity [Fig.6], as measured by the delta between actual WIP and targets, was improved by 17%

Where;

$$\text{Line balance} = 1 - \sum |DELTA\ WIP| / \sum TarGet\ WIP\ [5]$$

## IV. CONCLUSION

We have developed autonomous line balance control algorithm in scientific way and the test result shows 17% Line Balance improvement in the real production. This translates into higher equipment utilization, more outs, less cycle time and reduction of SBY no WIP tool state time, which leads to maximize the investment efficiency of the factory.

## REFERENCES

[1] Hirofumi Tsuchiyama, David Wiselman, "Smart Integrated Metrology Sampling" in International Symposium on Semiconductor Manufacturing 2022, YD-017

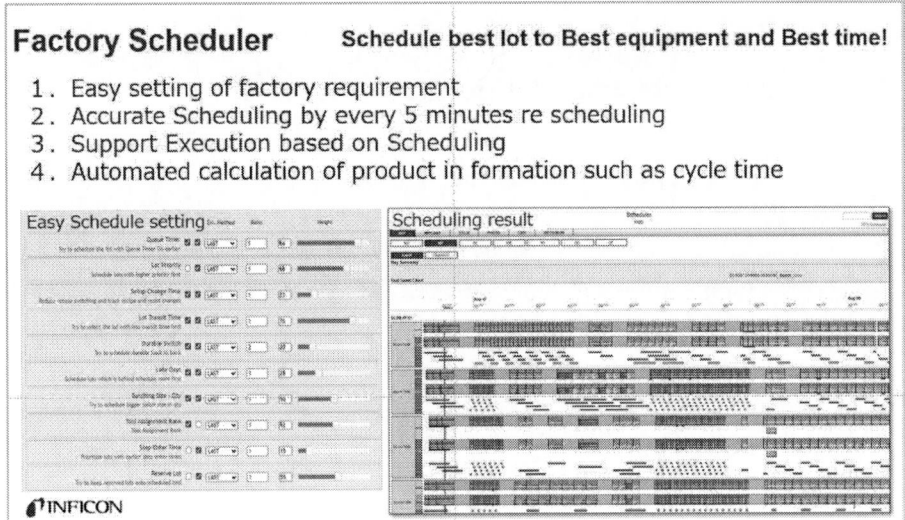

Fig.1 Factory Scheduler

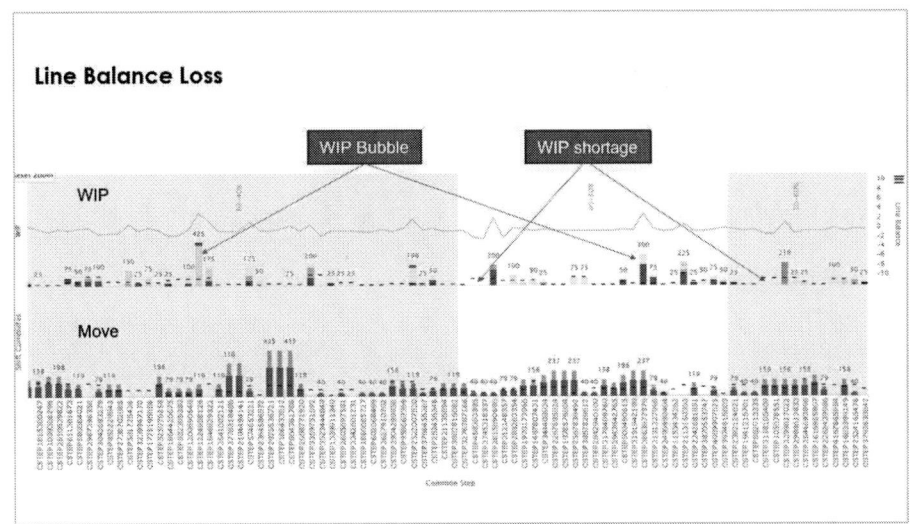

Fig.2 Line Balance Loss by WIP bubble and WIP shortage

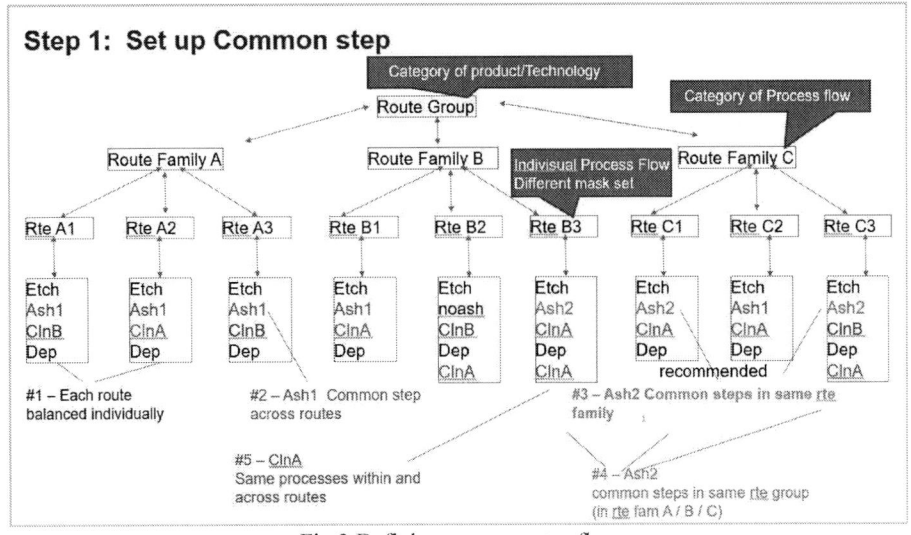

Fig.3 Defining common step flow

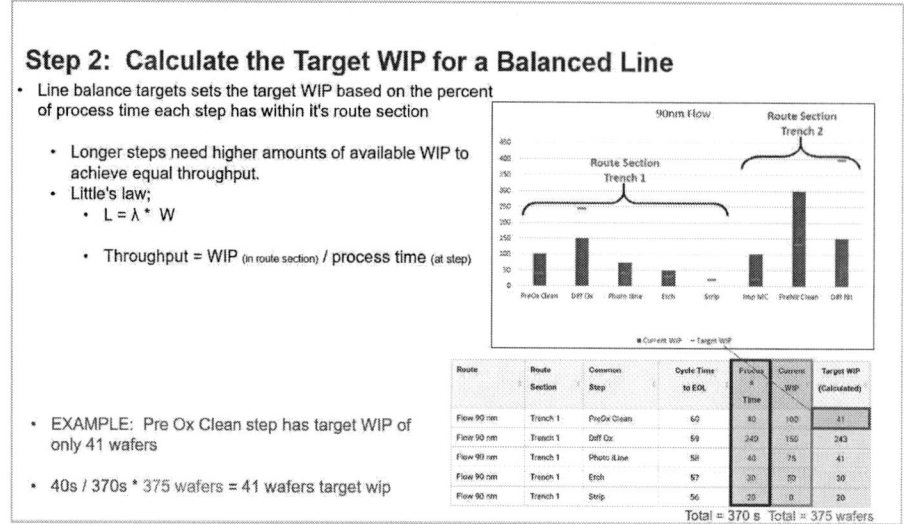

Fig.4 Calculate the Target WIP and Actual WIP

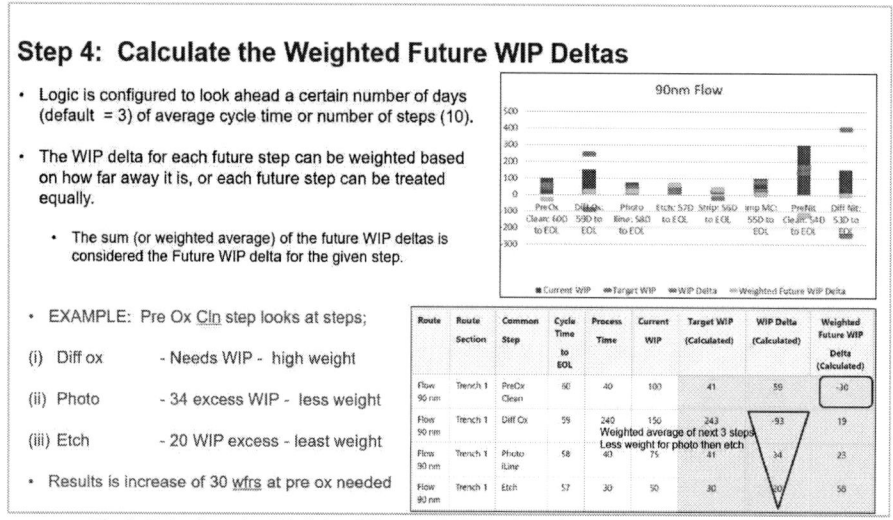

Fig.3 Calculate the Weighted Future WIP Delta with considering future step

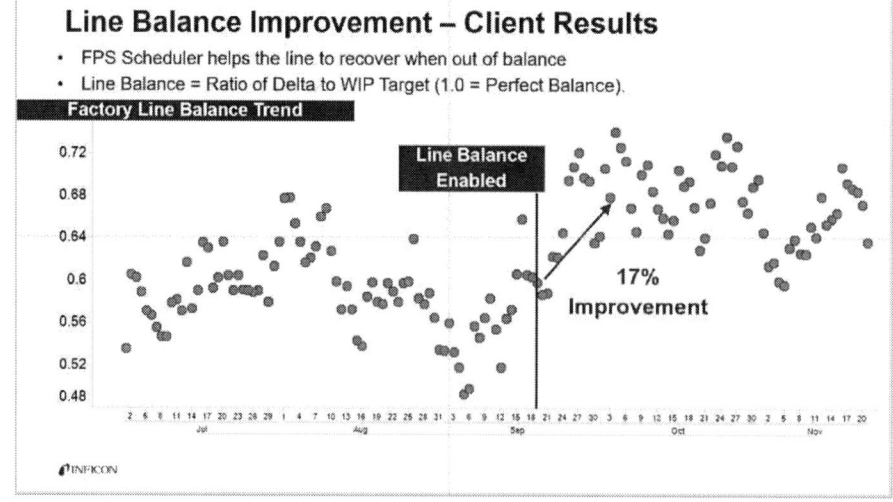

Fig.3 Line Balance control result with using Factory scheduler

# Data-driven Modeling for Production Dynamics

Sumika ARIMA
University of Tsukuba
Institute of Systems and Information
Engineering
Tsukuba, JAPAN
arima@sk.tsukuba.ac.jp

Yu SASAKI
University of Tsukuba
Institute of Systems and Information
Engineering
Tsukuba, JAPAN
s2120507@u.tsukuba.ac.jp

Sho MORIE
University of Tsukuba
Institute of Systems and Information
Engineering
Tsukuba, JAPAN
s2120516@u.tsukuba.ac.jp

Yuto KATAOKA
University of Tsukuba
Institute of Systems and Information
Engineering
Tsukuba, JAPAN
s1911210@u.tsukuba.ac.jp

Chending MAO
University of Tsukuba
Institute of Systems and Information
Engineering
Tsukuba, JAPAN
s2020480@u.tsukuba.ac.jp

Jia LIN
University of Tsukuba
Institute of Systems and Information
Engineering
Tsukuba, JAPAN
s1930142@u.tsukuba.ac.jp

*Abstract*— **This study introduced the application of VAR-LiNGAM, and Backpropagation Neural Network with node2vec for feasible data-driven modeling of dynamics of semiconductor production system in which the scale and complexity increase more and more. Open testbed SMT2020 is used evaluations.**

*Keywords—data-driven modeling, VAR-LiNGAM, node2vec, Backpropagation Neural Network (BPNN), factory dynamics*

## I. INTRODUCTION

This study will introduce two feasible approaches of data-driven modeling for dynamics of the semiconductor production systems (SPS) in which the scale and the complexity increase due to advances of product and process technologies responding to needs of markets and societies.

Now, the problem is that we still do not have a clear answer of methodologies to achieve the target throughput of final product in the huge and complex SPS. On the other hand, global customers of semiconductor industry watch not only good products but also operation efficiency and technology of makers severely to choose corporative suppliers in the future. Operation efficiency is not only for the company itself but also for customer competitiveness. However, how many semiconductor companies can clearly answer to the question?: when and how many products can be supplied better than competitors? That is the serious problem of the semiconductor industry regardless of the leading-edge or the legacy.

In the past, a lot of case studies of individual companies have piled for modeling and optimization of SPS for its management and control [1]. Particularly, the scheduling matter has become very important to handling the automated lot transportation. From the first phase around 2000s, event-driven models and simulation has been the most common and simple direction to control SPS and the multi-cluster tools mainly because the model is easily understandable to respond well to the physical production resource. Queueing theory [2][3], Petri-Net [4][5], mathematical formulations [5][6], and so on are used together with the simulation.

As more simple way, dispatching rule is used for high frequent real-time control, however, it leads a low-level solution of a trade-off of throughput and waiting time (and over Q-time) between process steps in high utilization level

of mass production [13]. For more sophisticated optimization, methods of Operations Research (i.e. mathematical optimization) such as the Linear Programing [6][7], Mixed Integer Programming [5][8], and other multi-criteria methods [9] have been applied to a partial problem. However, those have not covered the scope the whole SPS management. Moreover, there is issues of model maintenance in common.

As additional background, for a long time, one of the most terrible barriers to develop the common approach of modeling and optimization has been super-high security level of semiconductor factories. These days, a research group at the University of Hagen in Germany has released valuable open testbed (open frontend model) SMT2020 [10]. The number of views and citations of [10] has greatly increased in a few years, contributing to subsequent studies and industrial development. This study will utilize SMT2020 for PSP modeling because it covers almost the 10 difficulties of SPS scheduling mentioned in [11].

Still now, we cannot confirm the result of accurate modeling and control of the whole SPS, except of a study of [12] which achieved plan-actual move within 2% error. The result is wonderful, but the way of [12] was a discrete event simulation-based modeling which needs a large manpower, with psychology conscious UX for lean operations, highly-skilled operators, and all perseverance. It is not easy to imitate and diffuse. And there is also the model maintenance issue. Therefore, our challenge here is the common and feasible way of modeling without big efforts and resource.

There are two types of approaches, Hybrid flow-shop modeling and data-driven modeling. HFS is positioned as the most feasible and direct way of a common modeling of a small-middle size problem such as 30 workstations at most if a virtual capacity separation problem for hybrid flows can be solved like re-entrant flows smoothing [13]. It is supportive that HFS has rich literatures [14] as well as some multi-factory scheduling [15]. Furthermore, our previous research presented multi-criteria common optimization algorithms of HFS to balance the trade-offs of delivery, lead

978-1-6654-7134-3/22 $31.00 © 2022 IEEE

time, and workstations efficiency for Make-to-stock and Make-to-order systems [9][16].

On the other hand, the data-driven approach is more challenging but more flexible and scalable for bigger SPS problem of increasing scale and variability. Therefore, this research will mainly focus on the data-driven approach and discuss its possibility and limit. Very few is the previous research of the data-driven modeling of SPS. C.J.Kuo (2008) tried to build a virtual reduced SPS model to learn and judge of conditions to achieve the target throughput of final product by machine learning (ML) and mass production historical data [17]. However, usable ML methods of mid 2000s is too simple to perform high in complex and huge SPS model. In addition, the virtual model assumption was also too simple for sophisticated optimization [13] only with one bottleneck and others (non-bottleneck) for the whole of SPS. Now, some newer ML methods are applied in this study. VAR-LinGAM of Spatio-temporal causal search [18], node2vec [19] which is an extended version of word2vec to consider flow network, will be applied as the automated data-driven SPS modeling.

## II. LITERATURE REVIEW

### A. Data-driven SPS Modeling

Very few is the previous research of the data-driven modeling of SPS. C. J. Kuo (2008) tried to build a virtual SPS model to learn and judge of conditions to achieve the target throughput of final product by machine learning (ML) and mass production historical data [17]. For instance, single layer Backpropagation Neural Network (BPNN) is used to clarify the standard number of Work-in-Process of a bottleneck station, and Classification and regression trees (CART) is used to the condition of detect important features to influence the WIP.

Well-build input features based on the queueing theory introduced BPNN modelling [17] is unignorable technique for the model compaction and accuracy. Concretely, 6 features are used to estimate the daily workstation output (Fig.1). The status of the workstation is well expressed compactly. Average WIPs ($L$), Coefficient of variation of interarrival time of jobs ($C_a$). Coefficient of variation of service time (i.e. process time) ($C_s$), the number of machines ($m$), service rate ($\mu$), and available rate ($v$) of a workstation.

The workstation is set to bottleneck or others. 3 ML model is learnt before dispatching as following steps 1-3 [17]

1. Estimate the standard WIP of a bottleneck workstation to achieve the target throughput of final product by 1-layer BPNN (Fig.1).
2. Condition analysis of the bottleneck workstation (Fig.2).
3. Estimate the standard WIP of a non-bottleneck workstation to achieve the throughput of bottleneck station by 1-layer BPNN (Fig.1).
4. Dispatch WIPs in the bottleneck workstation based on 2 step priori of MIVS (Minimum Inventory Variability Scheduler) and X-factor (Fig.3).

In step 4, the standard WIP levels of feeder workstation (bottleneck) and Bleeder workstation estimated by Step 1 and 3 are used in the WIPs prioritization. Because there are many

WIPs have the same priority by MIVS portfolio prioritization (1-4), and thus the X-factor ($X_{accu}$) prioritize more precise.

Now, there are some issues. About modeling, usable ML methods of mid 2000s is too simple to perform high in complex and huge SPS model. In addition, the bottleneck is given as unique and deterministic in [17] though the bottleneck often dynamically moves due to tool down time and sometimes multiple stations consist of bottleneck in actual mass production (Fig.4)[20]. The influence of that should be considered in. As the process model, for better judgement and action, it is better to be clarified that influence of status of each process workstations to the final product throughput, besides the model simplification.

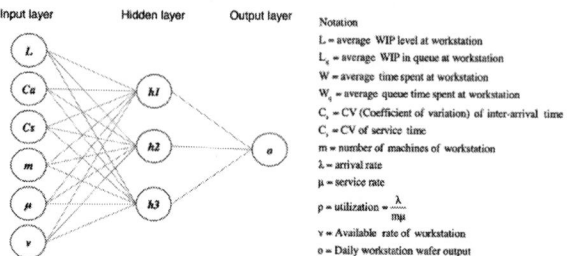

Fig. 1.   6 features used in daily output estimation/forecast of BPNN [17].

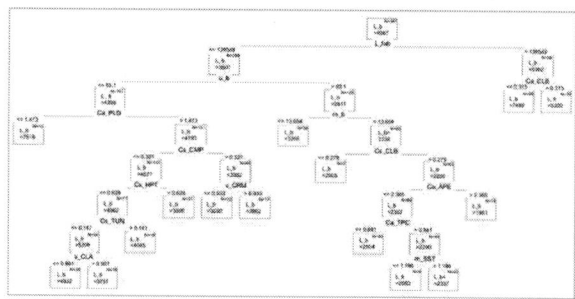

Fig. 2.   Decision Tree analysis (CART) [17].

| | | Bleeder | |
|---|---|---|---|
| | | WIP≧Standard | WIP＜Standard |
| Feeder | WIP≧Standard | Priority① | Priority② |
| | WIP＜Standard | Priority③ | Priority④ |

$$X_{accu} = \frac{Accu.\,cycle\,time}{Accu.\,raw\,process\,time}$$

Fig. 3.   Dispatching by MIVS and X-factor [17].

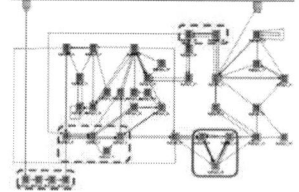

Bottlenecks in parallel production network (e.g. a half of frontend, source: [20]): Node: tool group, Red solid/dashed circle: permanent/temporal BNs with Q-time restriction.

### B. node2vec

Many of real problems have network structures in which many nodes are connected. For example, a production process consists of many steps ordered and workstation to be used, and so its network analysis will give useful information. A graph-based neighborhood sampling method node2vec [19] is extended version of word2vec which is often used natural language analysis. Figure 5 shows node2vec algorithm [19]. Here, node2vec consists of the two steps: LearnFeatures and nove2vecWalk. In LearnFeatures step, A flexible sampling strategy (biased random walk) is proposed to incorporate the characteristics of two major sampling strategies of extremely different for generating a neighborhood set, BFS (Breadth-first Sampling) and DFS (Depth-first Sampling). Furthermore, a quadratic random walk with two parameters p,q are defined to explore the network structure, and a random walk passes through edges $(t, v)$ and reaches source node $v$ are considered.

---
**Algorithm 1** The *node2vec* algorithm.

---
**LearnFeatures** (Graph $G = (V, E, W)$, Dimensions $d$, Walks per node $r$, Walk length $l$, Context size $k$, Return $p$, In-out $q$)
  $\pi = \text{PreprocessModifiedWeights}(G, p, q)$
  $G' = (V, E, \pi)$
  Initialize $walks$ to Empty
  **for** $iter = 1$ **to** $r$ **do**
    **for all** nodes $u \in V$ **do**
      $walk = \text{node2vecWalk}(G', u, l)$
      Append $walk$ to $walks$
  $f = \text{StochasticGradientDescent}(k, d, walks)$
  **return** $f$

**node2vecWalk** (Graph $G' = (V, E, \pi)$, Start node $u$, Length $l$)
  Initialize $walk$ to $[u]$
  **for** $walk\_iter = 1$ **to** $l$ **do**
    $curr = walk[-1]$
    $V_{curr} = \text{GetNeighbors}(curr, G')$
    $s = \text{AliasSample}(V_{curr}, \pi)$
    Append $s$ to $walk$
  **return** $walk$

---

Fig. 4.  Algorithm of node2vec [19].

### C. VAR-LinGAM

VAR-LiNGAM [19] is a model for spatio-temporal causal discovery method for time series data with unknown causal structure. It considers chronological causality by combining two models, VAR (Vector AutoRegression) and LinGAM (Linear Non-Gaussian Acyclic Model). The VAR model is one of the representative models frequently used in time series analysis. The current vectorized time series x(t) is represented by a linear sum of the past time series x(t-τ) (τ=1,···,k). A general formulation is as in Eq.(1)

$$x(t) = \sum_{\tau=1}^{k} B_\tau x(t - \tau) + e(t) \quad (1)$$

Here, $B_\tau$ is a coefficient representing the direct influence from the past time t−τ to the current value x(t) of t. The VAR model includes time series with different indices.: $x(t - \tau)$, τ=1,2···.

LiNGAM is one of the semiparametric approach methods for causal search. A general formulation is as in Eq.(2)

$$X = BX + e \quad (2)$$
$$X = (x_1, \cdots, x_p)^T, e = (e_1, \cdots, e_p)^T$$

where X and $e$ represent the observed and exogenous variables, respectively, and $B$ represents the causal coefficient of the p×p matrix.

Now, A general formulation of VAR-LinGAM is as in Eq.(3)

$$x(t) = \sum_{\tau=0}^{k} B_\tau x(t - \tau) + e(t) \quad (3)$$

Here, the difference from the VAR model is τ, which represents the time difference, starts from 0. $B_0$ is a coefficient representing causality from other variables at the present time, and $B_\tau$ (τ=1, ···, k) is a coefficient representing the influence from $x(t - \tau)$ at the past time. This makes it possible to estimate the value x(t) of the current vector

variable using not only other variables at the present time but also vector variables at the past time, and so it is possible to create a causal graph based on time series.

### III. PROPOSED METHOD

This study will introduce the application of BPNN with node2vec and VAR-LiNGAM for data-driven SPS modeling.

### A. BPNN with node2vec

In this research, we vectorize the connection of each node (workstation, tool group) using node2vec. After that, nodes similarity to the target node is calculated by the cosine similarity from the vectors. Here, for example, the target node is a final step workstation, bottleneck workstation, and so on. Feature values of higher similarity node is added to BPNN learning while MSE (Mean Squared Error) and MAPE (Mean Absolute Percentage Error). We aim to improve prediction accuracy by introducing the features extracted by node2vec.

In addition, single layer and multi-layer BPNN are be compared in accuracy and computational time. Simpler BPNN and node2vec will have advantages to analyze a large network rather than Graph Neural Network [21] and so on.

### B. Utilize VAR-LinGAM

Still there is few of VAR-LinGAM application but one applied it to data generated by a vinyl acetate plant simulator [22]. In this study, the definition of time period $t$ is set by minimum AIC or BIC. In addition, parameter τ will be evaluated by accuracy.

### IV. NUMERICAL EXPERIENCE

As mentioned in Section I, open testbed SMT2020 is used in this study. In this study, the simulated result of SMT2020 is used instead of the historical data of mass production. 2 years (720 days) run is summarized after 180 days warmup.

### A. SMT2020

Fig.6 shows overview of scale and complexity of SMT2020.

| Area | Population | | Availability | | Utilization | |
| --- | --- | --- | --- | --- | --- | --- |
| | Number of tool groups | Number of tools | Average Availability | Proportion of SDT | Average Utilization | Maximum Utilization |
| Dielectric | 10 | 59 | 80.3% | 72.0% | 85.0% | 89.8% |
| Diffusion | 10 | 75 | 92.5% | 89.3% | 84.1% | 90.0% |
| Dry Etch | 23 | 362 | 88.5% | 80.5% | 86.0% | 91.8% |
| Implant | 9 | 35 | 80.4% | 71.2% | 58.2% | 89.9% |
| Lithography | 11 | 203 | 88.4% | 51.7% | 86.6% | 97.6% |
| Planarization | 6 | 34 | 89.3% | 90.1% | 71.2% | 87.2% |
| Thin Films | 11 | 90 | 65.2% | 71.9% | 82.5% | 90.5% |
| Wet Etch | 14 | 117 | 89.0% | 80.3% | 79.4% | 89.7% |
| Defectivity Metrology | 7 | 16 | 96.5% | 90.0% | 52.6% | 84.0% |
| Litho Metrology | 4 | 53 | 96.5% | 90.1% | 90.1% | 91.7% |
| Thin Films Metrology | 2 | 4 | 96.5% | 90.2% | 46.1% | 74.9% |
| Total | 105 | 1043 | | | | |

| Product | RPT (days) | FF | |
| --- | --- | --- | --- |
| | | Regular lots | Hot Lots |
| 1 | 23.76 | 2.10 | 1.43 |
| 2 | 25.46 | 2.06 | 1.42 |
| 3 | 27.18 | 2.07 | 1.41 |
| 4 | 15.96 | 2.02 | 1.45 |
| 5 | 10.99 | 2.17 | 1.54 |
| 6 | 14.16 | 2.12 | 1.48 |
| 7 | 16.85 | 2.08 | 1.43 |
| 8 | 17.47 | 2.17 | 1.46 |
| 9 | 18.27 | 2.14 | 1.46 |
| 10 | 18.98 | 2.11 | 1.46 |

Fig.6 SMT2020 overview [10] Left table: 10 areas of 105 tool groups consisting of 1043 tools with availability information used to 341-643 process steps with reentrant job flow. Right table: 10 product-mix of different RPT and slack ratios of Regular & Hot Lots.

### B. Evaluatio of BPNN with node2vec

In case the modeling of daily throughput of Bottleneck workstation (40), the best degree of node2vec is determined as 3 both in MSE and MAPE, and 3 highest similarity nodes (Table.I) are used in BPNN.. Figure 7 shows the relation of the node 40 and nodes {57,12,24}. Not only node 40 itself but also features of 3 other nodes are influenced to the throughput. As summary, MAPE of BPNN with node2vec (average of 10 replications of random walk) is 9.29% and improved by 8.6% than MAPE 17.9% of original BPNN. For computation, node2vec takes only 0.343 [sec] and BPNN takes 0.147 [sec] for 1 layer and 0.297 [sec] for 2 layers by using a general PC (Apple MacbookPro M1(2020), 8GB memory) Python 3.8.5.

TABLE I.  BOTTLENECK AND HIGH SIMILARITY NODES (DEGREE=3).

| id | station group | station family | #step used | #tool | cascading | #step used/#tool | Bottleneck |
|----|---------------|----------------|-----------|-------|-----------|------------------|-----------|
| 40 | Diffusion | Diffusion_FE_100 | 9 | 3 | NO | 3.00 | permanent |
| 57 | Implant | Implant_91 | 45 | 7 | YES | 6.43 | temporal |
| 12 | Dry_Etch | DE_FE_54 | 5 | 3 | NO | 1.67 | temporal |
| 24 | Def_Met | DefMet_FE_106 | 7 | 1 | NO | 7.00 | temporal |

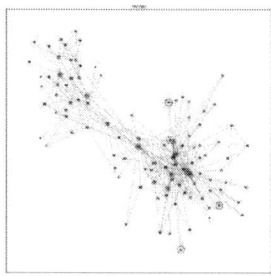 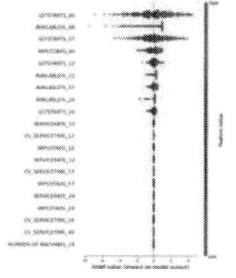

(a) workstation network by node2vec  (b)SHAP value of 6 features of 4 nodes

Fig.7 Numerical results of BPNN with node2vec applied to SMT2020 data.

*C. Evaluation of VAR-LinGAM*

We applied VAR-LinGAM for analyses A1-A4 as follows: Discovery of fab final throughput cause by the failure rate of tool groups (A1), the preventive maintenance rate of tool group (A2), the throughput of tool groups(A3), or the final throughput of tool area (A4) to which the tool group obtained in A3 belongs. As examples of the numerical results (Analysis 1,2), VAR-LiNGAM discovers the influence of the final throughput (e.g. regular lot) at t from the unavailable time of the critical tool groups (some of 105) at t-1 (Fig.8). The condition of discovery here is frequency over 80%, and It is numerical results when t is weekly. As summary, the final throughput is influenced by unavailable time of tool groups of smaller number of tools and/or of larger usage (#step/#tool). That is reasonable. Note that "STNGRP" in Table II means tool area which the same type of tool groups belongs.

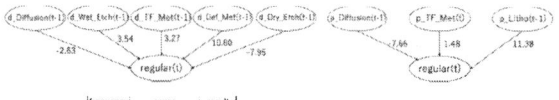

(a) causal network of Analysis1    (b) causal network of Analysis2

Fig.8 Numerical results of VAR-LinGAM applied to SMT2020 data.

*Highlight rows whows Significance level 1% for (a) and 5% for (b)

TABLE II.  INFORMATION OF DETECTED WORKSTATION.

| STNGRP | #step | #tool | #step /#tool |
|--------|-------|-------|--------------|
| Dry_Etch | 527 | 312 | 1.69 |
| Def_Met | 268 | 15 | 17.87 |
| Dielectric | 214 | 54 | 3.96 |
| Diffusion | 145 | 73 | 1.99 |
| Implant | 223 | 36 | 6.19 |
| Litho | 518 | 170 | 3.05 |
| Litho_Met | 575 | 45 | 12.78 |
| Planar | 91 | 26 | 3.5 |
| TF | 218 | 79 | 2.75 |
| TF_Met | 126 | 4 | 31.25 |
| Wet_Etch | 1067 | 99 | 10.78 |

CONCLUSION

This study introduced the application of VAR-LiNGAM and BPNN with node2vec for data-driven modeling of semiconductor production system. As the numerical results,

node2vec improves the accuracy of the model. VAR-LiNGAM detected the down time influence of the permanent/temporal bottleneck process to the final throughput.

REFERENCES

[1] L. Mönch, et.al.,"A survey of problems, solution techniques, and future challenges in scheduling semiconductor manufacturing operations," J. Sched., 14:583-599, 2011.

[2] K. SAITO, S. ARIMA, et.al., "Application of Resource Planning System for VLSI Assembly Facility," Proceeding of ISSM1999, pp. 345-348, IEEE, 1999.

[3] J. Asmundsson, et.al., "Tractable nonlinear production planning models for semiconductor wafer fabrication facilities," IEEE TSM, 19 (1), pp.95-111, 2006.

[4] S. ARIMA et.al., "Petri Net Simulator of Manufacturing Systems Considering Multiple Machines, Operators, and Maintenance for a Product-Mix," Proceeding of ISSM2000, IEEE, pp.1-6, 2000.

[5] C. Jung, et.al., "An Efficient Mixed Integer Programming Model Based on Timed Petri Nets for Diverse Complex Cluster Tool Scheduling Problems" IEEE TSM, 25(2), IEEE, pp186.-199, 2012.

[6] S. ARIMA et.al., "Operator Allocation Planning for a Product-Mix VLSI Assembly Facility," IEICE Transactions on Electronics. Vol.E84-C, No.6, IEICE, pp.832-840, June 2001.

[7] J.-Y. Bang, et.al., "Hierarchical Production Planning for Semiconductor Wafer Fabrication Based on Linear Programming and Discrete-Event Simulation," IEEE Trans. Automation Science and Engineering, 7(2), pp.326-336, 2010.

[8] A. Klemmt, et.al., "A. Scheduling jobs with time constraints between consecutive process steps in semiconductor manufacturing," Proceedings of the 2012 Winter Simulation Conference, No.194/pp.1-10, 2012.

[9] J. Lin, et.al., "Optimization of Multi-objective Function of n-step Hybrid Flow shop Scheduling," Proceedings of ISSM2020, IEEE, WC-51/pp.1-4, 2020.

[10] D. Kopp, et.al., "MT2020-A Semiconductor Manufacturing Testbed," IEEE Transaction on Semiconductor manufacturing, 33(4), pp.522-531, 2020.

[11] M. Ham et.al., "Dynamic photo stepper dispatching/scheduling in wafer fabrication," in Proc.ISSM2005, IEEE, pp. 75–79, 2005.

[12] Y. Ishii, et.al., "Highly Accurate Management in Dynamically Changing Fab," IEEE TSM, 22(4), pp.482‾490, 2009.

[13] S. ARIMA, et. al., "OPTIMIZATION OF RE-ENTRANT HYBRID FLOWS WITH MULTIPLE QUEUE TIME CONSTRAINTS IN BATCH PROCESSES OF SEMICONDUCTOR MANUFACTURING," IEEE TSM, 28(4), IEEE, pp528.-544, 2015.

[14] R.Ruiz, et.al., "The hybrid flow shop scheduling problem." European Journal of Operational Research. 2010.

[15] J. Lohmer, et. al., "Production planning and scheduling in multi-factory production networks: a systematic literature review," International Journal of Production Research, Volume 59, 2028-2054, 2021.

[16] J. LIN, et.al., 2022, "Multi-criteria optimization of n-step hybrid flow-shop scheduling," Proceedings of the 2022 International Symposium on Flexible Automation/ISFA, pp.1-8,2022.

[17] C.-J. Kuo, C.-M. Liu, and C.-Y. Chi, "Standard WIP determination and WIP balance control with time constraints in semiconductor wafer fabrication," J. Quality, vol. 15, no. 6, pp. 409–423, 2008

[18] HYVÄRINEN, Aapo, et al. "Estimation of a structural vector autoregression model using non-gaussianity," Journal of Machine Learning Research, 2010, 11.5.

[19] A. Grover, et.al., "node2vec: Scalable feature learning for networks," Proceedings of the 22nd ACM SIGKDD international conference on Knowledge discovery and data mining, pp. 855-864, 2016.

[20] T. SAITO, Y. Sato, S. Arima, "Novel Dispatching Rule of Phase-mix and Product-mix Wafer Factory for Cost-free Productivity Improvement on Inventory Reduction and Short Cycle Time," Proceedings of ISSM2007, IEEE, pp. 261-264, Oct.2007.

[21] J. Zhou, et.al., "Graph neural networks: A review of methods and applications," AI Open, Vol.1, pp.57-81, 2020.

K. Koyama, et. al., "Evaluation of Time Series Causal Discovery Method Using Plant Simulator," IEICE Technical Report PRMU2021-34, IEICE, pp.57-60, 2021-12.

# Recent Status of EUV Lithography, What is the Stochastic Issues ?

Toru Fujimori
*Electronic Materials Research Laboratories*
*FUJIFILM Corporation*
4000 Kawashiri, Haibara-Gun,
Shizuoka 421-0396, Japan
toru.fujimori@fujifilm.com

*Abstract*—The performance of EUV resist materials are still not enough for the expected HVM requirement, even by using the latest qualifying materials. One of the critical issues is the stochastic issues, which will be become defectivity, like nano-bridge or nano-pinching. The analyzing summary and the resulting classification the stochastic issues in lithography was described in this paper.

*Keywords—EUV, resist materials, stochastic, defect*

## I. INTRODUCTION

In 2019, finally, extreme ultraviolet (EUV) lithography has been applied to high volume manufacturing (HVM) for preparing advanced semiconductor devices. That was very important year for EUV enthusiasts and semiconductor industry. Because it takes for a long time, more than 30 years, to study EUV lithography for realizing HVM. However, the performance of EUV resist materials are still not enough for the expected HVM requirement, even by using the latest qualifying EUV resist materials. One of the critical issues is the stochastic issues, which will be become 'defectivity', like nano-bridge or nano-pinching [1].

## II. EXPERIMENTAL

### A. Materials

Each polymer materials were synthesized according to the conventional polymerization methods [2]. A series of photoresist were prepared by mixing organic solvents, the protected co-polymers, optimized amount of PAG and organic quencher or the related compounds. The resulting solution was filtered using 0.03 or 0.02 μm polyethylene filter or other filter before lithographic performance evaluation.

### B. The evaluation conditions of lithographic performance

The photoresist solution was spin-coated on a silicon wafer that was treated with an organic under layer (UL) or spin on glass (SOG) stacked with spin on carbon (SOC), and the resulting film was pre-baked at appropriate temperature for 60 sec to give a specific film thickness for each patterning features. The wafer was exposed with EUV light (13.5 nm) from an ASML NXE:3300 / 3400 series with 0.33 NA or Small Field Exposure Tools (SFET) with 0.33 NA. After exposure, the wafer was baked (PEB) at moderate temperature for 60 sec. and developed with 2.38% TMAH aq. or organic solvent developer for moderate time. CD-SEM measurements were performed with a Hitachi CG4100 or CG5000.

## III. RESULTS AND DISCUSSION

### A. Challenging of EUV resist materials

The pattern shrinkage has been driven by shorter exposure wavelength. The wavelength of EUV source is 14 times shorter than ArF source, however, the photon numbers are also 14 times fewer [3]. This is the most critical issues of EUV lithography realization. The resolution of line and space patterning observed 13 nm with 42 mJ/cm$^2$, looks not bad. However, stochastic error, which are nano-bridge and/or nano-pinching were observed, which becomes an obstacle for high volume manufacturing (Fig. 1).

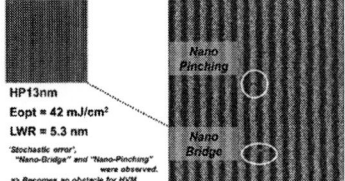

Fig. 1. The stochastic error in 13 nm line and space patterning.

### B. Analyzing of the stochastic issues

In the past, speaking of the stochastic issue of EUV lithography was basically considered from low photon number from EUV light source, which means 'photon shot noise', called 'Photon stochastic'. It was still critical concerning point of the stochastic issue, even with recent progress on source power improvement. However, the stochastic issue is not only from them but also from EUV materials and processes, called 'Chemical stochastic'. The 'Chemical stochastic' means caused from resist materials and processes for lithography, materials uniformity in the film, catch the photon efficiency, reactive uniformity in the film, and dissolving behavior with the developer (Fig. 2) [4].

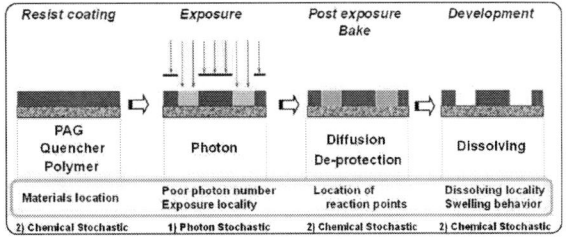

Fig. 2. The classification of stiochastic issues in lithography

Each step must be reduced each stochastic issue to improve the quality of lithographic performance.

## CONCLUSION

The analyzing summary of the stochastic issues in EUV lithography was reported. According to the article, 2 (two) major stochastic issues, which are 'Photon stochastic' and 'Chemical stochastic', were observed in the lithography steps. Each step must be reduced each stochastic issues to improve the quality of lithographic performance.

## REFERENCES

[1] P. De Bisschop, J. Van de Kerkhove, J. Mailfert, A. Vaglio Pret, J. Biafore, *Proc. SPIE*, **9048** (2014) 904809.

[2] G. Odian, "Principles of polymerization" *John Wiley* (2004).

[3] John J. Biafore, Mark D. Smith, Chris A. Mack, James W. Thackeray, Roel Gronheid, Stewart A. Robertson, Trey Graves, David Blankenship, *Prpc., SPIE*, **7273** (2009) 727343.

[4] T. Fujimori, *International Microprocesses and Nanotechnology Conference, the 34th, MNC* (2021).

# Technology Trends and Characteristics of Patent Information Disclosure in Advanced Semiconductor Photoresist

Kosuke Watahiki
Graduate School of Sciences and
Technology for Innovation
Yamaguchi University
Yamaguchi, Japan
c008wcw@yamaguchi-u.ac.jp

Yoshihiro Midoh
Graduate School of Information
Science and Technology
Osaka University
Osaka, Japan
midoh@ist.osaka-u.ac.jp

Kazuya Okamoto
Graduate School of Innovation and
Technology Management
Yamaguchi University
Yamaguchi, Japan
kokamoto@yamaguchi-u.ac.jp

*Abstract*—**This paper focuses on advanced semiconductor photoresists and analyzes the patent data from the following four perspectives: relationship between semiconductor trends and resists; profit gains of the photoresist industry; advanced extreme ultraviolet lithography trends and sustainable development goals (SDGs); and patent quality and filing trends for photoresists. As a result, we obtain the trend of semiconductor photoresists accurately reflecting miniaturization and its relation to SDGs. In addition, a guideline for patent quality and information disclosure characteristics is obtained. Photoresist is a crucial component of the semiconductor industry in miniaturization, and patents are an effective application tool for the trend investigation.**

*Keywords*— *semiconductor, photoresist, patent, correlation coefficient, natural language processing (NLP)*

## I. INTRODUCTION

In recent years, the COVID-19 pandemic has caused global economic stagnation. However, the spike in the Internet of Things (IoT) has led to a worldwide semiconductor shortage. To accelerate the semiconductor industry's growth, transistor miniaturization based on Moore's law is crucial [1]. Photolithography is one of the most dominant processes in semiconductor miniaturization (geometric scaling). Photoresist is a fundamental material for this process, for which Japanese manufacturers have an industrial competitive edge. One of the key factors in maintaining this advantage is the R&D capability of commercial products, and companies disclose crucial technological information in patents as explicit knowledge [2]. In this study, we investigate an inherent, intrinsic basis of the competitive edge by patent investigation compared with other information based on technology and trend analysis of advanced photoresists.

Photolithography utilizing photon (boson in quantum theory) is a mainstream technique to form fine patterns on photoresists. The pattern resolution follows the Rayleigh formula, as shown in Equation 1.

$$Pattern\ resolution\ =\ k_1 \cdot \frac{\lambda}{NA} \qquad (1)$$

where k1, λ, and NA denote the process coefficient, exposure light source wavelength, and numerical aperture equal to n·sinθ (n is the ambient refractive index, and θ is the maximal half-angle of the imaging lens), respectively. In addition to the reduction of k1 due to resolution enhancement technologies, such as new process developments and evolution of pattern design technologies, λ has been shortened from the classical g-line (436 nm) and i-line (365 nm) to Krypton Fluoride (KrF, 248 nm) and Argon Fluoride (ArF, 193 nm) excimer lasers. With regard to NA, tremendous efforts have been made in exposure optics to increase NA to approximately 0.92 and further to 1.35 (n=1.44 under a pure-water environment) through immersion technology. Since 2020, extreme ultraviolet lithography (EUVL, 13.5 nm) has been applied in mass production [3,4]. Because photoreaction is used in the patterning process, technological changes in the material design of photoresists have occurred for each exposure light source wavelength.

## II. EXPERIMENTAL

We used "CyberPatent Desk" as the patent database for the search period from 1990 to 2022, and extracted the patent metadata and specific information such as the title, assignee, abstract, application date, and classification codes such as international patent classification (IPC) from the documents. The IPC system already exists for patent classification and has limitations in terms of novel technologies and detailed classification. In contrast, Japanese patent identifiers consist of the file index (FI) and file-forming term (F-term), which contain a new domain that allows a broader, more comprehensive range of technological classifications compared to the IPC, especially with regard to the F-term, which includes over 300,000 detailed taxonomies. Moreover, the F-term describes the technology in more detail, and plays a critical role in patent research. Several studies have evaluated product innovation patterns using the F-term classification system [2]. Regarding FI, each classification consists of a symbol, such as "A23K20/184" (Accessory food factors for animal feeding-stuffs—hormones). The first letter is the "section symbol" and consists of a letter such as "A" (Human necessities), "G" (Physics), and "Y" (New technologies). Next, is the "class symbol," consisting of a two-digit number ("A23") which represents "Foods or foodstuffs: their treatment, not covered by other classes." Finally, the "subclass" ("A23K") represents "Fodder," followed by a one- to three-digit "group number," an oblique stroke, and a number consisting of at least two digits representing a "main group," "subgroup," or "lower subgroup."

We examined the G03F7/038_601, G03F7/039_601, and G03F7/004 FI codes, which represent a type of Japanese patent. We surveyed a total of 10,969 domestic patents and analyzed their technologies and trends by separating each light source using an F-term code, as shown in TABLE I.

978-1-6654-7134-3/22 $31.00 © 2022 IEEE

TABLE I. F-term indicating exposure light source.

| Light source | F-term | Theme | Detail |
|---|---|---|---|
| EUV | 2H125CB12 | Materials for photolithography | · · Extreme ultraviolet (EUV) |
| | 2H196EA07 | Processing of photosensitive resins and photoresists | · · X-ray, EUV |
| | 2H197CA10 | Exposure and alignment of photoresist sensitive materials | · · X-ray, EUV |
| | 2H197GA01 | | · Exposure wavelength is EUV |
| | 2H225CB14 | Materials for photolithography | · · Extreme ultraviolet (EUV) |
| ArF | 2H125CB09 | Materials for photolithography | · · ArF excimer(193nm) |
| | 2H125FA03 | | · Immersion exposure |
| | 2H197AA12 | Processing of photosensitive resins and photoresists | · · Immersion exposure |
| | 2H225CB09 | Materials for photolithography | · · ArF excimer(193nm) |
| | 2H225CB10 | | · · · Immersion exposure |
| KrF | 2H125CB08 | Materials for photolithography | · · KrF excimer(248nm) |
| | 2H225CB08 | Materials for photolithography | · · KrF excimer(248nm) |

In addition, natural language processing (NLP) was used to identify information disclosure characteristics from the examination status of patents and contents of abstracts. NLP is a machine learning technology that uses computers to interpret, manipulate, and understand text or speech in the human language [5]. In this case, bidirectional encoder representations from transformers (BERT) were used for NLP. BERT is designed to pretrain deep bidirectional representations from unlabeled text by jointly conditioning both the left and right contexts in all layers. Consequently, the pre-trained BERT model can be fine-tuned with only one additional output layer to create state-of-the-art models for a wide range of tasks, such as question answering and language inference, without substantial task-specific architecture modifications [6].

## III. RESULTS AND DISCUSSION

### A. Relationship between semiconductor trends and resists

Fig. 1 shows the cumulative number of patents since 1990 by light source and the result of fitting the logistic regression curve using Equation (2). The larger the number, the smaller the growth factor c along with shorter wavelengths, which indicates higher R&D expenses and longer product life cycles.

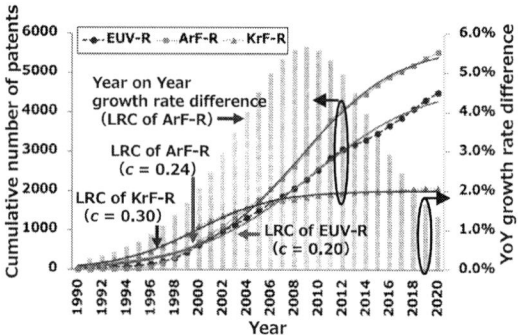

Fig. 1. Cumulative number of patents since 1990 by the light sources and the result of fitting the logistic regression curve (LRC).

$$y = a/(1 + be^{-cx}) \qquad (2)$$

The regression curve of EUV resists (EUV-R) shows an inflection point in 2012, and a growth trend is observed in contrast to the maturation of KrF resist (KrF-R) and the slowdown of ArF resist (ArF-R), confirming an acceleration toward the practical application of EUVL in 2020 (Fig. 2). In reality, EUVL has been deployed for practical mass production [4].

### B. Profit perspective of the resist industry

TABLE II shows the correlation coefficients (R) of the moving averages of sales revenue since 2008 and number of patents for each period from the previous year for ArF-R. A shift in the correlation from positive to negative was confirmed. However, for the period from 2009 to 2011, the correlation trend is considered to have been broken owing to the 2008 financial crisis. This indicates a slowdown in the growth rate of ArF-R patents corresponding to R&D investment since 2008, as shown in Fig. 1, and an increase in sales revenue. Thus, the return on investment (ROI) tends to be established over a long time period.

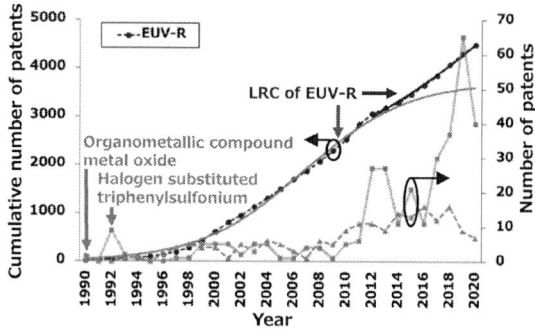

Fig. 2. Analysis of EUV-R patents in F-terms.

TABLE II. Correlation coefficients (R) of moving averages of ArF-R sales revenue since 2008 and the number of patents separated by each period from the previous year.

| Term | 1 year | 2 years | 3 years | 4 years | 5 years |
|---|---|---|---|---|---|
| 2008 | 0.99 | 0.83 | 0.99 | 0.92 | 0.92 |
| 2009 | 0.54 | 0.50 | 0.58 | 0.64 | 0.58 |
| 2010 | 0.73 | 0.71 | 0.75 | 0.78 | 0.75 |
| 2011 | 0.85 | 0.84 | 0.85 | 0.87 | 0.86 |
| 2012 | 0.90 | 0.90 | 0.90 | 0.91 | 0.90 |
| 2013 | 0.80 | 0.91 | 0.91 | 0.91 | 0.90 |
| 2014 | 0.28 | 0.53 | 0.78 | 0.87 | 0.89 |
| 2015 | 0.07 | 0.23 | 0.51 | 0.76 | 0.85 |
| 2016 | -0.03 | 0.05 | 0.23 | 0.47 | 0.70 |
| 2017 | -0.18 | -0.09 | 0.01 | 0.16 | 0.37 |
| 2018 | -0.34 | -0.27 | -0.17 | -0.08 | 0.05 |
| 2019 | -0.43 | -0.39 | -0.31 | -0.24 | -0.16 |
| 2020 | -0.47 | -0.47 | -0.44 | -0.39 | -0.33 |
| 2021 | -0.53 | -0.52 | -0.51 | -0.50 | -0.47 |

## C. Advanced EUVL trends and SDGs

Fig. 2 shows that the patent classification of halogen-substituted triphenylsulfonium has been on the rise since 2012, and those classified as "metal resists" have been on the rise since 2008, indicating that improving the sensitivity and line edge roughness (LER)/line width roughness (LWR) have become an issue, including in SDGs and ESG. In a major semiconductor lithography conference, SPIE, the risk of equipment contamination due to non-cleaning, along with the EUV light absorption efficiency of halogens with respect to EUV resists was reported in 2014 [7], which may be one of the reasons why the patent classification for halogens dropped once from 2014 to 2016. The photoresist patent trends clearly demonstrate the nature of semiconductor miniaturization.

## D. Patent quality and filing trends for resists

To understand the characteristics of patent information disclosure regarding photoresists, we first classified patents by examination status into four categories (Registered: Class-0, Not Requested: Class-3, etc.), as shown in Fig. 3. In this study, the number of forward citations (i.e., number of times the patent has been cited) for each patent was assumed to be the quality of the patent [8,9]. Fig. 4 shows the visualized distribution of the number of forward citations for each category, which corresponds to the quantification of the patent quality. This shows the importance of registered patents based on the average number of citations each year.

To take a closer look at the examination status of patents, we divided unexamined patents into two categories: dormant application and not requested patents; and registered patents into three categories: expiration of term, and abandonment of patent rights, and decision on opposition, resulting in eight categories (Fig. 5). Furthermore, to verify whether there were differences in disclosure characteristics among the eight categories, document classification by NLP (BERT) was applied to the specification abstracts of the patents in the eight categories. As a result, in the flow shown in Fig. 6, a valid text classification model was generated by training (accuracy in the confusion matrix: 0.957); further, the 768-dimensional BERT text vector was reduced to a two-dimensional vector using t-distributed stochastic neighbor embedding (t-SNE) and clustered, as shown in Fig. 7. It was found that clustering by examination status classification was possible, confirming that essential patents have a certain pattern in their descriptions. Using text analysis to create a co-occurrence network graph of the clusters in Fig. 7, keywords related to halogens, such as iodine and bromine; inorganic materials, such as tin, titanium, and ligands; and impurities, such as filtration, which did not appear in registered patents and rejected patents, were extracted as feature terms in the cluster of unexamined patents. It was suggested that one of the factors contributing to the categorization may have involved new keywords and sentences. In addition, the categories are not completely classified by the clusters; different categories exist in each cluster. Using text analysis on a group of patents in an actual category and a group of patents whose predictions differed from the actual category to create a co-occurrence diagram, the feature words matched more than 60%, suggesting that context analysis by BERT is important.

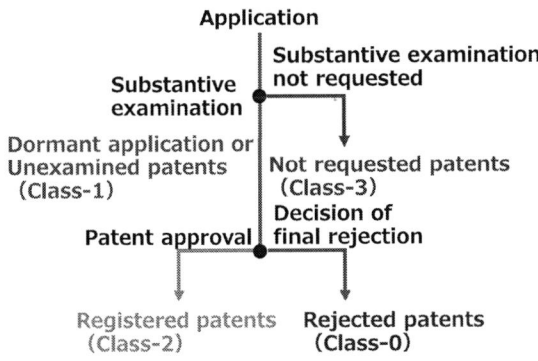

Fig. 3. Patent classification by examination status into four categories.

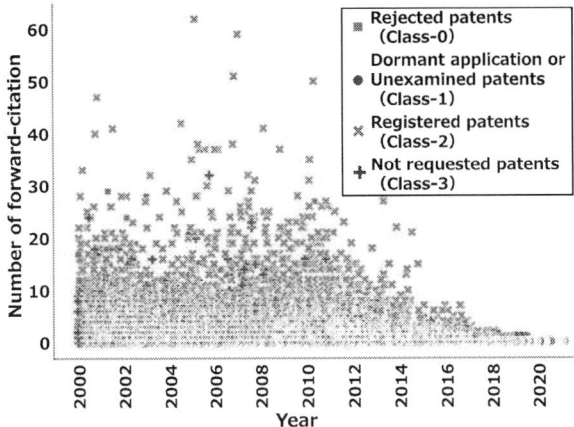

Fig. 4. Distribution of the number of each patent, classified by patent examination status.

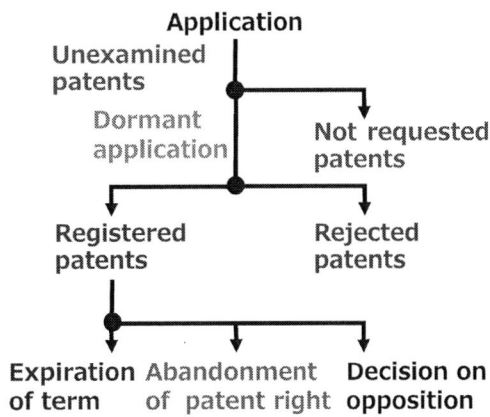

Fig. 5. Classified patents by examination status into eight categories.

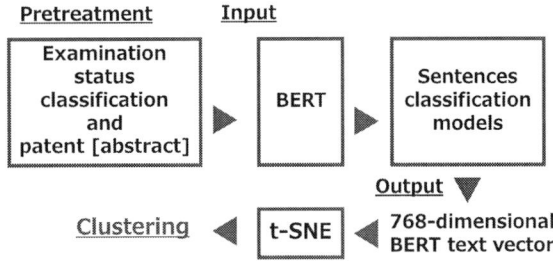

Fig. 6. Flow chart of patent clustering by NLP/BERT.

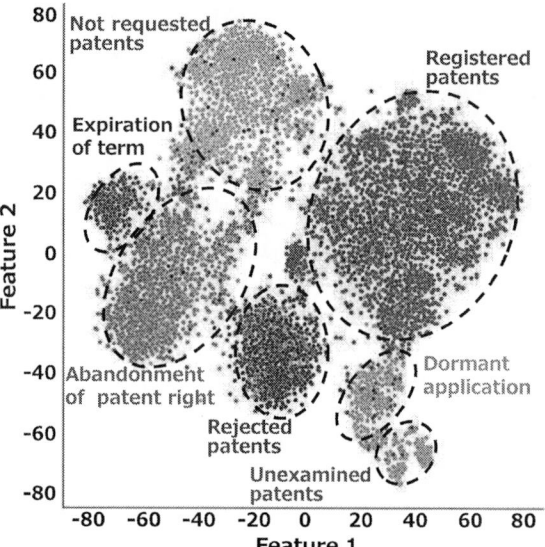

Fig. 7. Patent classification by examination status by NLP/BERT.

## IV. CONCLUSIONS

In this study, we used patent information to understand technology trends and characteristics in the advanced semiconductor industry. By focusing on photoresists, which are key enablers for miniaturization, we revealed a trend that matches the market, accompanying the transitor scaling of semiconductors and reflecting an SDGs/ESG background. By visualizing the quality of registered patents and the examination status, we obtained a guideline for the characteristics of information disclosure. In the future, we aim to clarify the relationship between the scope of patent rights and information disclosure by obtaining information disclosure characteristics from the patent specification in more detail.

## REFERENCES

[1] G. E. Moore, "Cramming more components onto integrated circuits," Electronics, vol. 38, no. 8, pp. 82-85, 1965.

[2] Y. Kanechika and K. Okamoto, "Advanced Thermal Materials and Systems: Technology and Trend Analysis for the Future," International Conference on Electronics Packaging (ICEP), pp. 51-52, 2021.

[3] K. Okamoto and, R. Sato, "Trends and Future Prospects on System-Design Integration based on Hyper-Miniaturization, 3D-Integration and Advanced Packaging," Journal of Japan Institute of Electronics Packaging, vol. 21, no. 6, pp. 531-541, 2018. (in Japanese)

[4] Alberto Pirati, Jan van Schoot, Kars Troost, Rob van Ballegoij, Peter Krabbendam, Judon Stoeldraijer, Erik Loopstra, Jos Benschop, Jo Finders, Hans Meiling, Eelco van Setten, Niclas Mika, Jeannot Dredonx, Uwe Stamm, Bernhard Kneer, Bernd Thuering, Winfried Kaiser, Tilmann Heil, Sascha Migura, "The future of EUV lithography: enabling Moore's law in the next decade", Proc. SPIE 10143, Extreme Ultraviolet (EUV) Lithography VIII, 101430G, March 2017.

[5] G. G. Chowdhury, "Natural Language Processing," Annual Review of Information Science and Technology, vol. 37, p. 51-89, 2003.

[6] J. Devlin, M.-W. Chang, K. Lee, and K. Toutanova, "BERT: Pre-training of Deep Bidirectional Transformers for Language Understanding," arXiv preprint arXiv:1810.04805, 2018.

[7] E. Shiobara, T. Takahashi, N. Sugie, Y. Kikuchi, I. Takagi, K. Katayama, H. Tanaka, S. Inoue, T. Watanabe, T. Harada and H. Kinoshita, "Contribution of EUV resist components to the non-cleanable contaminations," Proc. SPIE 9048, Extreme Ultraviolet (EUV) Lithography V, 904819, April 2014.

[8] M. Trajtenberg, "A penny for your quotes - patent citations and the value of innovations," Rand Journal of Economics, vol. 21, no. 1, pp. 172–187, 1990.

[9] B. H. J. A. &. T. M. Hall, "Market value and patent citations," Rand Journal of Economics, vol. 36, no. 1, pp. 16–38, 2005.

# Practical load impedance monitoring system externally installed in plasma etching equipment

Yuji Kasashima
*Sensing System Research Center*
*National Institute of Advanced*
*Industrial Science and Technology*
*(AIST)*
Tosu, Japan
kasashima-yuji@aist.go.jp

Shinji Kuniie
*Saitama R&D Center*
*Advantest Corporation*
Kazo, Japan
shinji.kuniie@advantest.com

Toshiyuki Sayama
*Saitama R&D Center*
*Advantest Corporation*
Kazo, Japan
toshiyuki.sayama@advantest.com

Tatsuo Tabaru
*Sensing System Research Center*
*National Institute of Advanced*
*Industrial Science and Technology*
*(AIST)*
Tosu, Japan
t-tabaru@aist.go.jp

**We have developed the load impedance monitoring method for plasma etching process, which can be externally installed in mass-production equipment. The monitoring system can detect micro-arc discharge and monitor the condition of the film deposited on inner wall of process chamber. In this study, we have upgraded the monitoring system to enhance precision, practicality, and versatility. The system can be used as an effective method for real-time and noninvasive monitoring of plasma etching process.**

*Keywords—plasma impedance, real-time monitoring, noninvasive monitoring, S-parameter, network analyzer*

## I. INTRODUCTION

Micro-arc discharge and particles generated in plasma etching process cause the decrease in the production yields and the overall equipment efficiency (OEE) at the mass-production line of LSI [1-8]. Micro-arc discharge occurs suddenly and rapidly during plasma process on the microsecond time scale. The arcing damages wafers and chamber parts, and induces unstable plasma discharge condition, which causes particle contamination [9]. These result in many scrapped wafers and the increase in manufacturing costs. The problem is that such arcing goes unnoticed at the plasma process until a flaw on a wafer is detected at the inspection process. Therefore the practical method for the in-situ detection of micro-arc discharge is highly required.

In mass-production lines, generation of particles depends on the number of processed wafers. Etching reaction products attach and gradually deposit on the inner walls as more wafers are processed. Then the particles are generated by flaking off of the films [9-11]. LSI circuit is shorted by the particles and the process must be stopped for periodical maintenance with cleaning of process chambers. When the number of particles, which falls on a Si wafer and is counted by the wafer surface inspection system at sampling inspection, or the total etching time reaches the management values, the process chambers are cleaned. However, these are indirect methods, so that the cycle of such the periodic maintenance is not necessarily optimized. Optimizing the maintenance schedule reduces maintenance and manufacturing costs, improving the OEE. In

other words, in addition to the monitoring method for micro-arc discharge, a continuous and in-situ monitoring method for the chamber inner wall to decide timing of the maintenance is also highly required. Developing such a practical method will also allow a predictive maintenance regime.

We have developed the load impedance monitoring method for plasma etching process [12-14]. Although impedance monitoring methods for plasma process have been studied, these are not practical for the mass-production equipment. The conventional methods have to install high voltage and current probes in a low impedance circuit line between the matching circuit and the powered electrode. However, it is usually difficult to modify the equipment at the mass-production line only for additionally installing the probes.

The developed impedance monitoring system can be externally installed in mass-production equipment and detect micro-arc discharge and monitor the condition of the film deposited on inner wall of process chamber [12-14]. In this study, we have upgraded the monitoring system to enhance precision, practicality, and versatility.

## II. EXPERIMENTAL METHOD

The experimental apparatus shown in Fig. 1 is mass-production plasma etching equipment with a capacitive rf discharge [11]. A bare Si wafer of 200 mm diameter is chucked onto the powered electrode by using an electrostatic chuck. The etching sequence and equipment parameters used are similar to those used at manufacturing facilities. The etching gas consists of $SF_6$ at a pressure of 18 Pa. The distance between the ground and powered electrodes is set at 60 mm. An rf power of 1000 W is supplied.

## III. UPGRADE OF LOAD IMPEDANCE MONITORIG SYSTEM

The impedance monitoring system consists of a directional coupler, the board network analyzer, and the software for the calculation of the load impedance. As shown in Fig. 1, the directional coupler is installed between the rf power supply and the matching circuit in the 50 Ω transmission line. That is, the system can be applied to mass-production equipment without remodeling. In the system, the attenuated forward and reflected rf powers containing phase information are measured with the directional coupler and the board network analyzer. The impedance at this measured point involves both the load impedance, which is the impedance of the load side from the output port of the auto-matching circuit, and the impedance of the matching circuit including the variable capacitances for impedance matching. Therefore, the system simultaneously monitors the capacitances to separate the load impedance from the entire impedance at measured point by calculating the circuit equation based on the equivalent circuit model.

In the upgraded monitoring system developed in this study, the load impedance is calculated by using not the equivalent circuit model but S-parameters of the circuit. Hence the component values of capacitances and inductances regarding to the capacitors and the inductors in the matching circuit, which are often not disclosed, are not needed for the impedance calculation. This improvement contributes to enhancing the precision because the calculation is not affected by the parasitic components in the matching circuit. The upgraded system monitors the load impedance by using a function of the circuit elimination of the network analyzer for S-parameter; i.e., the load impedance is calculated based on the S-parameter, which is obtained by eliminating the S-parameter of the matching unit from that of the entire circuit seen from the measured point at the directional coupler. In addition, the software for impedance calculation has not to be modified even if the type of the auto-matching unit is changed, whereas they have to be modified when the equivalent circuit method is used.

In addition, for the detection of micro-arc discharge a wave detector is employed at the measurement port of the reflected power in the board network analyzer. To install the wave detector the analyzer can detect the arcing independently from the time interval of the impedance measurement. Although the previous system can also detect the arcing, the system with the detector has been improved in terms of continuous monitoring. According to these improvements, the monitoring system enhances the precision, the practicality, and the versatility of the monitoring system.

## IV. RESULTS AND DISCUSSION

The validity of the upgrade of the monitoring system is investigated. The precision of the system based on the S-parameter method is compared with that based on the equivalent circuit method. Under the condition that the dummy load is connected to the output port of the matching circuit instead of the process chamber, the resistance and the reactance components of the load impedance are measured. Table 1 shows the measurement results of comparison between the S-parameter and the equivalent circuit method. The result indicates that the difference between the real and the measured values of the S-parameter method is much lower than that of the equivalent circuit one. This is due to the S-parameter method does not depend on the parasitic components in the matching circuit, whereas the equivalent circuit method is affected by them. The result demonstrates that the S-parameter method has higher precision.

To further confirm the effectiveness of the upgraded monitoring system, micro-arc discharge is tried to be detected. In the experiment, the arcing is induced at the backside of a wafer by applying higher voltage to the electrostatic chuck than usual condition. The measurement result shown in Fig. 2

indicates that the detected signal contains many impulsive changes with larger amplitude than normal level. These signals reveal that the plasma discharge became unstable condition instantaneously and repeatedly due to the micro-arc discharge [15, 16], and as a result the reflected power are generated by the impedance mismatch. Accordingly, the monitoring system has successfully detected the arcing by using a wave detector installed at the port of the reflected power in the network analyzer.

In near future works, we will also confirm that the monitoring system can detect the change in the condition of the inner wall of process chamber.

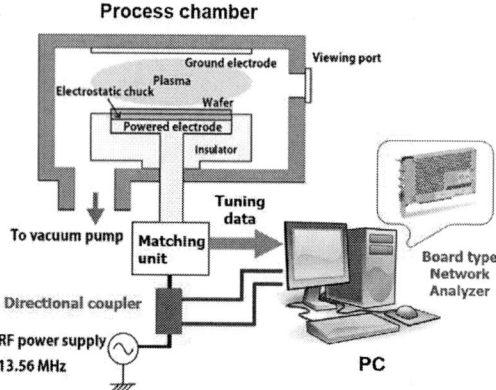

Fig. 1. Schematic view of plasma impedance monitoring system.

TABLE I. COMPARISON OF IMPEDANCE MONITORING RESULTS BETWEEN THE EQUIVALENT CIRCUIT METHOD AND THE S-PARAMETER METHOD WHEN THE DUMMY LOAD IS USED.

| Monitoring method | Difference between the real and the measured values | |
|---|---|---|
| | Resistance (Ω) | Reactance (Ω) |
| equivalent circuit method | +0.4 | +25.0 |
| S-parameter method | +0.1 | -0.4 |

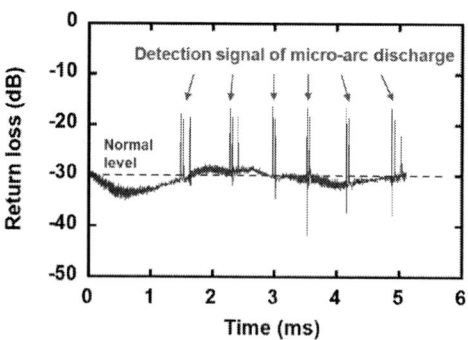

Fig. 2. Detection result of micro-arc discharge.

## V. CONCLUSIONS

The load impedance monitoring method for plasma etching process, which can be externally installed in mass-production equipment, has been upgraded. The results in this study indicates that the monitoring system employing the S-parameter method has higher precision and can detect micro-arc discharge. The improvements lead to enhance precision, practicality, and versatility. The upgraded monitoring system can be used as an effective method for real-time and noninvasive monitoring of plasma etching process at the mass-production line. In near future works, we will try to demonstrate that the upgraded system can also detect the change in the condition of the inner wall of process chamber.

## ACKNOWLEDGMENT

The authors would like to thank Hiroyuki Shiotsuka, Takashi Fujisaki, Satoru Aoyama, and Naoya Kimura (Advantest Corporation), Akikazu Kaneda, and Koichi Nakayama (Advantest Kyushu Systems Co., Ltd.), and Dr. Fumihiko Uesugi and Dr. Taisei Motomura (AIST) for their assistance and helpful discussions.

## REFERENCES

[1] H.-S. Jun, "Diffusive plasma dechucking method for wafers to reduce falling dust particles," Jpn. J. Appl. Phys., vol. 52, no. 6R, pp.1–5, Jun. 2013.

[2] N. Ito, T. Moriya, F. Uesugi, M. Matsumoto, S. Liu, and Y. Kitayama, "Reduction of particle contamination in plasma-etching equipment by dehydration of chamber wall," Jpn. J. Appl. Phys., vol. 47, no. 5R, pp. 3630–3634, May 2008.

[3] M. A. Hussein and R. B. Turkot, "Particle control in dielectric etch chamber," IEEE Trans. Semicond. Manuf., vol. 19, no. 1, pp. 146–155, Feb. 2006.

[4] T. Moriya, N. Ito, and F. Uesugi, "Capture of flaked particles during plasma etching by a negatively biased electrode," J. Vac. Sci. Technol. B, Microelectron. Nanometer Struct. Process. Meas. Phenomena, vol. 22, no. 5, pp. 2359–2363, Jul. 2004.

[5] T. Moriya, N. Ito, F. Uesugi, Y. Hayashi, and K. Okamura, "Generation of positively charged particles at an anode and transport to device wafers in a real radio frequency plasma etching chamber for tungsten etch-back process," J. Vac. Sci. Technol. A, vol. 18, no. 4, pp. 1282–1286, 2000.

[6] F. Uesugi, N. Ito, T. Moriya, H. Doi, S. Sakamoto, and Y. Hayashi, "Real-time monitoring of scattered laser light by a single particle of several ten of nanometers in the etching chamber in relation to its status with the equipment," J. Vac. Sci. Technol. A, vol. 16, no. 3, pp. 1189–1195, 1998.

[7] N. Ito, T. Moriya, F. Uesugi, H. Doi, S. Sakamoto, and Y. Hayashi, "Observation of the trajectories of particles in process equipment by an in situ monitoring system using a laser light scattering method," J. Vac. Sci. Technol. B, Microelectron. Nanometer Struct. Process. Meas. Phenomena, vol. 16, no. 6, pp. 3339–3343, 1998.

[8] G. Lapenta and J. U. Brackbill, "Simulation of dust particle dynamics for electrode design in plasma discharges," Plasma Sources Sci. Technol., vol. 6, no. 1, pp. 61–69, Feb. 1997.

[9] Y. Kasashima, T. Motomura, N. Nabeoka, and F. Uesugi, "Numerous flaked particles instantaneously generated by micro-arc discharge in mass-production plasma etching equipment," Jpn. J. Appl. Phys., vol. 54, no. 1S, Jan. 2015, Art. no. 01AE02.

[10] Y. Kasashima, N. Nabeoka, T. Motomura, and F. Uesugi, "Many flaked particles caused by impulsive force of electric field stress and effect of electrostriction stress in mass-production plasma etching equipment," Jpn. J. Appl. Phys., vol. 53, no. 4, Apr. 2014, Art. no. 040301.

[11] Y. Kasashima, N. Nabeoka, and F. Uesugi, "Instantaneous generation of many flaked particles by impulsive force of electric field stress acting on inner wall of mass-production plasma etching equipment," Jpn. J. Appl. Phys., vol. 52, no. 6R, Jun. 2013, Art. no. 066201.

[12] S. Kuniie S. Sato, N. Kimura, S. Wakamoto, Y. Kasashima, T. Motomura, T. Tabaru, and F. Uesugi, "Micro-Arc Discharge Detection Method in Plasma Process by Monitoring Load Impedance," J. Plasma Fusion Res., vol. 96, no. 3, pp. 117-121, Mar 2020.

[13] Y. Kasashima, T. Motomura, H. Kurita, N. Kimura, and F. Uesugi, "Detection of microarc discharge using a high-speed load impedance monitoring system," Appl. Phys. Express, vol. 7, no. 9, Aug 2014, Art. no. 096102.

[14] T. Motomura, Y. Kasashima, O. Fukuda, F. Uesugi, H. Kurita, and N. Kimura, "Practical monitoring system using characteristic impedance measurement during plasma processing," Rev. Sci. Instrum., vol. 85, no. 2, Feb 2014, Art. no. 026103.

[15] Y. Kasashima, T. Tabaru, M. Yasaka, Y. Kobayashi, M. Akiyama, N. Nabeoka, T. Motomura, S. Sakamoto, and F. Uesugi, "In-situ detection method for wafer movement and micro-arc discharge around a wafer in plasma etching process using electrostatic chuck wafer stage with built-in acoustic emission sensor," Jpn. J. Appl. Phys., vol. 53, no. 3S2, Mar. 2014, Art. no. 03DC04.

[16] Y. Kasashima, T. Motomura, N. Nabeoka, and F. Uesugi, "Numerous flaked particles instantaneously generated by micro-arc discharge in mass-production plasma etching equipment," Jpn. J. Appl. Phys., vol. 54, no. 1S, Oct. 2015, Art. no. 01AE02.

# Advanced Process Monitoring through Fault Detection and Classification for Robust Statistical Process Control of Tantalum Nitride Reactive Sputtering

Stephanie Y Chang
*Advance Process Technology*
*Skyworks Solutions, Inc.*
Newbury Park, USA
Stephanie.Chang@skyworksinc.com

Shiban Tiku
*Plant Management*
*Skyworks Solutions, Inc.*
Newbury Park, USA
Shiban.Tiku@skyworksinc.com

Lam Luu
*Advance Process Technology*
*Skyworks Solutions, Inc.*
Newbury Park, USA
Lam.Luu@skyworksinc.com

*Abstract*—This paper discusses Fault Detection and Classification (FDC) deployed in high-volume manufacturing for the Tantalum Nitride (TaN) reactive sputtering process. TaN thin film resistors (TFR) with an average sheet resistance ($R_s$) of 50 ohms/sq with high uniformity and negative temperature coefficient of resistance (TCR) were achieved. Optimizing interdiction capabilities for real-time prevention and detection of parameter excursions strengthened statistical process control (SPC) and improved yield.

*Keywords— TaN, TFR, SPC, PCM, FDC, Machine Learning*

## I. INTRODUCTION

TaN thin film resistors have favorable properties, such as physiochemical inertness, linear TCR, mechanical hardness, and self-passivation. Under optimized process manufacturing, high stability and accuracy can be achieved [1]. Ensuring a robust process control over the TaN thin film $R_s$ and uniformity is increasingly pertinent for microwave integrated circuits (MMIC) in applications of both Gallium Arsenide (GaAs) and Gallium Nitride (GaN) devices [2].

Reactive processes, such as sputtering TaN, are especially sensitive to the chamber and shielding conditions. For a stable process and to minimize tantalum (Ta) cathode poisoning, the process chamber must be adequately conditioned through a combination of "burn" and deposition sequences especially after the chamber is vented and the target is exposed to atmospheric conditions [3]. Optimization of the procedures for tool requalification after preventative maintenance (PM) activities has been addressed in a previous study [4].

Data-driven diagnosis of detected faults as well as effective analysis of historical data sets for trends and anomalies have emerged as powerful tools to improve quality and process control in a fast-paced manufacturing environment [5]. Instantaneous feedback through automated detection and classification of process parameters is a critical procedure that allowed further analysis of both the strengths and weaknesses within the TaN sputtering process. Solutions to several of the challenges encountered will be discussed in this work. Successful optimization and implementation of a third-party software's capabilities and features into the production environment substantially enhanced process visibility in the device fabrication flow.

## II. TaN REACTIVE SPUTTERING

### A. Inherent Challenges

The reactive deposition process in this single wafer sputtering system involves bombardment of argon (Ar) atoms against a Ta target, knocking off Ta atoms that subsequently react with nitrogen ions and form a deposited thin metal-nitride film on the wafer over time. As shown in Figure 1(a), the magnet behind the target intensifies the plasma discharge. The ratio and flow rates of these process gases directly affect the thin film quality and properties, which impact the resistor performance.

Thorough conditioning of the chamber is necessary to optimize both the process stability and intra-wafer, across-cassette, and run-to-run uniformity. Variation for the intra-wafer uniformity can be attributed to multiple factors, such as the plasma, magnetic field, atom bombardment, Ta target's wear pattern, etc. The deterioration of the Ta target's uniformity is most apparent at the end of its life cycle. Resembling that of a "sombrero," the degraded uniformity signature results in sputtered TaN thin films with alternating regions of high and low $R_s$ as shown in Figure 1(b). Therefore, it becomes increasingly pertinent for real-time process monitoring of both process trends and maintenance schedules to ensure consistent quality of all process runs.

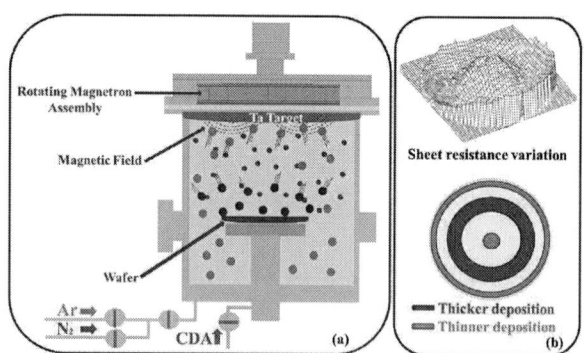

Fig. 1. (a) Reactive TaN sputtering has process gases, argon (Ar) and nitrogen ($N_2$), and platen cooling maintained by compressed dry air (CDA). (b) Target's wear pattern resulted in poor intra-wafer thin film uniformity.

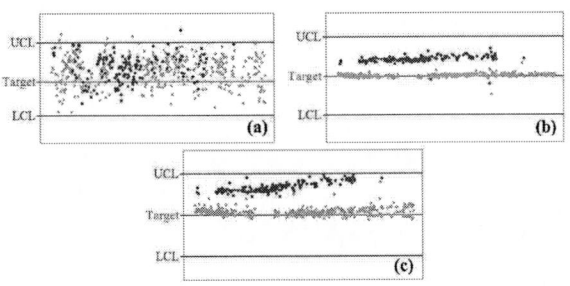

Fig. 2. A sputtering chamber with a high utilization percentage (nearing a target PM) has a relatively stable (a) inline average $R_s$ but shows poorer intra-wafer uniformity as indicated by higher (b) %sigma and (c) range values.

## III. FAULT DETECTION AND CLASSIFICATION

### A. Process Parameter Excursions

A major advantage of run-to-run monitoring of process parameters (e.g., cryogenic and platen temperatures, deposition power, voltage, chamber pressure, gas flow rates) is the ability to establish correlations between data of specific runs to wafers with yield issues (e.g., failed electrical parameters, poor uniformity). To prevent wafer scrap, the FDC system's interdicting capabilities halt the tool so that detected parameter excursions, which the tool may not alarm for, can be immediately troubleshot. FDC monitoring of process trends provided instantaneous insight into sources that contributed to erratic inline $R_s$ or a poorer intra-wafer uniformity signature. For instance, higher variability in Ar gas flow rates was attributed to a specific sputtering chamber's mass flow controller in Figure 3.

Wafers that were detected with out-of-control (OOC) process parameters were automatically flagged through FDC. These wafer runs were then collectively sorted into a list and reviewed for dispositioning and evaluated for risk assessment. Implementing a robust automated inline monitor of trending process parameters for all technologies and devices had significantly streamlined the down-selection of wafers that demonstrated higher process variation.

These selected wafers were then sent to a routine short loop sampling plan as shown in Figure 4 for earlier electrical characterization of critical resistor parameters. This reduced the average turnaround time by 4X, enabling faster feedback for earlier inline detection of shifts in TaN parameters as discussed in our previous work [6]. This short feedback loop was implemented after the discovery of the oxidative and thermal post lift-off ash treatment's stabilizing effects on the resistor parameters as shown in Figure 5. This step occurs downstream from the resistor layer stage.

As such, optimizing the FDC system and shortening the feedback loop for early electrical testing of TaN related parameters reduced process variation and tightened the distribution of the critical Process Control Monitor (PCM) resistor value for a specific resistor dimension by 25% as depicted in Figure 6.

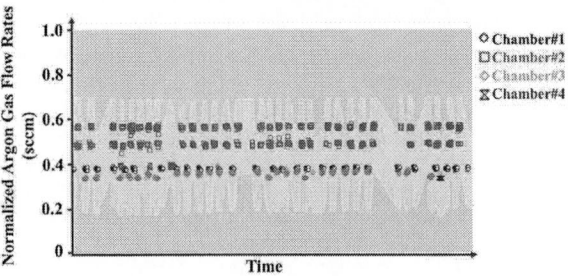

Fig. 3. Dynamic target hold x-bar set report tracked the standard deviation of charted residuals and revealed greater variability in Ar gas flow rates during TaN sputtering deposition for Chamber#2.

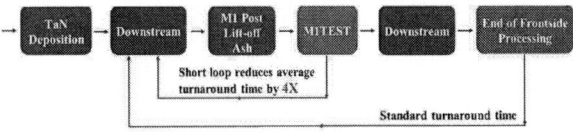

Fig. 4. A short loop (M1TEST) for earlier electrical characterization of TaN parameters was implemented to closely monitor for process drifts.

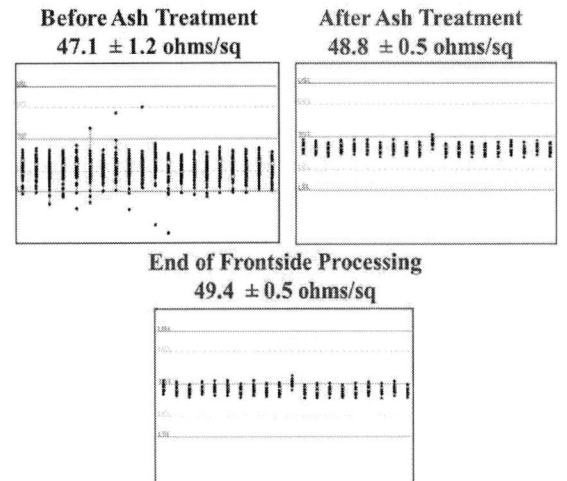

X-axis: each set of data corresponds to 1 wafer

Fig. 5. Oxidative and highly thermal ash treatment tightened the TaN $R_s$ distribution by 58% and promoted greater stability.

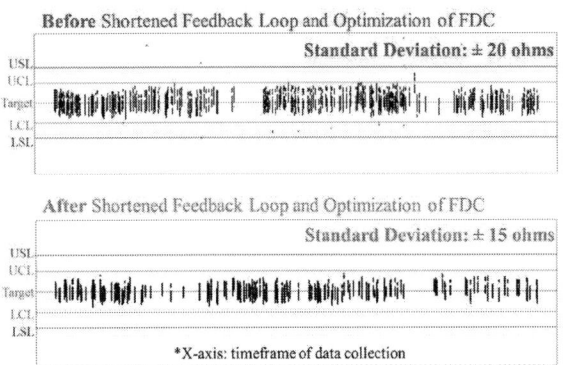

Fig. 6. Tightened SPC chart of several months' PCM resistor data for a specific resistor dimension to check the integrity of the resistor process. This was measured at the end of frontside processing for both before and after deployment of FDC monitoring and implementation of the early feedback testing of electrical parameters.

978-1-6654-7134-3/22 $31.00 © 2022 IEEE

## B. Automated Tracker and Scheduler for Preventative Maintenance

While some cases of detected parameter excursions were due to process variation, other cases aligned with approaching maintenance timelines. Performing PMs at a regulated basis is crucial not only for extending the Ta target life cycle but also serving as cost savings since Ta is a semi-precious metal. The online maintenance indicator in Figure 7 was implemented to schedule PMs by tracking the kWh-based target utilization percentages for every wafer run.

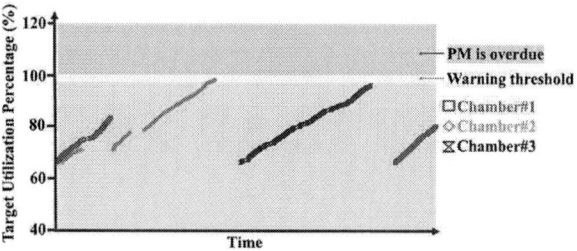

Fig. 7. Maintenance management tracks kWh-based target utilization percentages for scheduling routine preventative maintenance of each TaN sputtering chamber. Once the target utilization percentages enter the warning threshold, an alarm notification will signify the approaching PM.

Various benefits were achieved through ensuring a timely reassessment of the sputtering chamber condition (e.g., outgas, leak rates, cryogenic regeneration, bake-out) and equipment components (e.g., anode ring, shielding, shutter blade). As a result, unscheduled equipment downtime was significantly reduced, and risk was minimized for overloaded shielding or burning through the target into the backing plate which would result in product contamination and thus wafer scrap. After scheduled maintenance activities were completed, vigilant tracking of process parameters during requalification runs through FDC and inline SPC was necessary to ensure process stability and repeatability for consecutive wafer processing.

## C. Report and Model Types

Automated detection of these process parameters with immediate feedback provided the opportunity to identify areas of both strengths and weaknesses for optimizing tool capabilities. The high customization of the filtered model type enables tracking of statistical calculations most suited to the defined process parameter's trends.

While some reports have user-defined control limits, others determine limits from evaluating historical trending data for each unique sputtering chamber or process recipe. For instance, to remove the Ta target's surface contamination or oxidation, recipes used for conditioning the target after PMs reach higher platen temperatures than standard production recipes. The software's history splitting feature by recipe allows each process recipe to have its own historical target for tracking the platen temperature. After implementation of tool interdiction during conditioning runs with OOC platen temperatures caught by FDC, there was a considerable reduction in burnt or warped monitoring wafers that are shown in Figure 8. This is especially crucial since warped wafers can result in handling issues during wafer processing.

The robustness of automated detectability for fluctuations in platen temperature was further enhanced by setting up FDC reports that captured both the calculated range and average for the platen temperature during the duration of each wafer run.

This improved the platen temperature monitoring for both conditioning and standard production runs since additional data points were being accounted for rather than using a single point readout.

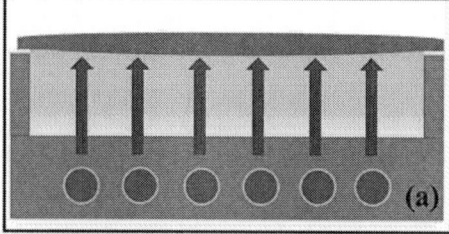

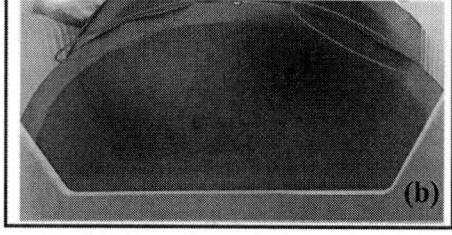

Fig. 8. (a) Wafer placed on the platen is heated during the sputtering process by radiation. (b) Extensively used silicon monitoring wafers for inline SPC were subjected to high thermal stress induced from burn-in conditioning of the Ta target.

## D. Tool Recipe Management and Interdiction

The system's central management of tool process recipes and revision control compares the current recipes against their respective golden recipes to ensure that recipe files remain uncorrupted and no unauthorized changes have been made. Mismatches for any parameters detected at a set time interval would interdict, thus preventing subsequent lots from running a nonstandard deposition recipe that would adversely affect TaN thin film composition and properties.

## E. Process Capability Timers

A capability timer was implemented to automatically switch off process capabilities that were no longer qualified or in usage for production. For instance, reload recipes, which contain different values for certain process parameters in comparison to the standard production recipe, are used for consecutive run-to-run processing in the same sputtering chamber. While this is favorable in the events of high work-in-progress (WIP), in which many wafers are queued at the resistor layer deposition step, it is crucial for vigilant monitoring of potentially erratic inline TaN $R_s$. OOC inline TaN data would disqualify the reload recipe to allow for recipe adjustments to retarget $R_s$.

Certain devices with nonstandard process specifications require a different targeted inline $R_s$. The FDC system would track when these nonstandard recipes were processed in the sputtering chamber and initiate a timer. If these nonstandard recipes were not run within a specified time interval, their corresponding process capabilities would be automatically switched off by the system's capability timers. This prevents accidental selection of unqualified process capabilities and signals that requalification is necessary to reenable those specific process capabilities.

### F. Process Tracking and Utilization for Maximizing Throughput

Reports tracking production's process utilization of each sputtering chamber provided greater insight that allowed the engineering and production teams to maximize throughput while abiding by process constraints (e.g., minimum time for stabilization between process runs on the same sputtering chamber). For instance, evaluations were conducted to assess if simultaneous wafer processing on two separate sputtering chambers as shown in Figure 9 from the same tool platform would affect FDC process parameters or TaN thin film characterization. Figures 10 and 11 are examples of FDC reports used to identify wafer runs that were processed at varied overlapping intervals. A comprehensive study allowed for the qualified release of simultaneous processing for the standard production recipe, thereby doubling the throughput at the resistor layer deposition step.

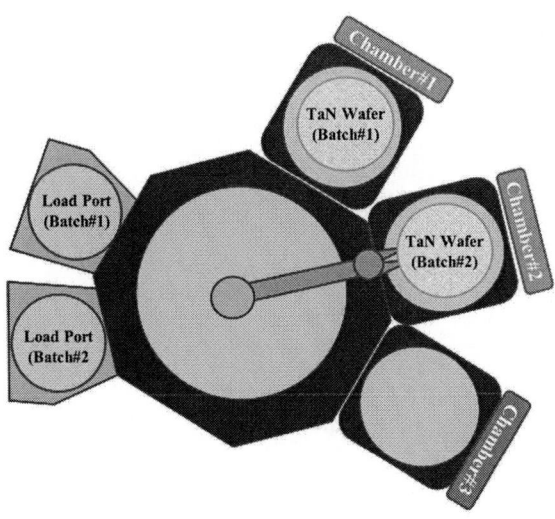

Fig. 9. Two chambers are actively in usage for simultaneously processed wafers from batch#1 and batch#2 for reactive TaN sputtering.

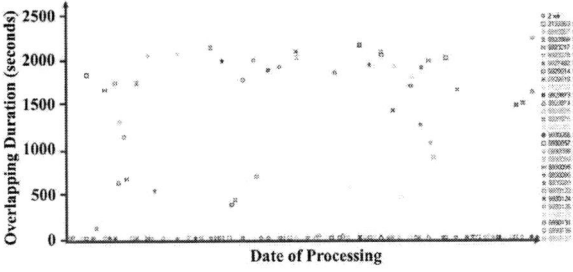

Fig. 10. Overlapping duration is shown for each pair of runs processed simultaneously on the same tool platform but in different sputtering chambers. Any data point with a non-zero overlapping value would indicate a simultaneously processed run.

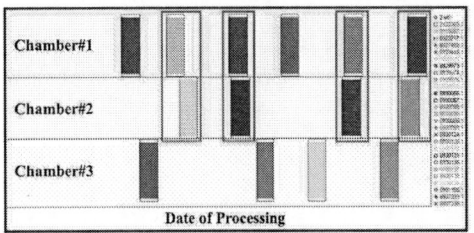

Fig. 11. Process tracking and utilization report allows for identification of simultaneously processed wafer runs. Each pair of simultaneously processed runs is boxed in red.

## IV. CONCLUSION

Data visualization and automated statistical analysis enhanced the visibility of process trends and equipment behavior for reliable monitoring of PM schedules, identifying sources of process variation and drifts, diagnosing tool alarms, and establishing correlations between process data and TaN thin film characterization. Substantial improvements to process monitoring were reflected in tightened TaN PCM resistor values, resulting in yield improvement. Additional engineering analytical tools and machine learning can be integrated into further developing the current third-party FDC system to perform automated training, create models, and flag anomalies. Enhanced capabilities would reduce risks to product manufacturing, efficiently address tool issues in a production environment, and comprehensively detect and analyze semiconductor processes through simplified deployment.

## ACKNOWLEDGMENT

The authors would like to acknowledge the contributions from colleagues across various cross-functional teams at Skyworks Solutions for their support to optimize the TaN resistor performance and strengthen the prevention and detection of process drifts in the fabrication flow through implementation of FDC. Special thanks to Nercy Ebrahimi for his continued support for this project as well as Ravi Ramanathan for his feedback and manuscript review.

## REFERENCES

[1] C. C. Lukose, G. Zoppi, and M. Birkett, "Thin film resistive materials: past, present and future," *IOP Conference Series: Materials Science and Engineering*, vol. 104. pp. 12003, 2016.

[2] X. Yao, Y. Pu, X. Liu, W. Wu, "Characterization and Reliability of Thin Film Resistors for MMICs Application Based on AlGaN/GaN HEMTs," *Journal of Semiconductors*, vol. 29, no.7, pp.1246-1248, 2008.

[3] T. Riekkinen, J. Molarius, T. Laurila, A. Nurmela, I. Suni, J.K. Kivilahti, "Reactive sputter deposition and properties of TaxN thin films," *Microelectronic Engineering*, vol. 64, no.1-4, pp.289-297, 2002.

[4] J. Sires, "Reactive Sputtering: TaN Process Characterization and Post PM Qualification Improvements," *2018 International Conference on Compound Semiconductor Manufacturing Technology (CS MANTECH) Technical Digest*, pp. 273-276, May 2018.

[5] G. A. Susto, M. Terzi, and A. Beghi, "Anomaly Detection Approaches for Semiconductor Manufacturing," Procedia Manufacturing, vol. 11, pp. 2018-2024, 2017.

[6] S. Chang, S. Tiku, L. Luu-Henderson, "Optimizing Process Conditions for High Uniformity and Stability of Tantalum Nitride Films," *2022 International Conference on Compound Semiconductor Manufacturing Technology (CS MANTECH) Technical Digest*, pp. 91-94, May 2022.

## Author Biography

Stephanie Chang is a process engineer at Skyworks Solutions, Inc.'s Advance Process Technology Group, specializing in thin films and photo processes. She graduated from University of California, Berkeley and previously worked at Lawrence Berkeley National Laboratory's Molecular Foundry where she researched liquid cell TEM, MEMS/NEMS, and metamaterials.

# Characterization of Light propagation loss in Photonics devices using High-Resolution CDSEM metrology

S. Levi[1], R. Le Tiec[1], C. Dupre[2], C. Vannuffel[2], T. Dewolf[2], S. Garcia[2], K. Millard[2,3], B. Meynard[2], Y. Lee[2], M. Colard[2,4], H. Al Dujaili[2], J. Faugier-Tovar[2] Shinsuke Mizuno[5]

1. PDC business group, Applied Materials, Rehovot 76705, Israel
2. CEA-LETI, MINATEC, 17 Rue des Martyrs, 38054 Grenoble, cedex 9, France
3. Univ. Grenoble Alpes, IMEP-LAHC, MINATEC – INPG, 38016 Grenoble, France
4. Institut de Recherche en Informatique, Mathématiques, Automatique et Signal (IRIMAS EA7499), Univ. Haute-Alsace, IUT Mulhouse, 61 rues A. Camus, F-68093 Mulhouse Cedex, France
5. AMJ – Applied materials Japan Yokoso Rainbow Tower 3-20-20 Kaigan, Minato-ku, Tokyo

*Abstract*— **Photonic devices manufactured on core materials (by example Si or SiN) wafers, using a semiconductor fabrication technique, was demonstrated to increase data transmission speed, consume less power, and generate less heat than conventional electronic circuits. Light propagation loss strongly dependents on the Waveguide edge roughness. To reduce roughness, three patterning methods are evaluated, dry vs wet lithography, OPC (Optical proximity correction) and Hydrogen anneal on two types of waveguides, RIB and STRIP, fabricated on SOI wafer. In this paper, we demonstrate innovative methods to measure edge roughness of straight and curved WGs with a CDSEM. CD roughness measurements are correlated with light propagation loss.**

*Keywords*— *Photonics, Roughness, LER, LWR, PSD CDSEM, Propagation losses*

## I. INTRODUCTION

Waveguides (WG), interconnects between photonic devices in the circuit, made from a silicon core (Silicon Photonics referred as SiPh) come with two structures RIB or STRIP. Given the material properties, optical signals can be transported without significant losses. Given the physics of photonics, older CMOS nodes may seem to be suitable to fabricate the photonic devices and circuits. Light propagation in WG takes place due to total internal reflection, therefore WGs are curved or tilted, to carry the light across the designed circuit. Patterning process variation such as CD uniformity or WG edge roughness can increase propagation losses. To enable tight process control, we developed a CDSEM that enables image grab in a large field of view with high resolution. It can measure CD uniformity on relatively large WG CD, and edge roughness with high sensitivity. Line Edge Roughness (LER) [1]is defined as the variation of the detected edge points by the CDSEM algorithm, from a known shape, for examples straight line. Since the WGs are curved or tilted across the design, edge roughness measurement can incorporate physical edge roughness and curvature deviation from design intent, for cases where patterning significantly deviates from design intent, it is possible to apply a polynomial fit. To deconvolute pattern curvature from edge roughness, new metrology approach was developed. This paper highlights the usage of high-resolution CDSEM for process characterization of Silicon. Line edge roughness (LER) is a well-known key detractor of the optical losses within the photonics devices [4]. To reduce the patterned WG roughness, three different techniques, applied and studied. Dry vs immersion lithography, Optical Proximity Correction

(OPC) applied on curved WGs and Hydrogen annealing of the Silicon WGs after Etch. To evaluate how roughness changes by each of one of these methods, the WG roughness is characterized with a CDSEM. New roughness algorithm, using PSD method [2] is reflecting roughness changes in the frequency domain. Roughness changes are correlated with optical losses measured on the same WGs.

## II. ROUGHNESS CHARACTERIZATION USING POWER SPECTRAL DENSITY ANALYSIS (PSD)

To detect the pattern edges bottom points with a CD-SEM on curved line-space pattern images, innovative metrology algorithms used. These topographical points, provides a set of two-dimensional coordinates, reflecting the edge bottom topography in a two-dimensional plane. Transforming the topographical points coordinates into the Fourier plane along the frequency domain, enables to characterize dominate roughness frequencies in the pattern edge roughness. Since the WG are curved, roughness measurement had to take this curvature into consideration, in this paper we present results via polynomial fit. Fitting roughness measurement to a straight line (Fig 1) results with increase of LF band.

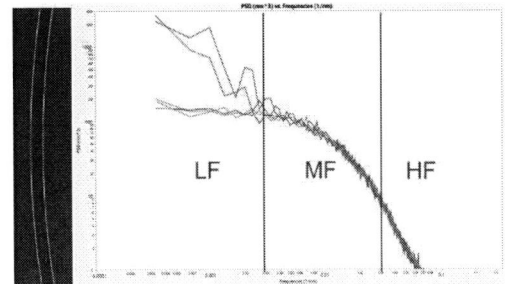

*Fig 1. PSD chart, linear vs polynomial fit*

As the roughness is mainly of a stochastic nature, to quantify roughness measured values the PSD calculation is separated into three sections, Low roughness frequencies (LF), Mid roughness frequencies (MF) and High roughens frequencies (HF). For this work, LF is all frequencies smaller than 1/200nm, MF all frequencies in between 1/200nm and 1/20nm and HF are smaller than 1/20nm. These are separated at the frequency domain where roughness value is calculated from the PSD histogram for each section and presented as a numeric value.

## III. PHOTONICS PROCESS

The photonics process is built of multiples blocks. On which we can have a non-exhaustive list like the P/N Junction, Photodetector formation and our main interest, the WG formation.

### A. RIB & STRIP WGs formation

Si Photonics devices may have multiple types of WGs. In this work two different types of WG, RIB and STRIP are studied. For Silicon, transition is tested with 1310 nm wavelength, using the TE emission mode, to have the following profiles (Fig 2).

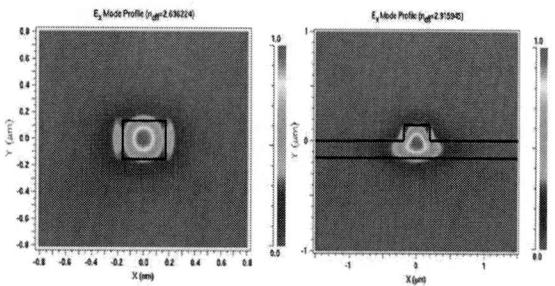

*Fig 2: STRIP (left) and RIB (right) with light propagation density*

In this study for Silicon Photonics (SiPh), the base wafer is a SOI. With a bulk Silicon, thin layer of buried oxide, and a Silicon layer with thickness that varies, pending the characteristics of the device. The first process steps are, deposition of Hard Mask (HM) material, which consist of SiO2 and SiN. Then Lithography steps, to define the RIB and the STRIP structures. In this work ArF dry vs wet lithography are compared, for LER and light propagation loss. Next is a partial Etch step, marked in this work as "Etch 01". It consists of the HM etch followed by a certain Etch into the Silicon layer. The Etch depth is defined by light propagation density and impact device characteristics. Photoresist is stripped. To finalize the RIB & STRIP structures, a second litho step covers the RIB pattern. Last step is to etch the Silicon till the buried oxide layer marked in this work as "Etch 02", forming the final RIB & STRIP structures:

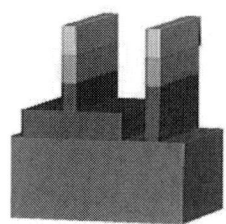

*Fig 3: RIB (Left) and STRIP (right) 3D illustration*

Hydrogen anneal can be added after Etch 02 to smooth the WG edge roughness. In this work pre and post anneal LER is measured and correlated with propagation loss. To continue the integration, the WGs are encapsulated.

### B. SMO and OPC principles applied to Photonics

In lithography, Source-Mask Optimization (SMO) techniques is used to maximize image contrast printed with scanner into the photoresist. For CMOS, this technique has been optimized mainly for Line/Space & Contact patterns. Therefore, SMO was simulated and studied to optimize pattern fidelity. Optical Proximity Correction (OPC), is another methodology used for CMOS manufacturing that was simulated and applied to the WG patterning process, aiming to improve pattern fidelity of smaller features and reduce line edge roughness of curved WG.

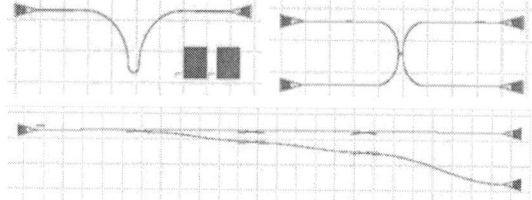

*Fig 4: Lightguides designed with various curvatures*

The first approach of OPC model, is to Manhattanize the designed feature. It alters pattern writing on the mask as a function of small steps. The optimizations of SMO and OPC were studied for CD control and pattern fidelity. In this work we measure LER pattern pre and post SMO & OPC for comparison.

## IV. EXPERIMENTAL RESULTS

### A. Dry lithography 193 vs immersion lithography 193i, Edge roughness vs propagation losses

Comparison of LER of Dry vs Wet lithography was reported in previous studies [5], LER significantly reduced with immersion lithography. These wafers where etched partially into the Silicon "Etch 01". Roughness measurement of RIB WG (Fig 5), shows a reduction of LER mainly at the Mid frequency band.

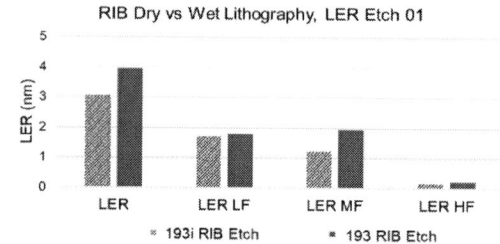

*Fig 5. Roughness of RIB immersion vs dry lithography*

Although wet lithography measurements showed a significant improvement of edge roughness over dry lithography, these were not fully reflected in the after Etch 01 of the RIB pattern. It is due to the partial etch that forms a foot at the pattern bottom. After "Etch 02" the STRIP is formed, LER of the STRIP WG was measured. For the STRIP, LER is reduced at all frequency bands (Fig 6).

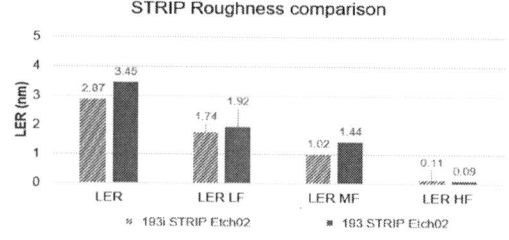

*STRIP Roughness comparison*

*Fig 6. Post Etch Roughness of RIB, immersion vs dry lithography*

The etch significantly influenced by the lithography process. The main reason for the different roughness characteristics between pattern types is linked with the different structure. The STRIP is etched down to the Oxide layer. Etch with stop layer, is characterized with sharp edges and a very clear bottom.

To evaluate the impact of LER on light propagation loss, the fabrication process continued the wafers, to the optical tests.

| Target | Propagation loss @1310nm (dB/cm) | 193I LER (nm) | Propagation loss @1310nm (dB/cm) | 193 LER (nm) |
|--------|------|------|------|------|
| RIB | 1.2 | 3.06 | 1.5 | 3.96 |
| STRIP | 2.9 | 2.87 | 3.5-4 | 3.45 |

Comparing Line Edge Roughness (LER) measurements vs propagation's losses, demonstrate a clear correlation between the two, using immersion lithography significantly contribute to reduce LER and propagation loss.

### B. SMO & OPC methodology on controlled curved WG

SMO and OPC influence on WG LER was studied on curved WG. "Low angle" and "High angle" WGs (Fig 7). Same design was printed on photomask, one with OPC and one without OPC. LER measurements performed on the two patterns, using the same measurement conditions.

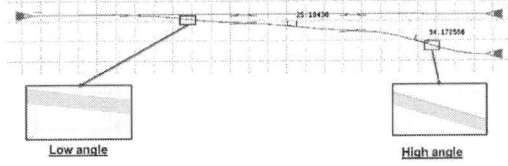

*Fig 7. Designed layout of curved WG*

LER measured and characterized with Power Spectral Density (PSD), at the beginning the calculation was to fit to a straight line. From the PSD spectra we noticed a similar fingerprint difference, between the patterns. The curved lines treated with SMO & OPC demonstrated a systematic peak around the 200nm wavelength.

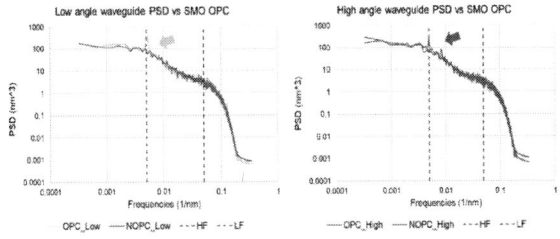

*Fig 8. PSD fingerprint of OPC vs No OPC WG*

Similar phenomena with slightly larger amplitude were measured on the high tilt angle WG. A singular peak on PSD chart suggests a stronger curvature of the patterned edge. Comparing the extracted the topographical points, these points are the exact pixels that the algorithms selected to define the line edge points. A change of the curvature in WGs while applying the SMO & OPC it is visually observed. The change is minor, since PSD results are evaluated on a large statistical sample the change is repeatable across wafer. With naked eyes, no differences are visible between the two patterns.

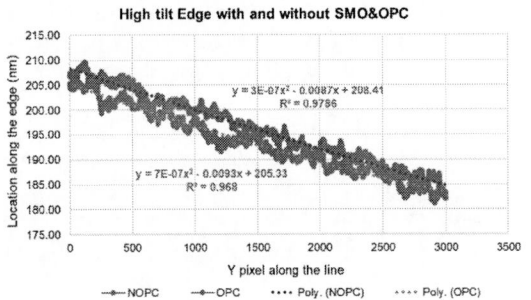

*Fig 9. Topographical points with, without OPC*

To conclude this study, LER measurements suggest that the SMO & OPC treatment, slightly increase the curvature of the printed line. This change can only be quantified using PSD methodology. LER measurements coupled with frequential analysis, Reflect valuable information on roughness characteristics. When fitting the PSD to a polynomial, to mimic the edge curvature, the peak around 200 nm was lost and roughness appeared to be similar between pre and post SMO & OPC treatment.

### C. Hydrogen anneal impact on WG edge roughness

Hydrogen Anneal, reorganized the materials lattice to move it to its crystalline state by applying a thermal stress. It results in significant improvement of LER. For this study wafers of the split with the 193nm dry lithography (after "etch02"), were treated with hydrogen anneal. LER measurements of the RIB pattern (Fig 10) showed LER reduction on the low frequency band only. The root cause is similar to the results obtained when comparing the Dry vs Immersion lithography.

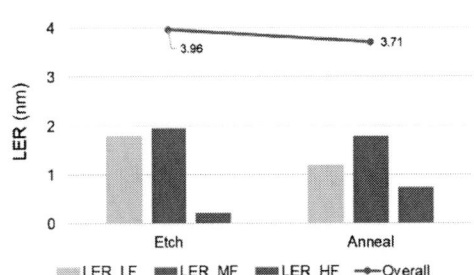

*Fig 10. RIB WG pre and post H2 Anneal*

For the STRIP a major reduction of LER is measured after hydrogen anneal treatment at all frequency bands.

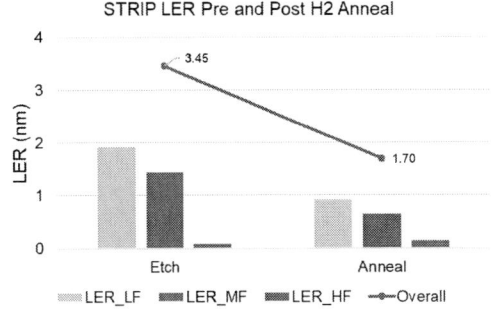

*Fig 11. STRIP WG pre and post H2 Anneal*

From the Top view CDSEM images (Fig 12), it is noticed that for the STRIP, edges become sharper after a anneal treatment and the RIB the edges become wider.

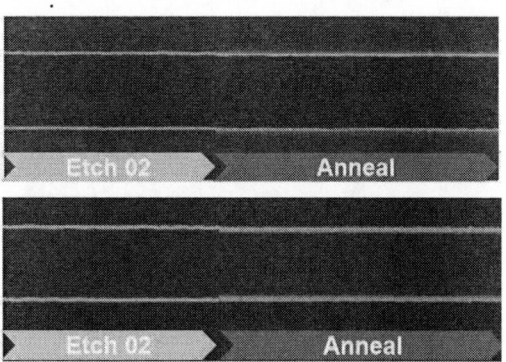

*Fig 12. RIB (upper pattern) & STRIP (lower pattern) image gallery per process steps*

Cross section images of the two WG, reveal similar phenomenon as top view images (fig 13). The RIB structures, prior the anneal treatment, form a rectangle shape, as intend with the design.

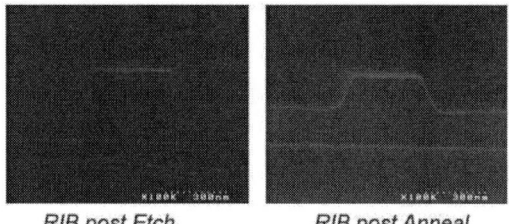

RIB post Etch          RIB post Anneal

*Fig 13 Cross section of RIP pattern pre and post Anneal*

After annealing, the pattern shape is more trapezoidal. This explains the LER measurement obtained by CDSEM Top view measurements. The foot formed by anneal process, results with a higher roughness measurement. Particularly at mid and low frequency bands. TEM images (Fig 14) reflecting the crystallin structure of the STRIP WF after annealing, contributes additional information.

On the STRIP structure, we are observing this:

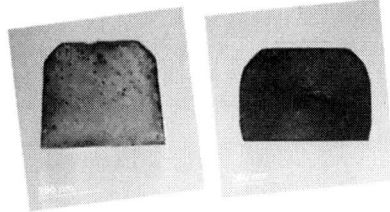

STRIP post Etch          STRIP post Anneal

*Fig 14 TEM images of STRIP pre and post anneal*

Post Etch 02, the STRIP shape looks like a standard rectangular line. However, after anneal process, the shape transformed into an octagonal shape. It translates into a much smoother edge. The TEM images reflect why STRIP LER is reduced after anneal process

To correlate LER measurements with propagation loss, the wafers were processed till propagation loss testing.

| Waveguide | Propagation lose @1310nm (dB/cm) | Etch LER (nm) w/o Anneal | Propagation lose @1310nm (dB/cm) | Etch LER (nm) With Anneal |
|---|---|---|---|---|
| RIB | 1.5 ⬆ | 3.96 ⬆ | 0.2 ⬇ | 3.71 ⬇ |
| STRIP | 3.5 ⬆ | 3.45 ⬆ | 1.1 ⬇ | 1.70 ⬇ |

As was already observed, the LER measurement is in correlation with the propagation loss.

Roughness measurements performed on SiN WG, were correlated with propagation loss. In the following work, we plan to apply these three treatments to demonstrate reduction in SiN propagation loss.

## CONCLUSION

This study demonstrates a correlation between WG edge roughness and light propagation loss. By using high resolution CDSEM, innovative PSD based algorithms and large statistical sample across wafer, we could better understand the changes cause by the patterning process to the WG. Immersion lithography, significantly reduced LER and therefore propagation loss. SMO & OPC models that we tested demonstrated low impact on LER mainly effecting line curvature but less contributing to mid and high frequency bands. Hydrogen anneal treatment to Silicon WG, significantly contributed to the reduction of LER. Correlation to light propagation loss suggest a high sensitivity of LER measurements and significant improvement of patterning process performance.

## ACKNOWLEDGMENTS

We would like to thank CEA-LETI for a strong collaboration on this study. Manufacturing the wafers and contributing with data characterization for better understanding of LER measurements and correlation to propagation loss. A special thank you to Quentin Wilmart that provided the RIB and STRIP profile mode and to Tair Duvdevani-Gabbay for the recreating of the process and the illustrations.

## REFERENCES

[1] S.Levi, I.Schwarzband, R.Kris, O.Adan, "Roughness characterization of gate all around Silicon NanoWire fabrication", IEEE 27th Convention of Electrical & Electronics Engineers in Israel (IEEEI), 2012

[2] Shimon Levi, Ishai Schwarzband, Roman Kris, Ofer Adan, Elly Shi, Ying Zhang, and Kevin Zhou "Edge roughness characterization of advanced patterning processes using power spectral density analysis (PSD)", Proc. SPIE 9782, Advanced Etch Technology for Nanopatterning V, 97820I (28 March 2016)

[3] Shimon Levi and al. "A holistic metrology sensitivity study for pattern roughness quantification on EUV patterned device structures with mask design induced roughness", Proc. SPIE 10585, Metrology, Inspection, and Process Control for Microlithography XXXII, 1058511 (2 August 2018)

[4] A. Fay and Al. "Mask grade impact on optical transmission losses" at EMLC 2019

[5] C. Dupré and Al. "Immersion lithography introduction in Si Photonics platform" at SPIE Photonics West 2020

[6] Kyllian Millard, Basile Meynard, Yann Lee, Matthias Colard, Jonathan Faugier-Tovar, Haidar Al Dujaili, Olivier Lartigue, Daivid Fowler, Elise Ghibaudo, and Christophe Martinez "Dense silicon-nitride PIC design and manufacturing on transparent substrate for a retinal projector evaluation", Proc. SPIE 12004, Integrated Optics: Devices, Materials, and Technologies XXVI, 120040J (5 March 2022)

# Plasma Process Classification using Causal Discovery Technique

KOBAYASHI Dai, KITSUNEZUKA Masaki, KATAOKA Yuki
AI Development Department, System Development Center
Tokyo Electron Ltd.
Sapporo, Japan
responsible author: dai1.kobayashi@tel.com

SHINAGAWA Jun
Tokyo Electron America, Inc.
Austin, U.S.A.
jun.shinagawa@us.tel.com

*Abstract*—The plasma etching process for semiconductor fabrication is too complex to specify the causal structure of the mechanism especially of process variation. Therefore, prediction of etching performance is affected by correlation but not actual causal relationship to process variation. Such correlation is called pseudo correlation. In this research, we introduced the causal discovery technique to clarify the causality of the parameters in process. This method has been applied for experimental process data with consumed parts. The causal structure has been estimated reasonable and a model based on the structure have been achieved better prediction precision for process performance and parts consumption.

*Keywords—machine learning, virtual metrology, process classification, causal analysis, causal discovery, dry etching, plasma process*

## I. INTRODUCTION

In the plasma process for semiconductor fabrication, process performance are varied by the several root-causes: recipe modifications, consumable parts conditions, machine individual differences, and other chamber parts variations. Such variation of process must be detected with the sensors, and the affected performance also must be predicted precisely. Based on the prediction, correction for the process variation also must be applied immediately. These processes significantly affect the stability and productivity of the fabrication. To achieve more effective control, we need to improve the prediction model to describe the mechanism of the process properly.

## II. EXISTING APPROACHS

Combination of Virtual Metrology (VM) and Advanced Process Control (APC) is a promising approach for detection and correction of the process variation. However, the coverage of the application highly depends on the precision and universality of the VM model.

### A. Virtual Metrology

A simple way to build such VM models is to adopt parameters highly correlated with process performance. Another approach is to apply machine-learning-based parameter selection to determine key explanatory parameters[1]. However, there are many cases where no causality relationship has been established between explanatory or correlated parameters and process variations,

which are called "pseudo correlation". Such pseudo correlation makes the VM models unreliable because the performance is predicted by unsubstantial parameters. Such a VM model can be applied only for limited cases in which pseudo correlation cause only negligible effect. To improve the universality of the VM model application, causality of the parameters must be taken into account during the VM model building sequence.

### B. Causal Discovery

Recently, causal discovery methods such as LiNGAM [2] and DAGs with NO TEARS [3] are presented to learn the causal structure from data. Generally, these methods supposed the Structural Equation Modeling(SEM) for the causal structure, in which parameters are represented as nodes and relations between parameters are represented as directed edges. The combination and weights of edges are optimized under acyclic requirement, i.e. to set constrain to prevent looping of the edges. This approach has been expected to solve the pseudo-correlation problem by specification of the causal structure. Nevertheless, they are slow in calculation and low in precision with a large number of parameters having strong collinearity, which is common for parameters in the plasma process.

## III. PROPOSED CAUSAL DISCOVERY METHOD

In this work, we have proposed a causal discovery method with a correction to enhance model robustness and applied it to build a VM model which can take the effect of consumable parts into consideration. Fig 1 shows the flow chart of the proposed method and the application for model construction, which employs an iteration loop to correct the unreliable causal relation due to multi-collinearity by removing problematic edges and re-training the causal structure.

This approach works very well on the complicated plasma process where numerous sensors are deployed to monitor the chamber and plasma conditions. It should be noted that the learned causal structures make a VM model robust, as well as identifies a root cause of process variation.

## IV. EXPERIMENTAL RESULT

The process performance variation by parts consumption is one of the significant known issues on the plasma semiconductor process. Therefore, we applied the proposed method to experimental data with variation of consumed parts condition as well as recipe parameters.

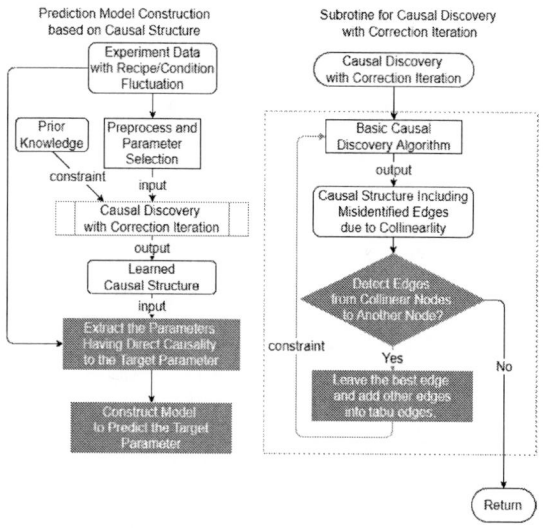

Fig 1. Flow chart of causal structure learning by the proposed method and VM model construction based on the learned causal structure.

### A. Experiment Data

For this analysis, a data set with 25 process runs were acquired with variation of 4 consumable parts conditions (new /consumed) and skew of 5 recipe parameters (+/-10%). Time traces were also recorded by sensors for electrical, gaseous, temperature, and other condition monitoring. As a process performance parameter, Etching Rate(E/R) was also measured for each process. Summary of the data set is shown in Table 1.

The recipe and sensor parameters were normalized as the fractional variations from the standard values, while the parts conditions were quantified as 1 for fully consumed and 0 for new. Hence, the coefficient between parameters is defined as the propagation degree of variation.

Table 1. Data set summary.

| Number of Parts with Consumption Condition | 4 |
| | (Top Electrode, Bottom Electrode, Cover Ring, and Depo shield) |
| Number of Recipe Parameters for Scans | 5 |
| | (Gas Pressure, Total Gas Flow, RF Power for low and high frequency, and Direct Current Source) |
| Number of Sensor Parameters | 25 |
| Number of Process Performance Parameters | 1 |
| | (ER) |
| Number of Experiments | 52 |

### B. Learned Causal Structure

The causal structure for the E/R variations was generated by proposed causal discovery method as shown in Fig 2. The nodes shown are recipe parameters(cyan frame), consumable

parts(cyan frame and green filled), sensor(black frame or magenta frame and white filled), and performance(magenta frame and orange filled). Nodes with cyan frame are root cause parameters, while nodes with magenta frame are terminal parameters. The edges between nodes represents the regression coefficient values. The causal structure visualizes parameterized causal relationships over the E/R variations as a flow from top to bottom.

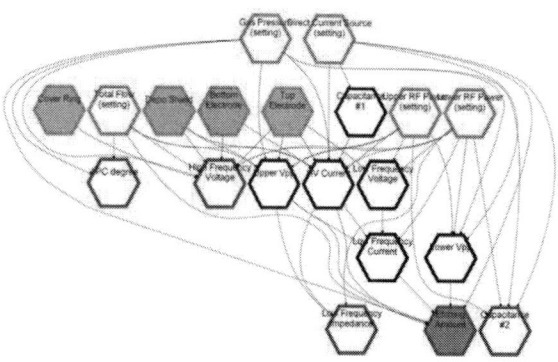

Fig 2. Learned causal structure from experiment data. Parameters without any edges or removed by preprocess are not shown in this figure.

### C. Etching Rate Prediction

VM models for E/R prediction were constructed by two parameter selection methods – one based on the casual structure ("causality-based model") and the other based on the correlations("correlation-based model"). Same number of explanatory parameters were used for both VM models in order to benchmark two parameter selection methods independent of its impact on the prediction accuracy. For the first model, the parameters having causality to the E/R variations were selected as explanatory parameters. While for the other model, the parameters having high correlation with the E/R variations were selected.

Prediction results are shown in Fig 3 and model parameters and performance metrics are summarized in Table 2. The causality-based model shows better fitting than the correlation-based model especially for smaller variations of E/R with narrower confidential interval. The accuracy of the models were evaluated by r-squared score: 0.979 for the causality-based model and 0.937 for the correlation-based model which are shown in Table 2. The better accuracy of the causality-based model can be attributed to removal of the pseudo-correlation parameters included in the correlation-based model. The prediction model suffers extra smearing when pseudo-correlation parameters are included in the model. Furthermore, causality-based explanatory parameters include parameters with effective and independent components even if their correlations to the E/R variations are low. Meanwhile, the correlation-based parameter selection method drops these parameters with low correlations.

Thus, the VM model based on causal structure is constructed using causal upstream parameters properly. Hence, variations of conditional parameters like parts consumption are also be taken account to the prediction properly. This characteristic feature of the approach expects to make a model coverage more universal.

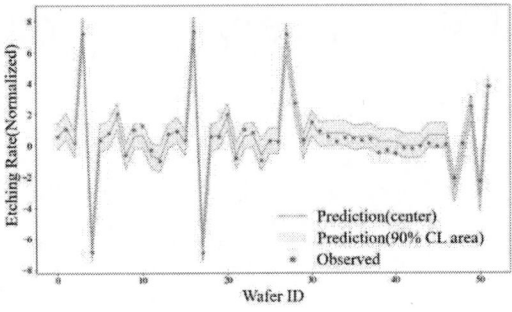

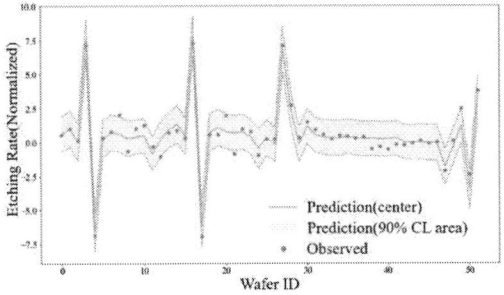

Fig 3. Prediction results for E/R by the VM model with selected parameters based on causal structure(top) and correlation(bottom). Observed values(cyan dots) and prediction values(green line) with 90% confidence limit interval(green shaded area) are sown.

Table2. Summary of VM model parameters and results of the prediction performances.

|  | Based on Causal Structure | Based on Correlation |
|---|---|---|
| Number of parameters | 7 | 7 |
| Model | Multiple Linear Regression | Multiple Linear Regression |
| Bare R2 score | 0.983 | 0.945 |
| Adjusted R2 score | 0.979 | 0.937 |
| RMSE | 0.323 | 0.572 |

### D. Degree of Consumption Prediction

It is also important to determine the root cause of the process variation for fast plasma process tool recovery. The causal structure expect to make it possible to estimate the degree of consumption for consumable parts. The parts consumption is one of the causes of such process variations, and it is important to determine which consumable part is causing the process variations. In this case, the degree of

consumption for each consumable part have to be monitored. The causal downstream parameters are affected by the parameter variation including parts consumption. Therefore, prediction model for each consumable part was constructed using downstream sensor parameters and causal upstream parameters excluding parts consumption parameters.

Fig 4 shows the prediction results for the degree of top electrode consumption as an example. The prediction model was constructed with 4 downstream parameters; High and Low Frequency RF Voltage, Direct Current, Upper peak-to-peak voltage (Vpp), and their 6 upstream parameters; setting values of Total Gas Flow, Gas Pressure, Direct Current(DC) Source, Upper RF Power, Lower RF Power, and measured value of one of the Capacitors, shown in Fig 2. Degree of consumption with new and consumed electrodes are quantified as 0 and 1, respectively. The setting consumption values are well predicted by using setting and measured sensor values only. This approach expects to be utilized for

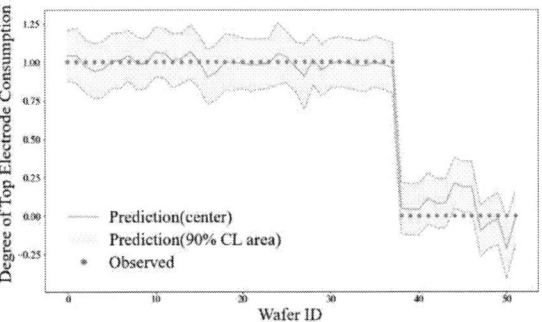

Fig 4. Prediction results for degree of top electrode consumption by the model with selected parameters based on causal structure(top). The color conventions are the same as for Fig 3.

determination of parts maintenance timing or recipe tuning considering consumption.

## V. CONCLUSION

The causal structure learned by proposed causal discovery method is useful to VM model construction and root cause identification for the process performance variation. This technique allows to improve model versality and time spent for root-cause investigation and resolution of process variation.

### REFERENCES

[1] J. Shinagawa, et al., "Successful pairing of plasma domain knowledge and machine learning techniques for virtual metrology model with root-cause analysis capabilities", 2021 APCSM conference.

[2] S. Shimizu, et al., "A linear non-gaussian acyclic model for causal discovery. Journal of Machine Learning Research, 7: 2003--2030, 2006.

[3] X. Zheng, et al., "DAGs with NO TEARS: Continuous Optimization for Structure Learning", NeurIPS 2018.

# Plasma Diagnostics and Characteristics of Hydrofluorocarbon Films in Capacitively Coupled CF$_4$/H$_2$ Plasmas

Shih-Nan Hsiao
Center for Low-temperature Plasma
Sciences, Nagoya University
Nagoya, Japan
hsiao@plasma.engg.nagoya-u.ac.jp

Yusuke Imai
School of Engineering
Nagoya University
Nagoya, Japan
imai.yusuke.d5@s.mail.nagoya-u.ac.jp

Nikolay Britrun
Center for Low-temperature Plasma
Sciences, Nagoya University
Nagoya, Japan
britun@plasma.engg.nagoya-u.ac.jp

Takayoshi Tsutsumi
Center for Low-temperature Plasma
Sciences, Nagoya University
Nagoya, Japan
tsutsumi@plasma.engg.nagoya-u.ac.jp

Kenji Ishikawa
Center for Low-temperature Plasma
Sciences, Nagoya University
Nagoya, Japan
ishikawa@plasma.engg.nagoya-u.ac.jp

Makoto Sekine
Center for Low-temperature Plasma
Sciences, Nagoya University
Nagoya, Japan
sekine@plasma.engg.nagoya-u.ac.jp

Masaru Hori
Center for Low-temperature Plasma
Sciences, Nagoya University
Nagoya, Japan
hori@nuee.nagoya-u.ac.jp

*Abstract*—**Plasma diagnostics including electron density, temperature, neutral atomic densities of the CF$_4$/H$_2$ plasmas were performed in a capacitively-coupled reactor using surface-wave probe, Langmuir probe and vacuum ultraviolet absorption spectroscopy. The plasma density increased monotonically with varying H$_2$ content from 30 to 90 %. The electron temperature first decreased with H$_2$ up to 50 % and then increased at higher H$_2$ concentration. The HF concentration reached a maximum value at a H$_2$ of approximately 50 %, which is probably due to balance between H and F radicals from the plasma. Increasing the H$_2$ content resulted in a higher H concentration and a less cross-linked structure of the amorphous hydrofluorocarbon films, analyzed by using *in situ* Fourier transformation infrared spectroscopy.**

*Keywords—Capacitively-coupled plasma, Hydrofluorocarbon, Plasma diagnostics, Hydrogen, CF$_4$*

## I. INTRODUCTION

In the recent decades, plasma processes with high density plasma sources have been continuously developed to meet the requirements from the sustainable growth in manufacture, the etch process for the dielectric materials is getting more challenging to meet the requirements, for instance, precise control of etching amount, high aspect ratio contact etching for deep holes, and etch selectivities among materials.

To etch the Si-based materials, typically, the fluorocarbon gases have been used for over decades. In these cases, the deposition of fluorocarbon film by CF$_x$ radicals and ions not only provide the etchants but also would protect the sidewall as if ion energy distribution can be well-controlled [1]. The addition of other gases, such as O$_2$, Ar can be used to change the features of the plasma and improve the etch performance as well [2]. The addition of hydrogen to fluorocarbon plasmas has been reported to manipulate the etch selectivities among the SiO$_2$, SiN and Si. [3, 4]. The hydrofluorocarbon gases such as (C$_2$H$_3$F$_3$, C$_3$H$_2$F$_6$, etc) have also been demonstrated their superiority for Si-based material etching [5, 6]. It turns out that hydrogen is one of key factors to affect the etching

behavior and mechanism, especially for SiN films [7-9]. Many diagnostics on the CF$_4$/H$_2$ plasmas have been reported for monitoring gas phase species by using a quadrupole mass spectroscopy (QMS) and optical emission microscopy (OES) [10]. Although the relative concentration of radicals as a function of H$_2$ percentage has been reported, the absolute amount of radical density is still lack. Furthermore, the relationship between plasma properties and characteristics of the polymer deposition by CF$_4$/H$_2$ plasmas is still unknown, which is also significant for the fluorocarbon-assisted cyclic etching, especially for the ALE process .

In this study, the plasma diagnostics and film structure of the hydrofluorocarbon deposition using CF$_4$/H$_2$ gas mixture in a capacitively-coupled plasma reactor as a function of H$_2$ percentage was investigated.

## II. EXPERIMENTAL

### A. Capactively-coupled Plasma Reactor

A capacitively-coupled plasma reactor with a very high frequency of 100 MHz at the upper electrode was used, as illustrated in Fig. 1. Before introducing a feedgas for plasma discharge, a background vacuum below $5 \times 10^{-4}$ Pa was kept. The feedgas mixture of CF$_4$/H$_2$ were supplied through a showerhead at the upper electrode with a diameter of 150 mm. The power input for the plasma discharge was fixed at 300 W. The working pressure was fixed at 4 Pa. A dummy Si wafer with a diameter of 100 mm was set at the bottom electrode without applying bias voltage. The electrode gas between top and bottom is approximately 95 mm. Substrate temperature of Si wafer was controlled by a circulant cooling system at 20 °C. To improve the thermal conductivity between the Si wafer and bottom electrode a back He flow with a pressure of over 700 Pa was introduced.

### B. Plasma Diagnostics and Polymer Film Characterization

The surface wave probe was used to measure the plasma density ($N_e$), which is capable of high accuracy even when the probe surface is contaminated by a deposited film, typically in

978-1-6654-7134-3/22 $31.00 © 2022 IEEE

a fluorocarbon plasma [11]. The electron temperature ($T_e$) of the plasmas was analyzed with a Langmuir probe (Impedans Ltd.). The H radical in the plasmas with different percentage of hydrogen was measured with a vacuum ultraviolet absorption spectroscopy (VUVAS). The film structure of the polymeric films deposited by the $CF_4/H_2$ plasmas was characterized with an attenuated total reflection Fourier Transformation infrared spectroscopy (ATR-FTIR). In the ATR-FTIR measurement, a Ge prism with diameter of 80 × 20 × 1 mm$^3$ (45° bevel edges) was used (iS50, Thermofisher).

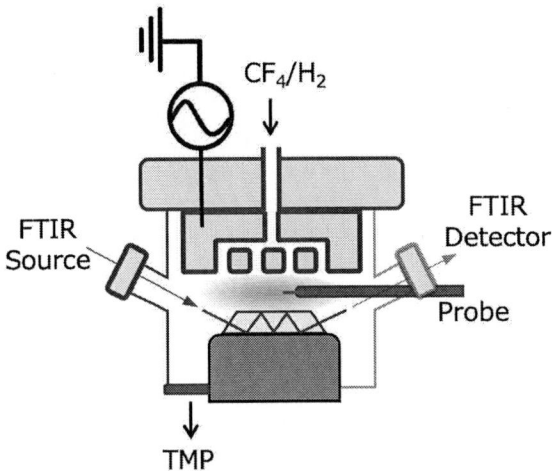

Fig. 1. Schematic of the capacitively-coupled plasma system.

### III. PLASMA DIAGNOSTICS

Fig. 2 shows the $N_e$ as a function of $H_2$ content in the $CF_4/H_2$ plasmas. The $N_e$ monotonically increased from approximately 1.06 to $1.64 \times 10^{11}$ cm$^{-3}$, as the $H_2$ content was increased. This can be attributed to the decrease of $CF_4$ percentage in the plasma due to its nature of electronegative property.

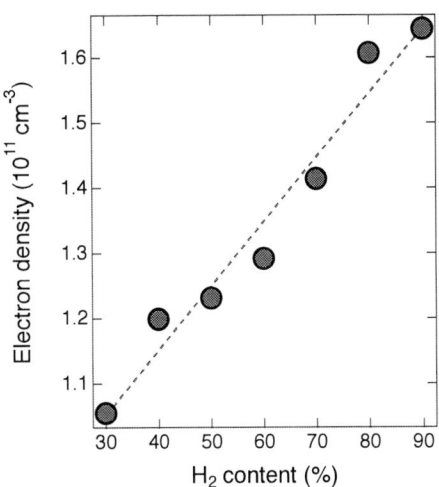

Fig. 2. Dependence of electron density as a function of $H_2$ content.

The dependence of $T_e$ on $H_2$ percentage in the $CF_4/H_2$ plasmas is illustrated in Fig. 3. For the plasma with $H_2$ = 30 %, the $T_e$ was turned out to be ~3.66 eV. It decreased to ~2.75 eV as $H_2$ content was increased to 50 %, and then increased to

3.75 eV for the $H_2$ content of 90 %. This kind of U-shape dependence of $T_e$ on $H_2$ content has been also reported before; however, none of any explanations has been addressed [12]. The followings are our concerns: I) When the $H_2$ content is increased, one can expect that the particle mean free paths will be shorten, which lead to a high collision frequency between species that, in turn, cause a decrease of $T_e$ due to a higher chance of collision between electron and other species, similar to the pressure effect ; II) As the $H_2$ percentage is increased, effective ion mass can be expect to be increase, which may lead to an increase of $T_e$, based on the ion species particle conservation model [13].

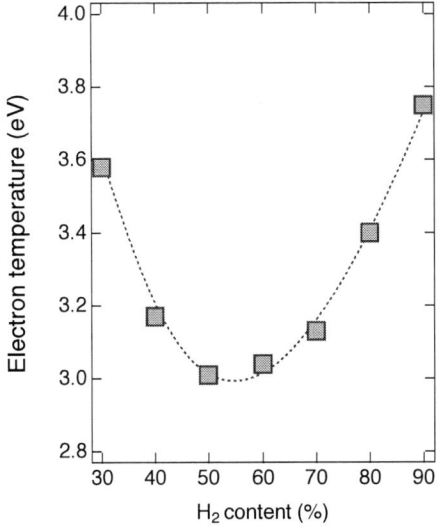

Fig. 3. Dependence of electron temperature as a function of $H_2$ content.

The absolute density of hydrogen fluoride (HF), analyzed with the FTIR, in the gas phase from the plasmas has a maximum magnitude of ~ $6 \times 10^{14}$ cm$^{-3}$ for the $H_2$ content of approximately 50 % in the feedgas (not shown). Combined with the results from the absolute amount of H radical obtained by the VUVAS, it reveals that the H radical increased rapidly for the mixtures with > 50 % of $H_2$ additive to $CF_4$. Since the H radical scavenges F which lead to formation of HF molecules, it infers that the F radical should also decrease with increasing $H_2$ % in $CF_4/H_2$ plasmas, similar to the previous literatures [10].

### IV. THIN FILM CHARACTERIZATION

Fig. 4 illustrates the *in situ* monitoring of absorbance change of the ATR-FTIR spectra for the hydrofluorocarbon polymer films deposited by $CF_4$ plasma with different percentages of $H_2$ additive. The four main bands in the spectra are assigned to $C–F_x$ stretching at 1230 cm$^{-1}$, C–H deformation at 1460 cm$^{-1}$, C=C stretching at 1700 cm$^{-1}$ and C–H stretching at 2900 cm$^{-1}$. The absorbance area of the C–H bands increased as the $H_2$ content was increased; in the meanwhile the $C–F_x$ and C=C decreased with $H_2$ content being increased. The decrease of C=C band indicates a less cross-linked fluorocarbon polymer for the condition with greater $H_2$ content, i.e., a hydrated diamond-like amorphous carbon. Furthermore, the C–H and $C–F_x$ bonds shifted toward lower and higher wavenumber (frequency), respectively, as the $H_2$ content was decreased. This frequency shift phenomenon is attributed to changes in back-bonds that incorporate more fluorine atoms with high electronegativity into C–H and C–F$_x$ moieties [14]. Thus, it also indicates that the H (F) concentration in the hydrofluorocarbon films increased (decreased) with increase of H$_2$ contents.

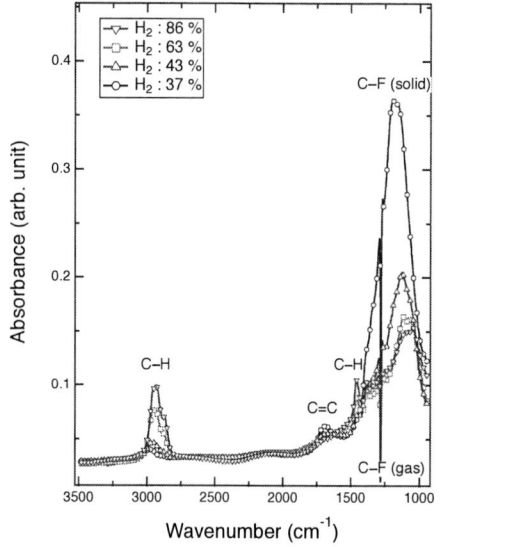

Fig. 4. Absorbance change of FTIR spectra in the hydrofluorocarbon polymer films deposited on a Ge prism by CF$_4$/H$_2$ plasmas

## V. SUMMARY

The electron density, electron temperature and neutrally atomic density in the capacitively coupled CF$_4$/H$_2$ plasmas were investigated as a function of H$_2$ percentage in the feedgas using probe-based and spectral absorption techniques. The $N_e$ increased monotonically with increasing H$_2$ content due to the change of electronegativity in the gas mixture. The $T_e$ first decreased with increasing H$_2$ up to 50 %, and increased for H$_2$ > 50 %. The additive of H$_2$ is expected to increase the mean free path of the species in the gas phase, which causes the higher probability of the collisions between electrons and gas species that may decrease of $T_e$. On the other hand, the effective ion mass decreases with increasing H$_2$ may be responsible for the $T_e$ increase, based on the ion species particle conservation model [13]. The magnitude of neutral HF molecules reached to a maximum as H$_2$ was increased to 50 %, which probably due to a balance between H and F radicals in the plasmas. By using *in situ* ATR-FTIR technique, it was found that with increasing H$_2$ content the structure of the polymeric film deposition became less cross-linked and its chemical composition was also less fluorinated.

## VI. REFERENE

[1] M. Hori, "Radical-controlled plasma processes," *Rev. Mod. Plasma Phys.,* vol. 6, p. 36, 2022.

[2] M. Matsui, T. Tatsumi, and M. Sekine, "Relationship of etch reaction and reactive species flux in C$_4$F$_8$/Ar/O$_2$ plasma for SiO$_2$ selective etching over Si and Si3N4," *J. Vac. Sci. Tech. A,* vol. 19, p. 2089, 2001.

[3] D. C. Marra and E. S. Aydil, "Effect of H$_2$ addition on surface reactions during CF$_4$/H$_2$ plasma etching of silicon and silicon dioxide films," *J. Vac. Sci. Tech. A,* vol. 15, p. 2508, 1997.

[4] Y. Zhang, G. S. Oehrlein, and F. H. Bell, "Fluorocarbon high density plasmas. VII. Investigation of selective SiO$_2$-to-Si$_3$N$_4$ high density plasma etch processes," *J. Vac. Sci. Tech. A,* vol. 14, p. 2127, 1998.

[5] S. N. Hsiao *et al.*, "Selective etching of SiN against SiO$_2$ and poly-Si films in hydrofluoroethane chemistry with a mixture of CH$_2$FCHF$_2$, O$_2$, and Ar," *Appl. Surf. Sci.,* vol. 541, p. 148439, 2021.

[6] H.-W. Tak *et al.*, "Effect of hydrofluorocarbon structure of C$_3$H$_2$F$_6$ isomers on high aspect ratio etching of silicon oxide," *Appl. Surf. Sci.,* vol. 600, p. 154050, 2022.

[7] S. N. Hsiao *et al.*, "Effects of hydrogen content in films on the etching of LPCVD and PECVD SiN films using CF$_4$/H$_2$ plasma at different substrate temperatures," *Plasma Proc. Polym.,* vol. 18, p. e210078, 2021.

[8] S. N. Hsiao, T.-T.-N. Nguyen, T. Tsutsumi, K. Ishikawa, and M. Hori, "On the etching mechanism of highly hydrogenated SiN films by CF$_4$/D$_2$ plasma: Comparision with CF$_4$/H$_2$," *Coatings,* vol. 11, p. 1535, 2021.

[9] S. N. Hsiao, K. Nakane, T. Tsutsumi, K. Ishikawa, M. Sekine, and M. Hori, "Influence of substrate temperatures on etch rates of PECVD-SiN thin films with a CF$_4$/H$_2$ plasma," *Appl. Surf. Sci.,* vol. 542, p. 148550, 2021.

[10] H.-H. Doh, J.-H. Kim, K.-W. Whang, and S.-H. Lee, "Effect of hydrogen addition to fluorocarbon gases (CF$_4$, C$_4$F$_8$) in selective SiO$_2$/Si etching by electron cyclotron resonance plasma," *J. Vac. Sci. Tech. A,* vol. 14, p. 1088, 1996.

[11] H. Sugai and K. Nakamura, "Recent innovations in microwave probes for reactive plasma diagnostics," *Jpn. J. Appl. Phys.,* vol. 58, p. 060101, 2019.

[12] H. Shindo, M. Konishi, and Y. Horike, "Emissive probe study of CF$_4$/H$_2$ etching plasma," *Jpn. J. Appl. Phys.,* vol. 30, p. 578, 1991.

[13] M. Sode, T. Schwarz-Selinger, and W. Jacob, "Quantitative determination of mass-resolved ion densities in H$_2$-Ar inductively coupled radio frequency plasmas," *J. Appl. Phys.,* vol. 113, p. 093304, 2013.

[14] X. Wang, H. R. Harris, K. Bouldin, H. Temkin, and S. Gangopadhyay, "Structural properties of fluorinated amorphous carbon films," *J. Appl. Phys.,* vol. 87, p. 621, 2000.

# *In Situ* Measurement and Analysis of Low Pressure Gas Concentration Distribution Using 70-dB SNR 1,000 Frame-per-second Absorption Imaging System

Yushi Sakai, Yoshinobu Shiba, Takafumi Inada, Tetsuya Goto, Tomoyuki Suwa, Akihito Sutoh, Tatsuo Morimoto, Yasuyuki Shirai, and Shigetoshi Sugawa

*New Industry Creation Hatchery Center*
*Tohoku University*
Sendai 980-8579, Japan.
E-mail: yushi.sakai.c7@tohoku.ac.jp

Tetsu Oikawa and Aoi Hamaya
*Graduate School of Engineering*
*Tohoku University*
Sendai 980-8579, Japan.

Rihito Kuroda
*Graduate School of Engineering*
*and New Industry Creation*
*Hatchery Center*
*Tohoku University*
Sendai 980-8579, Japan

*Abstract*—We are developing a 70-dB SNR 1,000 frame-per-second absorption imaging system, and it enables us to visualize gas concentration distribution and its dynamic behavior in a chamber for semiconductor manufacturing. In the steady-state gas flow condition, the spectrometry of $NO_2$ gas shows a good linear relationship between the absorbance and the partial pressure above approximately 0.1 Pa with the system. Such low partial pressure measurement enables the visualization of beginning of jet out of $NO_2$/Ar mixed gases from a nozzle. Furthermore, 1,000 frame-per-second imaging enables analyses of the gas velocity distribution and its width.

*Keywords*—*Absorption imaging, CMOS image sensor, global shutter, SNR, in situ, gas concentration monitoring, semiconductor manufacturing.*

## I. INTRODUCTION

The manufacturing industry of electronic devices, such as semiconductors, power devices, LEDs, and displays, have made tremendous technological progress from the latter half of the 20th century to the 21st century. In order to keep up with the progress, semiconductor manufacturing processes have been also constantly required technological advances, such as improvement of dimensional accuracy of nanometer-order structures, yield enhancement, contamination countermeasures, and so on. While such process technologies are being advanced, atomic layer deposition (ALD) and etching (ALE) are attracting attention as next-generation semiconductor manufacturing process. It is necessary for ALD and ALE to control a material surface at the monoatomic layer level precisely, so gas concentration control/monitoring technologies in the process chamber have been more important. Until now, a lot of studies about gas control in chamber for semiconductor manufacturing process have been undertaken, for example, fast gas replacement and concentration control technology in the process chamber [1]-[4], and *in situ* monitoring in the chamber by infrared spectroscopy [5]-[7]. About in-line gas concentration monitoring, although it is not *in situ*, a compact gas concentration sensor unit with a single photodiode using ultraviolet light absorption was developed and it realized high sensitivity, fast response, and compatible with various organometallic gases in electronic device manufacturing process [8], [9]. To realize next-generation ALD and/or ALE processes with high accuracy, high uniformity and high throughput, it is essential to understand distribution of process gas concentrations in the chamber which changes significantly over time during cyclic sequences of alternative introductions of process and purge gases [10]. For this purpose, we are now developing a high-speed multi-dimensional absorption imaging technology to obtain temporal and spatial distribution of process gas concentration in the chamber using 70-dB SNR high-speed global shutter CMOS image sensors [11]. In this paper, measurement results and analysis of $NO_2$ gas injected into the chamber from the gas nozzle are presented.

## II. DEVELOPED SPECTROPHOTOMETRY SYSTEM

Fig. 1 shows (a) a photograph and (b) a schematic illustration of the developed spectrophotometry system installed into a chamber for 300 mm diameter wafer used in this work. The chamber inner diameter was 430 mm. The chamber has three pairs of quartz windows placed each in opposite sides to perform absorption imaging. From one window, a light with 405 nm wavelength was injected into the chamber using LED. A cross section of the light was approximately 32 × 32 mm. The wavelength of 405 nm corresponds to the absorption peak of $NO_2$ gas. The developed camera module with 1/10 magnification telecentric lens was set just outside of the window at the opposite side, to capture the absorption image. Here, telecentric lens emit and receive only rays of light collimated to the optical axis. It is useful to improve spatial

PM–41

2022 International Symposium on Semiconductor Manufacturing (ISSM).
December 12–13, 2022

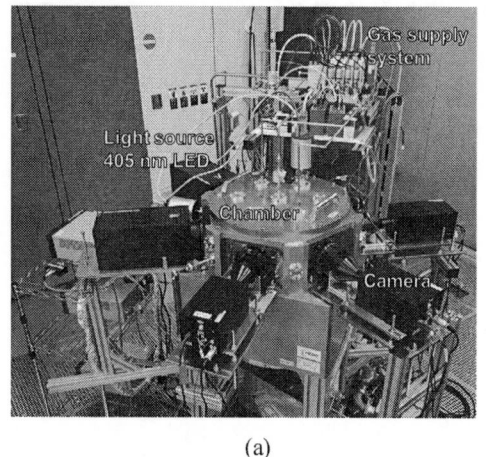
(a)

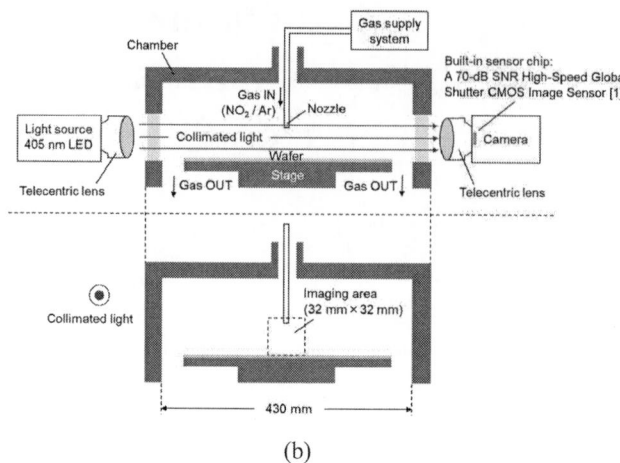
(b)

Fig. 1. (a) A photograph and (b) a schematic illustration and imaging area of the developed spectrophotometry system installed into a chamber used in this work.

resolution as well as for the 3-D image construction of gas concentration distribution. The employed global shutter CMOS image sensor was developed by the group of authors for the high-speed high precision *in situ* fluid absorption imaging. It has 140 × 140 pixels with 22.4 μm pitch with maximum 1,000 frames/s. The unprecedented high full well capacity of over 27 Me⁻ results in maximum 70 dB SNR, which makes the low concentration absorption imaging possible [11].

III.    MESUREMENT RESULT AND DISCUSSION

Fig. 2 shows the relationship between light absorbance and partial pressure of $NO_2$ gas in the low concentration region in the steady-state gas flow condition. Here the absorbance was obtained at the center of pixel array. The partial pressure of $NO_2$ was controlled by changing the flow ratio of $NO_2$/Ar gases and the pressure in the chamber. The maximum flow ratio and pressure were 1.5 % and 86 Pa respectively. The minimum flow ratio and pressure were 0.5 % and 5.6 Pa respectively. Thanks to the high SNR of the developed camera, the obtained results show good linearity above approximately 0.1 Pa. Here the partial pressure of 0.1 Pa is equivalent to several-ppm under atmospheric pressure, indicating the high precision of the developed absorption imaging system.

In Figs. 3 (1a) – (1c) and (2a) – (2c), 20 % $NO_2$/Ar gas jet from the nozzle with an inner diameter of 1 mm was measured in the range of absorbance below 0.005 obtained by different SNR conditions of 70 dB (this work) and 43 dB (conventional cameras), respectively. The flow rate of inlet gas was 20 sccm and the pressure in the chamber was 1.1 kPa. Under these conditions, the nozzle exit velocity was 4.2 m/s. Here 4.2 m/s is the same as exit velocity from a hole of

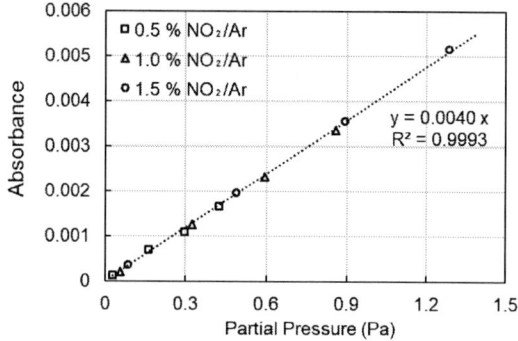

Fig. 2. Acquired calibration curve extracted from the obtained images. Partial pressure is the product of gas flow ratio and pressure in chamber, and it corresponds to the $NO_2$ gas concentration in vacuum chamber.

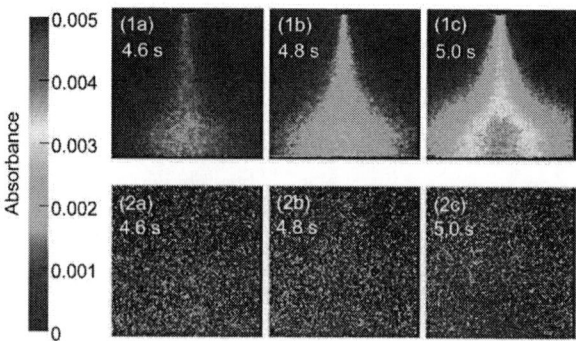

Fig. 3. Captured absorbance images of jet of $NO_2$ and Ar mixed gases from Φ1 mm nozzle at 1,000 frames/s by different SNR conditions, (1a) – (1c) 70 dB and (2a) – (2c) 43 dB. $NO_2$ flow ratio was 20 %, the pressure in the chamber was 1.1 kPa, and the calculated nozzle exit velocity was 4.2 m/s.

978-1-6654-7134-3/22 $31.00 © 2022 IEEE          70

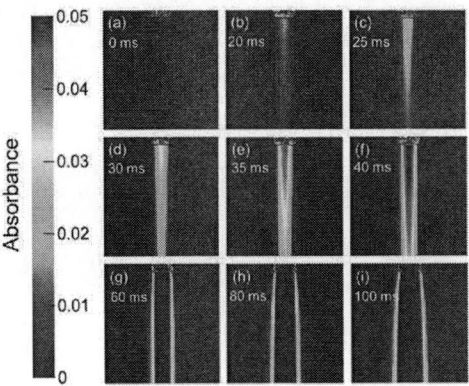

Fig. 4. Captured images of jet of 100 % NO₂ gas from Φ4.35 mm nozzle at different times captured at 1,000 frames/s. The pressure in the chamber was 22.3 kPa and the calculated nozzle exit velocity was 4.2 m/s.

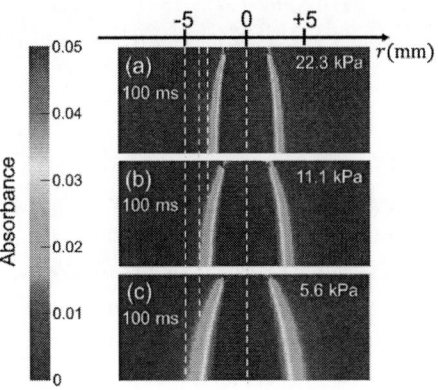

Fig. 5. (a)-(c) Captured images of jet of NO₂ gas from Φ4.35 mm nozzle at 100 ms for different pressures, (a) 22.3 kPa, (b) 11.1 kPa, and (c) 5.6 kPa. The calculated nozzle exit velocity was 4.2 m/s in all the pressure cases.

a shower plate with 4,218 holes, the flow rate 1 SLM, and the pressure in chamber 133 Pa. So the exit gas velocity 4.2 m/s is reasonable for semiconductor manufacturing process. Here, the absorbance of 0.005 corresponds to the NO₂ partial pressure of approximately 1.3Pa calibrated from Fig. 2. Therefore, Fig. 3 (1a) – (1c) shows that the developed spectrophotometry system is able to visualize the absorbance distribution clearly in the low partial pressure under 1.3 Pa, it is impossible with conventional cameras in Fig. 3 (2a) – (2c).

Fig. 4 (a) – (i) shows temporal and spatial absorbance distribution of 100 % NO₂ gas injected into the chamber from the nozzle with an exit hole diameter of 4.35 mm for 100 ms, where the inlet gas was changed from Ar to NO₂ at time 0 ms, while the chamber pressure was kept at 22.3 kPa. The gas flow rate was 756.8 sccm and the hole exit velocity was 4.2 m/s. Thanks to 1,000 frames/s imaging system, the dynamic change of gas flow distribution was successfully obtained.

Fig. 5 (a) - (c) shows the absorbance distribution of NO₂ gas flow at 100 ms at various pressures of (a) 22.3 kPa, (b) 11.1 kPa, and (c) 5.6 kPa from the nozzle with an exit hole diameter of 4.35 mm. We set the gas flow rate to keep the gas velocity the same at the hole exit (4.2 m/s). It was found that the width of gas jet distribution broadened as the pressure decreased.

To analyze the width of the gas jet in Fig. 5, we defined the vertical pseudo gas flow velocity at arbitrary position $k$ as follows:

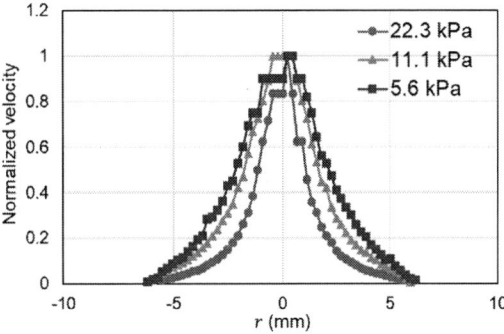

Fig. 6. Radial distribution of normalized gas-flow velocity at the height just above of the wafer substrate. The vertical distance from the nozzle exit is approximately 29.3 mm. The velocity was derived from absorbance transition for equation (1).

$$v_k = \frac{z_k}{t_{k,A} - t_{0,A}}, \qquad (1)$$

where $z_k$ is the vertical distance from nozzle exit to position $k$, $t_{0,A}$ and $t_{k,A}$ are the time when the absorbance first exceeds over $A$ at nozzle exit and position k respectively. In this study, threshold absorbance $A$ was set to 0.05 because sufficient resolution was guaranteed at 0.05. Fig. 6 shows radial distribution of normalized pseudo velocity calculated from equation (1) at the height just above of the wafer substrate, $z_k$ is 29.3 mm. It was found that the full width half maximum (FWHM) of the flow velocity distribution increased as the pressure decreased which was qualitatively consistent with the theoretical prediction [12]. The analysis of gas jet velocity distribution and the FWHM are expected to be useful in designing the hole-diameter and the hole-pitch of a shower plate, and the gap between the shower plate and wafer.

## IV. CONCLUSION

Using the developed *in situ* high-speed and high precision absorption imaging system, we succeeded in visualizing partial pressure distribution of $NO_2$ gas jet at a very low concentration. In addition, the pseudo velocity distribution was able to be obtained by absorbance transition and the full width half maximum of the flow velocity distribution was predicted. The results of this study are expected to contribute to the improvement of semiconductor manufacturing process technology. In the future, we will measure organometallic gases used in ALD process.

## ACKNOWLEDGMENT

This work was supported by the New Energy and Industrial Technology Development Organization (NEDO) under Project JPNP19005.

## REFERENCES

[1] M. Nagase, M. Kitano, Y. Shirai, and T. Ohmi, "Precise Control of Gas Concentration Ratio in Process Chamber," *Jpn. J. Appl. Phys.* Vol. 48, No. 1, pp. 016003-1-5, January 2009.

[2] S. Morishita, T. Goto, I. Akutsu, K. Ohyama, T. Ito, and T. Ohmi, "Fast Gas Replacement in Plasma Process Chamber by Improving Gas Flow Pattern," *Jpn. J. Appl. Phys.* Vol. 48, No. 1, pp. 016003-1-5, January 2009.

[3] S. Morishita, T. Goto, M. Nagase, and T. Ohmi, "Precise and high-speed control of partial pressures of multiple gas species in plasma

process chamber using pulse-controlled gas injection," *J. Vac. Sci. Technol.* A, Vol.27, No.3, pp.423-429, May/June 2009.

[4] S. Morishita, T. Goto, Y. Shirai and T. Ohmi, "Gas Flow Characteristics in a Plasma Process Chamber and Proposal of New Pulse-Controlled Gas Injection Method Using Interference Matrix Operation for Rapid Stabilization of Gas Pressure," *Jpn. J. Appl. Phys.*, Vol.51, No.1, 016001, December 2011.

[5] J. A. O'Neill, M. L. Passow, T. J. Cotler, "Infrared absorption spectroscopy for monitoring condensible gases in chemical vapor deposition applications," *J. Vac. Sci. Technol.* A 12,839, February 1994.

[6] S. Salimal, C. A. Wangb, R. D. Driverc, K. F. Jensen, "In situ concentration monitoring in a vertical OMVPE reactor by fiber-optics-based Fourier transform infrared spectroscopy," *J. Crystal Growth*, Volume 169,3,443, January 1996.

[7] J. E. Maslar, W. A. Kimes, and B. A. Sperling "In Situ Gas Phase Measurements During Metal Alkylamide Atomic Layer Deposition," *J. Nanosci. Nanotechnol.*, 11, 8226, September 2011.

[8] H. Ishii, M. Nagase, N. Ikeda, Y. Shiba, Y. Shirai, R. Kuroda, and S. Sugawa, "A High Sensitivity and Compact Real Time Gas Concentration Sensor for Semiconductor and Electronic Device Manufacturing Process," *ECS Trans.*, Volume 85, Issue 13, pp.1399-1405, 2018.

[9] H. Ishii, M. Nagase, N. Ikeda, Y. Shiba, Y. Shirai, R. Kuroda, and S. Sugawa, "A high-sensitivity compact gas concentration sensor using ultraviolet light absorption with a heating function for a high-precision trimethyl aluminum gas supply system," *Jpn. J. Appl. Phys.*, vol. 58, no. SB, Apr. 2019, Art. no. SBBL04.

[10] Z. Li, K. Cao, X. Li, R. Chen, "Computational fluid dynamic modeling of spatial atomic layer deposition on microgroove substrates," *Int. J. Heat Mass Transf.* 181 (2021) 121854.

[11] T. Oikawa et al., "A 70-dB SNR High-Speed Global Shutter CMOS Image Sensor for *in Situ* Fluid Concentration Distribution Measurements," *IEEE Trans. Electron Devices*, vol. 69, no. 6, pp. 2965–2972, Apr. 2022.

[12] T. Goto et al., "Establishment of very uniform gas-flow pattern in the process chamber for microwave-excited high-density plasma by ceramic shower plate," *J. Vac. Sci. Technol.* A 27, 686 (2009).

# A Study on Robust Noninteracting Control System Design with Disturbance Feedforward for 6-DoF Active Vibration Isolation Platform

Thinh Huynh
*Department of Mechanical System*
*Engineering*
*Pukyong National University*
Busan, South Korea
huynhthinh@hcmute.edu.vn

Dong-Hun Lee
*Department of Mechanical System*
*Engineering*
*Pukyong National University*
Busan, South Korea
hun_control@pknu.ac.kr

Young-Bok Kim
*Department of Mechanical System*
*Engineering*
*Pukyong National University*
Busan, South Korea
kpjiwoo@pknu.ac.kr

*Abstract*—This paper proposes a novel noninteracting control strategy for a 6-degree-of-freedom (DoF) active vibration isolation system (AVIS) that carries high-precision machinery or measuring instruments. External vibrations need rejecting effectively, and at the same time, the internal interactions and parameter uncertainties are also worth consideration. For these objectives, the system model is first derived. Then, the proposed control law consists of state feedback, unity feedback from the output, and finally, feedforward from the measured floor motion. The feedback gains are formulated such that noninteracting performance is achieved in the sense that each DoF independently follows its corresponding reference. Moreover, robust stability is obtained from linear matrix inequality (LMI) techniques. Simulation studies have been conducted to validate the proposed control system.

*Keywords— active vibration isolation system, noninteracting control, feedforward, linear matrix inequality.*

## I. INTRODUCTION

A vibration isolation platform is one of the infrastructures necessary for high-precision works, such as semiconductor manufacturing, ultra-fine product processing, and microscope measurement. The nature of the works requires their machinery are isolated from vibration disturbances. A simple solution for this is the use of mechanical springs and dampers in passive isolation systems. They can naturally suppress high-frequency vibrations without any control intervention. Unfortunately, they are incapable of dealing with low-frequency disturbances. Moreover, the resonance phenomenon is unavoidable in these systems. In other words, the passive isolation system cannot provide an ideal vibration environment for machinery and instruments working with micro-scale tolerances. Hence the semi-active and active vibration isolation systems (AVIS) whose are equipped with controllable actuators. Pneumatic actuators, hybrid actuators comprising pneumatic actuators and giant magnetostrictive actuators, piezoelectric stack actuators, and voice coil motors (VCMs) are usually used. They are controlled not only to remove the resonance phenomenon of the passive system but also to enhance the vibration isolation in a large range of frequencies.

The problem of controlling an AVIS has been widely studied. The main control objective is to effectively isolate the system from external vibration disturbances and preserve the system's stability. Many control strategies have been deployed, from the simple proportional-integral-derivative (PID) [1], composite nonlinear feedback [2], and sliding mode

[3], to H∞ robust controls [4]. Feedforward control techniques are often considered to improve the rejection of measurable disturbances. They could work alone, as in [5], [6], but usually combine with a feedback controller to enhance the system robustness, for instance [3], [4]. Besides external vibrations, other factors are also worth consideration, such as actuator backslash and inaccurate parameters [7].

Moreover, difficulties arise from a large number of controllable degrees of freedom (DoF) and actuators, as well as the interaction between them. The coupled relations in the system result in unintended actuation in one direction while the system operates in the other direction and vice versa. A direct approach is to transform the system model into a diagonal structure of independent subsystems. In [8] and [9], the authors used the modal decomposition method to expand the 6-DoF AVIS into six input-output pairs. Individual controllers are then designed for each pair resulting in a significant number of controllers. The modal decomposition is a great structure analytic method, but not an effective control approach. Instead, noninteracting control should be considered in the sense that each system input controls independently a corresponding system output rather than decoupling the system structure. Diagonal decoupling by static-state feedback, decoupling with dynamic compensation [10], noninteracting constraints [11], robust noninteracting control [12], input-output energy decoupling [13], row-by-row decoupling [14], and so forth have been applied in various applications. However, the remaining issue with these control system designs lies in the analytical method to find the appropriate control matrices that not only decouple the interactions but also achieve the desired performance.

In this paper, a robust noninteracting feedback control law in combination with floor disturbance feedforward is proposed for the 6-DoF AVIS. The resulting configuration is able to effectively isolate the external vibration disturbance and preserve the system's robustness. The control inputs act in a way that each output follows its corresponding reference signal, and interactions in the system structure are minimized. A new method to obtain the optimal controller gains using linear matrix inequality (LMI) techniques is introduced. The efficiency of the proposed control system is validated through simulation studies. The remaining of the paper is organized as follows: Section 2 formulates the system model. The design process of the control law is introduced in section 3. Simulation results are discussed in section 4. Finally, conclusions are drawn in section 5.

978-1-6654-7134-3/22 $31.00 © 2022 IEEE

## II. SYSTEM MODELING

The controlled AVIS and its schematic drawing are shown in Fig. 1. The system carries its payload on the top plate, while the bottom one is fixed on the ground or support frame. Between two plates are four coil springs and eight VCMs that act as support and perform vibration isolation. The movement of the top plate, as well as the instrument carried on it, is captured by eight vibration sensors. Each sensor measures the motion in one specific direction. Three additional sensors are on the bottom place to measure the floor vibration. The allocations of the actuators and sensors are given in detail in Fig. 1.

a. The 6-DoF AVIS.

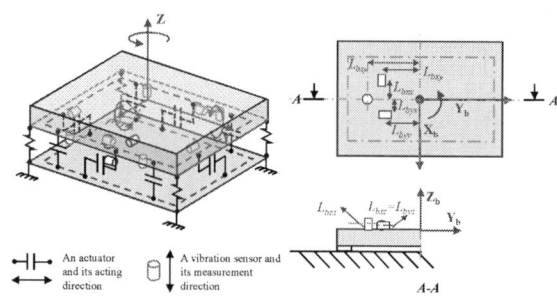

An actuator and its acting direction

A vibration sensor and its measurement direction

b. Schematic drawing of the AVIS and its bottom plate.

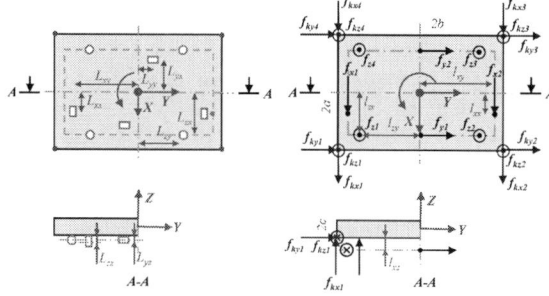

c. Springs, actuators, and sensors on the top plate.

Fig. 1 The 6-DoF AVIS and the allocation of actuators and sensors.

The system dynamics are represented by the following state-space model:

$$\dot{X} = AX + B_A U_a + B_D X_b,$$
$$Y = CX, \tag{1}$$

where the state vector $X \in R^n$, $(n = 12)$ contains the vector of top plate translation and rotation and their velocities,

respectively. The control input $U_a \in R^l$, $(l = 8)$ is the vector of the forces generated by the eight VCMs in $x$-, $y$-, and $z$-directions, respectively. $X_b \in R^n$ is the vectors of displacements and velocities of the bottom plate. Finally, $Y \in R^m$, $(m = 6)$ is the system output vector of translations and rotations of the top plate.

The matrices are listed as follows:

$$A = \begin{bmatrix} O_6 & I_6 \\ M^{-1}A_1 & M^{-1}A_2 \end{bmatrix}, \quad B_A = \begin{bmatrix} O_{6\times8}^T & M^{-1}B_a^T \end{bmatrix}^T,$$

$$C = \begin{bmatrix} I_6 & O_6 \end{bmatrix}, \quad B_D = M^{-1}\begin{bmatrix} O_6 & O_6 \\ D_1 & D_2 \end{bmatrix}, \quad M = \begin{bmatrix} mI_3 & O_3 \\ O_3 & J \end{bmatrix},$$

$$A_1 = 4\begin{bmatrix} -k_x & 0 & 0 & 0 & -ck_x & 0 \\ 0 & -k_y & 0 & -ck_y & 0 & 0 \\ 0 & 0 & -k_z & 0 & 0 & 0 \\ 0 & -ck_y & 0 & -(b^2k_z+c^2k_y) & 0 & 0 \\ ck_x & 0 & 0 & 0 & -(a^2k_z+c^2k_y) & 0 \\ 0 & 0 & 0 & 0 & 0 & -(b^2k_x+a^2k_y) \end{bmatrix},$$

$$A_2 = 4\begin{bmatrix} -c_x & 0 & 0 & 0 & -cc_x & 0 \\ 0 & -c_y & 0 & -cc_y & 0 & 0 \\ 0 & 0 & -c_z & 0 & 0 & 0 \\ 0 & -cc_y & 0 & -(b^2c_z+c^2c_y) & 0 & 0 \\ cc_x & 0 & 0 & 0 & -(a^2c_z+c^2c_y) & 0 \\ 0 & 0 & 0 & 0 & 0 & -(b^2c_x+a^2c_y) \end{bmatrix},$$

$$B_a = \begin{bmatrix} 1 & 1 & 0 & 0 & 0 & 0 & 0 & 0 \\ 0 & 0 & 1 & 1 & 0 & 0 & 0 & 0 \\ 0 & 0 & 0 & 0 & 1 & 1 & 1 & 1 \\ 0 & 0 & c+l_{yz} & c+l_{yz} & -l_{zy} & l_{zy} & l_{zy} & -l_{zy} \\ -c-l_{xz} & -c-l_{xz} & 0 & 0 & -l_{zx} & -l_{zx} & l_{zx} & l_{zx} \\ l_{xy} & -l_{xy} & l_{xz} & -l_{xz} & 0 & 0 & 0 & 0 \end{bmatrix},$$

$$D_1 = -A_1, \quad D_2 = -A_2 \tag{2}$$

In which, $O_{i\times j}$ and $I_k$ are the $i$-by-$j$ zero matrix and identity matrix of size $k$, respectively. $m$ and $J$ are the mass and inertia tensor of the top plate with its load. $k_i$ and $c_i$ ($i = x, y, z$) are the spring stiffness and damping ratio in the corresponding $i$-direction. It is worth noting that the stiffness and damping ratios are not constants but vary. In the other words, the system consists of uncertainties in addition to the model with nominal parameters. The other parameters are depicted in Fig. 1. Also from the allocations of sensors in Fig. 1, the measured motion of the top plate $Y_m$ and the bottom plate $Y_{mb}$ are calculated as follows.

$$Y_m = RY, \quad Y_m = \begin{bmatrix} x_1 & x_2 & y_1 & y_2 & z_1 & z_2 & z_4 & z_4 \end{bmatrix}^T,$$

$$Y_{mb} = R_b Y_b, \quad Y_{mb} = \begin{bmatrix} x_{1b} & y_{1b} & z_{1b} \end{bmatrix}^T,$$

$$R = \begin{bmatrix} 1 & 0 & 0 & 0 & -(c+L_{xz}) & L_{xy} \\ 1 & 0 & 0 & 0 & -(c+L_{xz}) & -L_{xy} \\ 0 & 1 & 0 & c+L_{xz} & 0 & L_{zx} \\ 0 & 1 & 0 & c+L_{xz} & 0 & -L_{zx} \\ 0 & 0 & 1 & -L_{zy} & -L_{zx} & 0 \\ 0 & 0 & 1 & L_{zy} & -L_{zx} & 0 \\ 0 & 0 & 1 & L_{zy} & L_{zx} & 0 \\ 0 & 0 & 1 & -L_{zy} & L_{zx} & 0 \end{bmatrix},$$

$$R_b = \begin{bmatrix} 1 & 0 & 0 & 0 & 0 & L_{bxy} \\ 0 & 1 & 0 & 0 & 0 & L_{byx} \\ 0 & 0 & 1 & L_{bzy} & 0 & 0 \end{bmatrix} \quad (3)$$

*Remark 1*: The AVIS is over-actuated with eight VCMs performing six forces and torques in 6-DoF motions. The force generated by each actuator affect not only its respective direction but also other directions.

*Remark 2*: There are internal interactions between the system states. That is motion in one direction results in redundant motions in the other.

## III. DESIGN OF ROBUST NONINTERACTING CONTROL WITH DISTURBANCE FEEDFORWARD

### A. Robust Tracking and Disturbance Rejection with Floor Feedforward

The schematic drawing of the proposed control system is given in Fig. 2. The feedback elements follow the classical robust tracking and disturbance rejection control law to calculate the required forces and torques in six DoFs. The control input for each actuator is chosen from the actuator allocation. They are given in Eq. (4)

$$U_a = B_a^+ U, \quad B_a^+ = B_a^T \left( B_a B_a^T \right)^{-1},$$
$$U = -FX + Gv, \quad \left( F \in R^{m \times n}, G \in R^m \right) \quad (4)$$

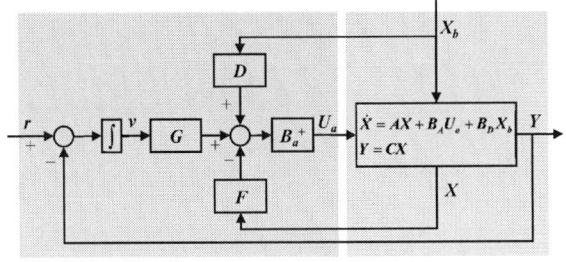

Fig. 2. Schematic drawing of the proposed control system.

Additionally, most external vibrations come from the floor where the AVIS is put on. From the measurements of the three sensors placed on the bottom plate, the feedforward law to reject these vibrations is suggested as follows:

$$U_a = DX_b,$$
$$D = -B_a^+ B_D \begin{bmatrix} R_b & O_{3 \times 6} \\ O_{3 \times 12} \\ O_{3 \times 6} & R_b \\ O_{3 \times 12} \end{bmatrix} \quad (5)$$

The system dynamics with the proposed feedback law can be written in the following form:

$$\begin{bmatrix} \dot{X} \\ \dot{v} \end{bmatrix} = \underbrace{\begin{bmatrix} A & O \\ -C & O \end{bmatrix}}_{\tilde{A}} \underbrace{\begin{bmatrix} X \\ v \end{bmatrix}}_{\tilde{X}} + \underbrace{\begin{bmatrix} O \\ I \\ O \end{bmatrix}}_{\tilde{B}} \begin{bmatrix} -FG \end{bmatrix} \underbrace{\begin{bmatrix} X \\ v \end{bmatrix}}_{\tilde{X}} + \underbrace{\begin{bmatrix} B_D \tilde{X}_b \\ r \end{bmatrix}}_{\tilde{D}}$$

$$(6)$$

$\tilde{X}_b$ is the mismatch between the 6-DoF floor disturbances and 3-translation measurements of the sensors. The following proposition introduces a useful method to obtain an optimal value of the matrix gain $F$ and $G$.

*Proposition 1* [15]: For the control system represented by Eq. (6), if there exist positive definite matrices $Z$ and $Q$, $(Z \in R^{m \times (m+n)}, Q \in R^{(m+n) \times (m+n)})$ and a positive value $\gamma$ such that the inequality

$$\begin{bmatrix} \tilde{A}Q + Q\tilde{A}^T + \tilde{B}Z + Z^T\tilde{B}^T & I & Q\tilde{C}^T \\ I & -\gamma I & O \\ \tilde{C}Q & O & -\gamma I \end{bmatrix} < 0, \quad (7)$$

holds, then the feedback gains chosen by

$$\begin{bmatrix} F & G \end{bmatrix} = ZQ^{-1} \quad (8)$$

ensures the transfer matrix $H$ of the system in Eq. (6) satisfies the bounded-real condition $\|H\|_\infty < \gamma$.

### B. Noninteracting performance condition

The gains obtained from Eq. (8) preserve the efficiency and stability of the control system. However, they cannot deal effectively with the interaction inside the system, as will be shown in the control results. A noninteracting control then should be considered such that the motion in each direction follows its corresponding reference signal and is independent of the other DoFs. A relationship between the time derivatives of the output and the states is obtained as follows:

$$Y^* = \left( A^* - \Delta F \right) X + \Delta Gv + g, \quad (9)$$

where

$$Y^* = \begin{bmatrix} Y_1^{(x_1)} & Y_2^{(x_2)} \cdots Y_m^{(x_m)} \end{bmatrix}^T,$$

$$A^* = \begin{bmatrix} C_1 A^{x_1} \\ C_2 A^{x_2} \\ \vdots \\ C_m A^{x_m} \end{bmatrix}, \quad \Delta = \begin{bmatrix} C_1 A^{x_1-1} B \\ C_2 A^{x_2-1} B \\ \vdots \\ C_m A^{x_m-1} B \end{bmatrix},$$

$$x_i = \begin{cases} \min \left\{ j \mid C_i A^{j-1} B \neq 0 \right\} & (j = 1, 2, ..., n) \\ n \quad \text{if} \quad C_i A^{j-1} B = 0 \quad \text{for all } j \end{cases},$$

$$g = \begin{bmatrix} C_1 BG v^{(x_1-1)} \\ C_2 BG v^{(x_2-1)} \\ \vdots \\ C_m BG v^{(x_m-1)} \end{bmatrix}, \qquad (10)$$

with $Y_i$ and $C_i$ the $i^{th}$ row of the output and the matrix $C$, respectively. $g$ is the term that contains higher-order time derivatives of the reference and the remaining disturbances.

*Proposition 2* [10]: A necessary condition for the system represented in Eq. (6) to achieve the noninteracting performance is that the matrix $\varDelta$ is invertible.

*Theorem 1*: The control system in Eq. (6) is both decoupled and stable with the feedback gains chosen by the following formulation:

$$F = \varDelta^{-1}\left(A^* + \bar{F} S_a\right), \quad G = \varDelta^{-1}\bar{G} \qquad (11)$$

In which:

- The nonsingular matrix $S_a = \begin{bmatrix} S^T & W^T \end{bmatrix}^T \in R^{n\times n}$, with $WB = O$ and

$$S = \begin{bmatrix} C_1 \\ \vdots \\ C_1 A^{x_1-1} \\ C_2 \\ \vdots \\ C_2 A^{x_2-1} \\ \vdots \\ C_m A^{x_m-1} \end{bmatrix} \in R^{\left(m+\sum_{i=1}^{m}(x_i-1)\right)\times n} \qquad (12)$$

- The feedback matrix $\bar{F}$ to be tuned such that the closed-loop system is asymptotically stable. The $i^{th}$ row of $\bar{F}$ satisfy

$$\bar{F}_i = \begin{bmatrix} \underbrace{f_{10} \cdots f_{1(x_1-1)}}_{\bar{F}_{i1}} & \underbrace{f_{20} \cdots f_{2(x_2-1)}}_{\bar{F}_{i2}} & \cdots & \underbrace{f_{m(x_m-1)}}_{\bar{F}_{im}} & f_w \end{bmatrix},$$
$$\begin{cases} \bar{F}_{ij} = 0 & \text{if} \quad j \neq i \\ \bar{F}_{ij} \neq 0 & \text{if} \quad j = i \end{cases}$$
$$(13)$$

- And finally, the unity feedback gain $\bar{G}$ is a diagonal matrix.

*Proof:*

Substituting the feedback gains chosen by Eq. (11) to (13) into Eq. (9) and neglecting the higher-order term g result in the following relation:

$$Y_i^{(x_i)} = -\left(-\bar{F} S_a\right) X$$
$$= -\bar{F}_{ii}\begin{bmatrix} Y_i \\ \vdots \\ Y_i^{(x_i-1)} \end{bmatrix} + \bar{G}_i v_i \qquad (14)$$

Therefore, the output in one direction only depends on the corresponding reference signal in that direction. Hence the proof is completed.

*Remark 3*: An appropriate matrix $\bar{F}$ can be obtained from the feedback gain $F$ in Eq. (8) of Proposition 1.

$$\bar{F}_s = \left(\varDelta F - A^*\right) S_a^{-1},$$
$$\begin{cases} \bar{F}_{ij} = 0 & \text{if} \quad j \neq i \\ \bar{F}_{ij} = \bar{F}_{s,ij} & \text{if} \quad j = i \end{cases} \qquad (15)$$

## C. Robust Stability Condition

As shown in Eq. (13), several elements of the state feedback matrix have been removed in order to achieve the decoupling performance. A similar concern may appear in the unity feedback as well. Then a problem to solve is to obtain an optimal $G$ such that the condition for noninteracting control is met, the system robustness is preserved, and the desired control performance is reached. For this, let the AVIS with noninteracting controller gains be written as follows:

$$\dot{\bar{X}} = \left(\bar{A} + \bar{B}\bar{K}\right)\bar{X} + \bar{B}_D \bar{D},$$
$$\bar{Y} = \bar{C}\bar{X}, \qquad (16)$$

where

$$\bar{X} = \begin{bmatrix} X \\ v \end{bmatrix}, \quad \bar{D} = \begin{bmatrix} \tilde{X}_b - \begin{bmatrix} Y_{mb}^T & O_{1\times3} & \dot{Y}_{mb}^T & O_{1\times3} \end{bmatrix}^T \\ r \end{bmatrix},$$
$$\bar{A} = \begin{bmatrix} A - B\left(\varDelta^{-1}A^* + \bar{F} S_a\right) & O_6 \\ -C & O_6 \end{bmatrix}, \quad \bar{B} = \begin{bmatrix} B\varDelta^{-1} \\ O_6 \end{bmatrix},$$
$$\bar{K} = \begin{bmatrix} O_{12\times6} & \bar{G} \end{bmatrix}, \quad \bar{B}_D = \begin{bmatrix} B_D & O_6 \\ O_{6\times12} & I_6 \end{bmatrix}, \quad \bar{C} = \begin{bmatrix} O_{6\times12} & I_6 \end{bmatrix}$$
$$(17)$$

*Theorem 2*: Facing the uncertainties and disturbance inputs, the control system represented by Eqs. (16) and (17) is robustly stable and decoupled if there exist positive matrices $\bar{Z}$ and $\bar{Q}$ in the forms

$$\bar{Z} = \begin{bmatrix} O_{6\times12} & Z_3 \end{bmatrix}, \quad Z_3 \in R^{6\times6},$$
$$\bar{Q} = \begin{bmatrix} Q_{11} & Q_{12} & Q_{13} \\ Q_{21} & Q_{22} & Q_{23} \\ O_6 & O_6 & Q_{33} \end{bmatrix}, \quad Q_{ij} \in R^{6\times6}, \qquad (18)$$

with $Z_3$ and $Q_{33}$ are 6-by-6 diagonal matrices, and if there exists $\sigma > 0$ such that the LMI

$$\begin{bmatrix} \bar{A}\bar{Q} + \bar{Q}\bar{A}^T + \bar{B}\bar{Z} + \bar{Z}^T \bar{B}^T & \bar{B}_D & \bar{Q}\bar{C}^T \\ \bar{B}_D^T & -\sigma I & O \\ \bar{C}\bar{Q} & O & -\sigma I \end{bmatrix} \leq 0, \qquad (19)$$

holds.

*Proof:*

Choosing $\bar{G} = Z_3 Q_{33}^{-1}$, the above LMI results in the bounded-real lemma for the system in Eq. (16). Then, the robust stability of the system is preserved. Since both $Z_3$ and $Q_{33}$ are diagonal, the resulting $\bar{G}$ is also a diagonal matrix.

Therefore, the noninteracting condition is met. The proof is completed.

## IV. RESULTS AND DISCUSSION

The proposed control strategy was simulated with the 6-DoF AVIS model derived in section 1. The parameters were identified from the experimental AVIS (Fig. 1) and their values are given in Table 1. To highlight the performance of the proposed system, two other control systems are taken into consideration. One is the feedback control obtained from Proposition 1, and the other is the noninteracting control without the disturbance feedforward. Their performances are accessed through the task of isolating 6-DoF floor disturbances. Isolation performances in frequency and time domains are shown in Figs. 3 and 4, respectively. In the time response, square pulse movements appear in each floor direction in turn.

TABLE I. SPECIFICATION OF THE CONTROLLED AVIS

| Parameter | Value | Parameter | Value |
|---|---|---|---|
| $m$ (kg) | 59.085 | $a$ (m) | 0.188 |
| $J_{xx}$ (kg.m2) | 0.1218 | $b$ (m) | 0.233 |
| $J_{yy}$ (kg.m2) | 0.1656 | $c$ (m) | 0.008 |
| $J_{zz}$ (kg.m2) | 0.2838 | $l_{zx}$ (m) | 0.156 |
| $k_z$ (N/m) | 22526.2 | $l_{zy}$ (m) | 0.147 |
| $k_x$, $k_y$ (N/m) | 12683 | $l_{xx}$ (m) | 0.069 |
| $c_z$ (Ns/m) | 26.64 | $l_{xy}$ (m) | 0.2 |
| $c_x$, $c_y$ (Ns/m) | 9.087 | $l_{xz}$ (m) | 0.013 |

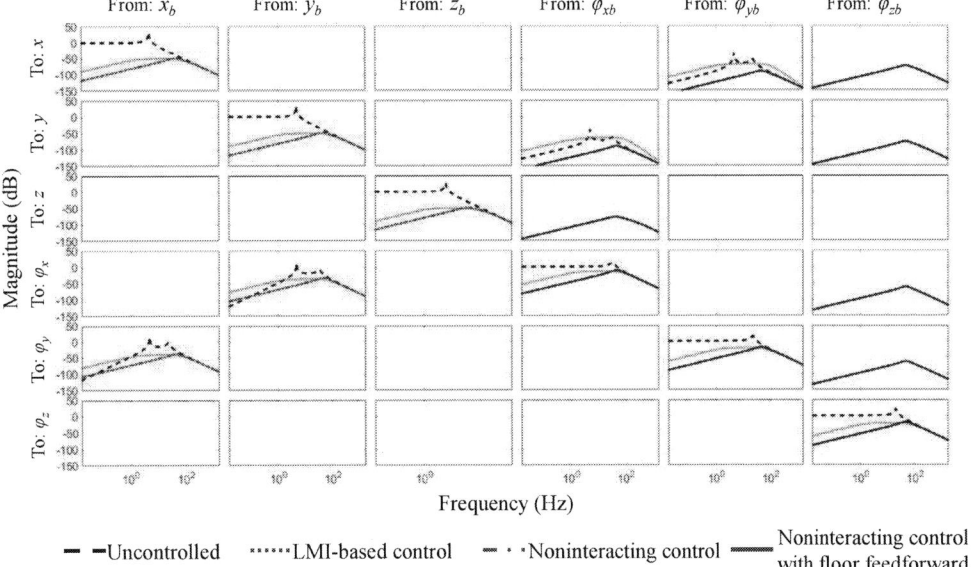

Fig. 3. Frequency response of the floor disturbances isolation performances.

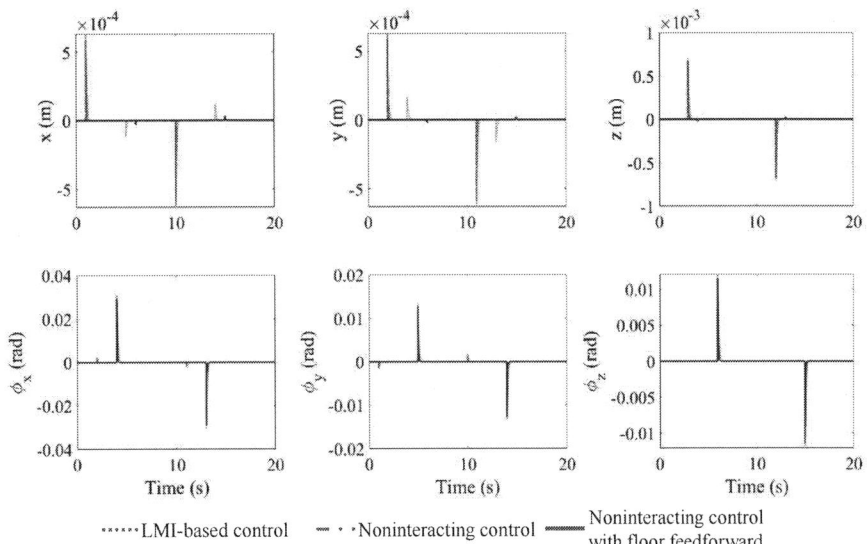

Fig. 4. Time response of the floor disturbances isolation performances.

As seen in Fig. 3, the mass-spring-damper system is capable of suppressing high-frequency vibrations from the floor. Unfortunately, the passive system does not work in the low-frequency range and there appears resonance phenomenon at the natural frequency of the system. Furthermore, internal interactions result in redundant movement. For instance, a vibration along the $x_b$-axis of the floor leads to a translation in the same direction and an additional rotation about the $y$-axis of the top plate.

On the other hand, three controllers have significantly improved the isolation performance, especially in the low-frequency range. Among the three, the feedback controller designed based on the LMI (7) barely copes with the internal interactions. This is seen in the time response in Fig. 4, where the redundant motions in $x$- and $y$-axes due to the rotation $\phi_{yb}$ and $\phi_{xb}$ cannot be effectively suppressed. The noninteracting controllers, on the other hand, decouple these interactions so their influence is barely seen in the time response. Their transmissibility is also lower than that of the LMI-based control in the low-frequency range. That is the better vibration isolation performance. Especially, with the feedforward controller, translational vibrations are almost suppressed completely. However, rotational vibrations, especially $\phi_{zb}$, result in unintended motion in the other directions. This is because of the mismatch between the sensor measurement and the true movement of the floor. Fortunately, these influences are small and hardly seen in the time response.

## V. CLUSIONS

The robust noninteracting control with disturbance feedforward has been proposed in this paper for the 6-DoF AVIS. The resulting closed-loop system is decoupled in the sense that each output DoF is independently controlled by the corresponding reference signal and the interactions between directions are effectively suppressed. The analytical method to obtain the noninteracting condition and optimal control value have been introduced. Simulation results validated the efficiency of the proposed system. The proposed control strategy can be applied to complex systems to enhance their performance and robustness. Additionally, experimental studies are going to be conducted in the near future for further evaluation.

## ACKNOWLEDGMENT

This work was supported by the National Research Foundation of Korea (NRF) grant funded by the Korea government (MSIT) (No. 2022R1A2C1003486).

## REFERENCES

[1] M. H. Kim, H. Y. Kim, H. C. Kim, D. Ahn, and D. G. Gweon, "Design and Control of a 6-DOF active vibration isolation system using a halbach magnet array," *IEEE/ASME Transactions on Mechatronics*, vol. 21, no. 4, pp. 2185–2196, 2016.

[2] Y. Kim, S. Kim, and K. Park, "Magnetic force driven six degree-of-freedom active vibration isolation system using a phase compensated velocity sensor," *Review of Scientific Instruments*, vol. 80, no. 4, p.

045108, Apr. 2009.

[3] H. Pu, Wenchuan Jia, and X. Chen, "Sliding mode control with adaptive feedforward compensator for ultra-precision active vibration isolation," in *The 8th Annual IEEE International Conference on Nano/Micro Engineered and Molecular Systems*, 2013, pp. 316–319.

[4] M. A. Beijen, M. F. Heertjes, H. Butler, and M. Steinbuch, "H ∞ feedback and feedforward controller design for active vibration isolators," *IFAC-PapersOnLine*, vol. 50, no. 1, pp. 13384–13389, Jul. 2017.

[5] M. A. Beijen, M. F. Heertjes, H. Butler, and M. Steinbuch, "Disturbance feedforward control for active vibration isolation systems with internal isolator dynamics," *Journal of Sound and Vibration*, vol. 436, pp. 220–235, Dec. 2018.

[6] M. A. Beijen, M. F. Heertjes, J. Van Dijk, and W. B. J. Hakvoort, "Self-tuning MIMO disturbance feedforward control for active hard-mounted vibration isolators," *Control Engineering Practice*, vol. 72, pp. 90–103, Mar. 2018.

[7] C. Chen, Z. Liu, Y. Zhang, C. L. P. Chen, and S. Xie, "Actuator Backlash Compensation and Accurate Parameter Estimation for Active Vibration Isolation System," *IEEE Transactions on Industrial Electronics*, vol. 63, no. 3, pp. 1643–1654, Mar. 2016.

[8] H. J. Kim, D. H. Lee, H. C. Park, and Y. B. Kim, "Direct Disturbance Suppression System Design for High Precision Fabrication Tables," *Transactions of the Korean Society of Mechanical Engineers - A*, vol. 44, no. 11, pp. 843–853, Nov. 2020.

[9] D. Jiang, J. Li, X. Li, C. Deng, and P. Liu, "Modeling identification and control of a 6-DOF active vibration isolation system driving by voice coil motors with a Halbach array magnet," *Journal of Mechanical Science and Technology*, vol. 34, no. 2, pp. 617–630, Feb. 2020.

[10] A. Morse and W. Wonham, "Status of noninteracting control," *IEEE Transactions on Automatic Control*, vol. 16, no. 6, pp. 568–581, Dec. 1971.

[11] M. Bonilla, L. Pallottino, and A. Bicchi, "Noninteracting constrained motion planning and control for robot manipulators," in *2017 IEEE International Conference on Robotics and Automation (ICRA)*, 2017, pp. 4038–4043.

[12] Y.-B. Kim, "An algorithm for robust noninteracting control of ship propulsion system," *KSME International Journal*, vol. 14, no. 4, pp. 393–400, Apr. 2000.

[13] Hongliang Liu and Guangren Duan, "Input-output Energy Decoupling Control for Linear Neutral Time-delay Systems," in *2006 6th World Congress on Intelligent Control and Automation*, 2006, vol. 1, no. 60374024, pp. 2353–2357.

[14] D. Chu and R. C. E. Tan, "Numerically Reliable Computing for the Row by Row Decoupling Problem with Stability," *SIAM Journal on Matrix Analysis and Applications*, vol. 23, no. 4, pp. 1143–1170, Jan. 2002.

[15] S. Boyd, L. El Ghaoui, E. Feron, and V. Balakrishnan, *Linear Matrix Inequalities in System and Control Theory*. Society for Industrial and Applied Mathematics, 1994.

# Impact of Cation Vacancies on Leakage Current on TiN/ZrO₂/TiN Capacitors Studied by Positron Annihilation

Akira Uedono
*Faculty of Pure and Applied Science*
*University of Tsukuba*
Tsukuba, Ibaraki 305-8573, Japan
ORCID: 0000-0001-6224-4869

Naomichi Takahashi
*Faculty of Pure and Applied Science*
*University of Tsukuba*
Tsukuba, Ibaraki 305-8573, Japan

Ryu Hasunuma
*Faculty of Pure and Applied Science*
*University of Tsukuba*
Tsukuba, Ibaraki 305-8573, Japan

Yosuke Harashima
*Graduate School of Science and Technology*
*Nara Institute of Science and Technology*
Ikoma, Nara 630-0192, Japan
ORCID: 0000-0003-0705-1583

Yasuteru Shigeta
*Faculty of Pure and Applied Science*
*University of Tsukuba*
Tsukuba, Ibaraki 305-8573, Japan
ORCID: 0000-0002-3219-6007

Zeyuan Ni
*S-Technology Development Center*
*Tokyo Electron Technology Solutions Ltd.*
Nirasaki, Yamanashi 407-0192, Japan
ORCID: 0000-0002-9815-8652

Hidefumi Matsui
*S-Technology Development Center*
*Tokyo Electron Technology Solutions Ltd.*
Nirasaki, Yamanashi 407-0192, Japan
ORCID: 0000-0003-1140-4361

Akira Notake
*SDC AI Development Department*
*Tokyo Electron Ltd.*
Sapporo, Hokkaido 060-0003, Japan
ORCID: 0000-0001-5761-882X

Atsushi Kubo
*S-Technology Development Center*
*Tokyo Electron Technology Solutions Ltd.*
Nirasaki, Yamanashi 407-0192, Japan

Tsuyoshi Moriya
*Advanced Data Planning Department*
*Tokyo Electron Ltd.*
Tokyo 107-6325, Japan
ORCID: 0000-0001-8049-2276

Koji Michishio
*Research Institute for Measurement and Analytical Instrumentation*
*National Institute of Advanced Industrial Science and Technology*
Tsukuba, Ibaraki 305-8568, Japan
ORCID: 0000-0003-1381-7856

Nagayasu Oshima
*Research Institute for Measurement and Analytical Instrumentation*
*National Institute of Advanced Industrial Science and Technology*
Tsukuba, Ibaraki 305-8568, Japan
ORCID: 0000-0002-5713-112X

Shoji Ishibashi
*Research Center for Computational Design of Advanced Functional Materials*
*National Institute of Advanced Industrial Science and Technology*
Tsukuba, Ibaraki 305-8568, Japan
ORCID: 0000-0002-4896-3530

*Abstract*—**TiN/ZrO₂/TiN capacitors were characterized using XRD, STEM, EDX, and monoenergetic positron beams. For an as-deposited ZrO₂ layer, an interlayer was formed in the layer. After post-deposition annealing at 550°C, the width of the interlayer expanded. The major vacancy-type defects in the ZrO₂ layer were determined to be a Zr-vacancy coupled with oxygen vacancies by the positron annihilation technique. After annealing, the size of these vacancies increased. The presence of the cation vacancies and their complexes in the ZrO₂ layer suggests that not only oxygen vacancies but also such defects play a role in the defect formation of the ZrO₂ layer and affect its electrical properties.**

*Keywords—ZrO₂, capacitor, vacancy, leakage current, positron annihilation*

## I. INTRODUCTION

High-dielectric constant ($k$) oxides such as $ZrO_2$ and $HfO_2$ have been extensively studied as an insulator for DRAM because of their high $k$ value, large bandgap, kinetic stability, their device process compatibility, etc. [1,2]. Because these oxides are fast ion conductors, oxygen vacancies ($V_O$s) are easily introduced during the device process. Such defects are one of the origins causing current leakage, and affect the stability of the crystal phase of high-$k$ films [3]. For high-$k$ based devices, TiN is often used as the top and bottom electrodes [4–9]. The capacitor with TiN electrode tends to show large leakage current, which is often attributed to the introduction of $V_O$ due to the scavenging effect of TiN.

Positron annihilation is a useful technique to detect vacancy-type defects and open spaces in solid state materials with non-destructive manner [10]. This technique has been successfully used to detect vacancy-type defects and open spaces in high-$k$ layers [11–14]. In the present study, we used monoenergetic positron beams to study the behaviors of vacancy-type defects in TiN/ZrO₂/TiN capacitors. The

978-1-6654-7134-3/22 $31.00 © 2022 IEEE

relationship between the leakage current and vacancies is discussed.

## II. EXPERIMENTAL –POSITRON ANNIHILATION–

When a positron is implanted into solids, it annihilates with an electron and emits γ quanta (Fig. 1). The energy distribution of the γ rays is broadened by the momentum component of the annihilating electron-positron pair $p_L$. A freely diffusing positron could be trapped by a vacancy because of Coulomb repulsion from ion cores. Because the momentum distribution of the electrons in defects differs from that of electrons in the bulk, the defects can be detected by measuring Doppler broadening spectra of the annihilation radiation. The change in the spectra due to the positron trapping is shown in Fig. 2. Because of the low electron density in vacancies, the lifetime of positrons is increased if positrons are trapped by vacancies. Thus, the positron lifetime is also the parameter used to detect vacancy-type defects.

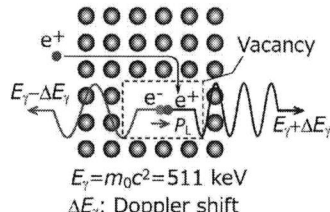

$$E_\gamma = m_0 c^2 = 511 \text{ keV}$$
$\Delta E_\gamma$: Doppler shift

Fig. 1. Positron annihilates with electron producing γ rays (511 keV). When positrons are implanted into solids, they could be trapped by vacancy-type defects. From energy distribution of annihilation γ rays due to the annihilation of positrons with electrons in the defect, one can detect vacancy-type defects.

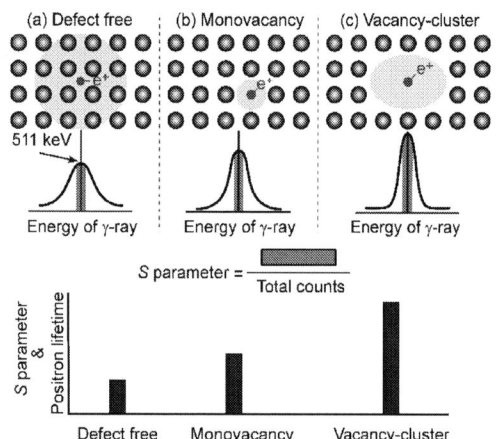

Fig. 2. Schematic drawing of Doppler broadening spectrum for (a) annihilation of positrons in free state and for annihilation of positrons trapped by (b) monovacancy and (c) vacancy cluster. Doppler broadening spectrum is characterized by $S$ parameter. Positron lifetime and the $S$ value are increased by the trapping of positrons by vacancy-type defects.

## III. EXPERIMENTAL –SAMPLE PREPARATION–

The sample structure used in the present experiments is $ZrO_2$(10 nm)/TiN(5 nm)/$SiO_2$(300 nm)/Si. TiN and $ZrO_2$ layers were deposited by using PVD technique. $SiO_2$ layers were thermally grown on Si substrates. A TiN bottom electrode (BE) was deposited on $SiO_2$/Si through the reactive sputtering of a pure Ti target under a $N_2$-plasma condition. Then, $ZrO_2$ layers were deposited on the BE-TiN layers. For some samples, a 5-nm thick TiN was deposited on the sample

as a top electrode. Forming gas annealing was performed at 550°C for 5 min in a $N_2$ atmosphere for the samples with and without the top TiN layer. Figure 3 shows the fabrication process and a cross section of the sample.

For these samples, additional 100-nm-thick TiN was deposited though a metal shadow mask with diameter holes varied from 250 μm to 350 μm. After cleaning the samples using wet etching with $H_2O_2$, the relationships between the leakage current density and applied voltage ($J$–$V$) were measured. After the removal of the top TiN layer, the microstructures of the samples were observed by using a scanning transmission electron microscope (STEM). Vacancy-type defects in the $ZrO_2$ layers were probed using the positron annihilation technique. Details on this technique are given elsewhere [10−14]. Before the measurements, the top TiN layer was removed by $H_2O_2$ wet etching. The Doppler broadening spectra of the annihilation radiation were measured by Ge detectors as a function of the incident positron energy $E$. The lifetime spectra of positrons were measured by using a vertical monoenergetic positron beam line. During the measurements, the value of $E$ was fixed at 0.6 keV. The positron lifetimes in monoclinic $ZrO_2$ were calculated using the QMAS (Quantum MAterials Simulator) code [15]. Details on this calculation procedure are given elsewhere [16,17].

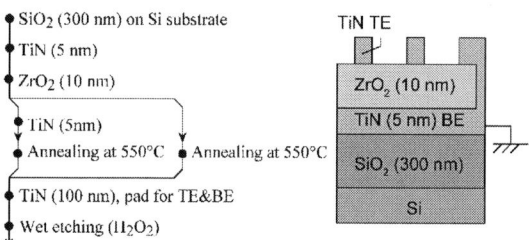

Fig. 3. Fabrication process of TiN/$ZrO_2$/TiN capacitor (left), and the side view of the sample (right)..

## IV. RESULTS AND DISCUSSION

Figure 4 shows $J$–$V$ relationships of the TiN/$ZrO_2$/TiN capacitors before and after post-deposition annealing. The bias voltage was applied to the top electrode while the bottom electrode was grounded. The results shown in the figure were obtained with the top electrode with a diameter of 260 μm. The $J$–$V$ relationships were measured with the top electrode with 40 different diameter sizes, and no clear dependence of the electrode size on the leakage current was observed The overall behavior of the obtained relationship was similar to those reported in the previous works [1]. The current density was increased by post-deposition annealing, and this tendency was suppressed by annealing with the top TiN layer. For the annealed samples, the current density at a positive voltage was higher than that at a negative voltage.

Figure 5 shows STEM images of the samples before and after post-deposition annealing. For the as-deposited sample [Fig. 5(a)], the formation of an interlayer was observed near the interface between the $ZrO_2$ layer and bottom-TiN layer. For the annealed sample without the top TiN layer [Fig. 5(b)], the interlayer was clearly observed. For the sample annealed with the top TiN layer [Fig. 5(c)], however, its width was shallower than that for the annealed sample without the top TiN layer. The observed interlayer is considered to have been introduced by the reaction between the $ZrO_2$ layer and the bottom TiN layer. Because the conductive interlayer could decreases the barrier-height of electrons from BE to the $ZrO_2$ layer, the observed asymmetry of the $J–V$ relationship can be associated with the formation of the interlayer near the bottom TiN layer.

Figure 6 shows the $S$ values as a function of incident positron energy $E$ for the samples before and after annealing. Above $E$=6 keV, $S$ increased as $E$ increased due to the positron annihilation in the Si substrate (the $S$ value corresponding to the positron annihilation in Si is 0.53). The $S$ value at $E$=3−5 keV can be attributed to the positron annihilation in the $SiO_2$ layer [18]. The $S$ value at $E$=1−1.5 keV can be associated with mainly the annihilation of positrons in the $ZrO_2$ layer. The $S$ value for a single crystal yttrium doped $ZrO_2$ was obtained to be 0.423, which is considered to be close to the defect-free $S$ value of $ZrO_2$. Thus, the observed high $S$ value for the $ZrO_2$ layer is due to the trapping of positrons by vacancies. The $S$ value for the annealed $ZrO_2$ layers was higher than that for as-deposited one, suggesting the size of vacancies increased by

annealing. This tendency was suppressed by the annealing with the top TiN layer.

The $S–E$ curves were analyzed by using the VEPFIT code [19], and the solid curves in Fig. 6 are fitting curves. The depth distributions of the $S$ value in the $ZrO_2$ film were derived by the analysis and the results are shown in the inset of Fig. 6. As shown in the inset, the $S$ value of the $ZrO_2$ film near the $ZrO_2$/bottom-TiN interface was higher than that near the surface (interface between top-TiN and the $ZrO_2$ film). The observed inhomogeneous distribution of $S$ can be associated with the formation of the interlayer in the $ZrO_2$ layer (Fig. 5). As shown in Fig. 6, the introduction of the defect-rich region near the $ZrO_2$/TiN interface was enhanced by the annealing without the top TiN layer.

The mean positron lifetimes for the $ZrO_2$ layers for the as-deposited and annealed $ZrO_2$ were obtained as 300 ps and 400-430 ps, respectively. The lifetimes of positrons trapped by vacancies in $ZrO_2$ were simulated, and the results are summarized in Table I. From a comparison between the experimentally obtained lifetimes and the ones calculated by using a computer simulation, the defect species in the as-deposited $ZrO_2$ layer was identified as Zr vacancy ($V_{Zr}$) coupled with $V_O$s, such as $V_{Zr}(V_O)_4$ (Fig. 7). For the annealed samples, the observed longer positron lifetimes suggest an introduction of vacancies larger than $V_{Zr}$-type defects.

The present experiments showed that that not only $V_O$s but also cation vacancies play an important role in the microstructure formation of the $ZrO_2$ layer, and affect electrical properties of the capacitors.

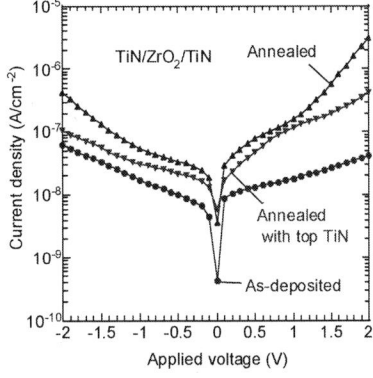

Fig. 4. $J–V$ curves for TiN/$ZrO_2$/TiN capacitors before and after post-deposition annealing. Annealing was done with and without top TiN layer.

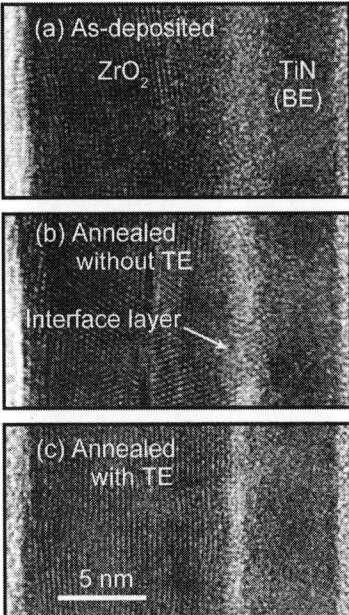

Fig. 5. STEM images of samples (a) before annealing and after annealing (b) without and (c) with top TiN layer. Measurements were performed after removing top TiN layer.

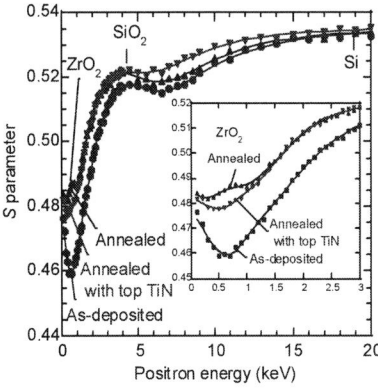

Fig. 6. S parameters as function of incident positron energy $E$ for samples before and after post-deposition annealing. Inset shows $S–E$ curves corresponding to subsurface region.

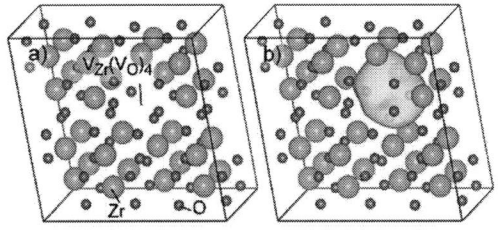

Fig. 7. (a) Atomic configuration of super cell of $V_{Zr}(V_O)_4$ and (b) distribution of the positron density around the defect. Green and red circles correspond to Zr and O atoms, respectively.

Table I. Lifetime of positrons annihilated from the free state and that of positrons trapped by Zn vacancy ($V_{Zn}$), $V_O$, and their complexes.

| Defect | Positron lifetime (ps) |
|---|---|
| Defect free | 172 |
| $V_{Zn}$ | 245 |
| $V_O$ | 175 |
| $V_{Zn}V_O$ | 251 |
| $V_{Zn}(V_O)_2$ | 260 |
| $V_{Zn}(V_O)_3$ | 276 |
| $V_{Zn}(V_O)_4$ | 294 |
| $V_{Zn}(V_O)_5$ | 315 |
| $V_{Zn}(V_O)_6$ | 339 |

## REFERENCES

[1] W. Jeon, "Recent advances in the understanding of high-k dielectric materials deposited by atomic layer deposition for dynamic random-access memory capacitor applications," J. Mat. Res. 35, pp. 775–794, 2019.

[2] J. Robertson and R. M. Wallace, "High-*k* materials and metal gates for CMOS applications", Mat. Sci. Eng. R 88, pp. 1–41, 2015.

[3] B. E. Park, I. K. Oh, J. S. Park, S. Seo, D. Thompson, and, H. Kim, "Simultaneous improvement of the dielectric constant and leakage currents of $ZrO_2$ dielectrics by incorporating a highly valent $Ta^{5+}$ element", J. Mat. Chem. C 6, 9794–9801, 2018.

[4] T. S. Böscke, J. Müller, D. Bräuhaus, U. Schröder, and, U. Böttger, "Ferroelectricity in hafnium oxide thin films", Appl. Phys. Lett. 99, 102903, 2011.

[5] D. S. Jeong, R. Thomas, R. S. Katiyar, J. F. Scott, H. Kohlstedt, A. Petraru, and, C. S. Hwang, "Emerging memories: resistive switching mechanisms and current status", Rep. Prog. Phys. 75, 076502, 2012.

[6] S. J. Kim, J. Mohan, S. R. Summerfelt, and, J. Kim, "Ferroelectric $Hf_{0.5}Zr_{0.5}O_2$" Thin Films: A Review of Recent Advances, JOM 71, 246–255, 2018.

[7] S. Oh, H. Hwang, and, I. K. Yoo, "Ferroelectric materials for neuromorphic computing", APL Mater. 7, 091109, 2019.

[8] Y. Kim, M.-J. Choi, and, H. W. Jang, "Ferroelectric field effect transistors: Progress and perspective", APL Mater. 7, 091109, 2021.

[9] T. Onaya, T. Nabatame, M. Inoue, T. Sawada, H. Ota, and, Y. Morita, "Wake-up-free properties and high fatigue resistance of $Hf_xZr_{1-x}O^{2-}$ based metal-ferroelectric-semiconductor using top $ZrO_2$ nucleation layer at low thermal budget (300°C)", APL Mater. 10, 051110, 2022.

[10] R. Krause-Rehberg and H. S. Leipner, "Positron Annihilation in Semiconductors: Defect studies", Springer Series in Solid-State Sciences, vol. 127, Springer-Verlag, Berlin, 1999, ISBN-10: 354064371.

[11] A. Uedono, K. Ikeuchi, K. Yamabe, T. Ohdaira, M. Muramatsu, R. Suzuki, A.S. Hamid, T. Chikyow, K. Torii, and, K. Yamada, "Annealing properties of open volumes in $HfSiO_x$ and $HfAlO_x$ gate dielectrics studied using monoenergetic positron beams", J. Appl. Phys. 98, 023506, 2005.

[12] A. Uedono, K. Ikeuchi, T. Otsuka, K. Shiraishi, K. Yamabe, S. Miyazaki, N. Umezawa, A. Hamid, T. Chikyow, T. Ohdaira, M. Muramatsu, R. Suzuki, S. Inumiya, S. Kamiyama, Y. Akasaka, Y. Nara, and, K. Yamada, "Characterization of HfSiON gate dielectrics using monoenergetic positron beams", J. Appl. Phys. 99, 054507, 2006.

[13] A. Uedono, T. Naito, T. Otsuka, K. Shiraishi, K. Yamabe, S. Miyazaki, H. Watanabe, N. Umezawa, T. Chikyow, Y. Akasaka, S. Kamiyama, Y. Nara, and, K. Yamada, Introduction of defects into $HfO_2$ gate dielectrics by metal-gate deposition studied using x-ray photoelectron spectroscopy and positron annihilation", J. Appl. Phys. 100, 064501, 2006.

[14] A. Uedono, T. Naito, T. Otsuka, K. Ito, K. Shiraishi, K. Yamabe, S. Miyazaki, H. Watanabe, N. Umezawa, T. Chikyow, T. Ohdaira, R. Suzuki, Y. Akasaka, S. Kamiyama, Y. Nara, and, K. Yamada, "Characterization of metal/high-*k* structures using monoenergetic positron beam", Jpn. J. Appl. Phys. 46, pp. 3214−3218, 2007.

[15] S. Ishibashi, T. Tamura, S. Tanaka, M. Kohyama, and, K. Terakura, "Ab initiocalculations of electric-field-induced stress profiles for diamond/c−BN(110) superlattices", Physical Review B 76, 153310, 2007.

[16] S. Ishibashi and A. Uedono, "First-principles calculation of positron states and annihilation parameters for group-III nitrides", J. Phys. Conf. Ser. 505, 012010, 2014.

[17] S. Ishibashi, A. Uedono, H. Kino, T. Miyake, and, K. Terakura, "Computational study of positron annihilation parameters for cation mono-vacancies and vacancy complexes in nitride semiconductor alloys", J. Physics. Conden. Matter 31, pp. 475401−475401, 2019.

[18] A. Uedono, W. Ueno, T. Yamada, T. Hosoi, W. Egger, T. Koschine, C. Hugenschmidt, M. Dickmann, and H. Watanabe, "Voids and vacancy-type defects in $SiO_2$/GaN structures probed by monoenergetic positron beams", J. Appl. Phys. 127, 054503, 2020.

[19] A. van Veen, H. Schut, M. Clement, J. M. M. de Nijs, A. Kruseman, and, M. R. IJpma, "VEPFIT applied to depth profiling problems", Appl. Sur. Sci. 85, pp. 216−224, 1995.

# Optimization of RF Frequencies in Dual-frequency Capacitively Coupled Plasma Apparatus Using Genetic Algorithm (GA) and Plasma Simulation

Shigeyuki Takagi
*Department of electrical and electronic Engineering, School of Engineering Tokyo University of Technology*
Hachioji, Japan
takagisgyk@stf.teu.ac.jp

Tatsuhiro Nakaegawa
*Department of electrical and electronic Engineering, School of Engineering Tokyo University of Japan*
Hachioji, Japan
samuraiblue.0210@gmail.com

Shih-Nan Hsiao
Center for Low-temperature Plasma Sciences
*Nagoya University*
Nagoya, Japan
hsiao@plasma.engg.nagoya-u.ac.jp

Makoto Sekine
*Center for Low-temperature Plasma Sciences*
*Nagoya University*
Nagoya, Japan
sekine@plasma.engg.nagoya-u.ac.jp

*Abstract*—*As a method to optimize the power frequency of dual-frequency plasma, we propose an optimization method that combines genetic algorithm and plasma simulation. A two-dimensional plasma simulation model of Ar plasma was constructed with a fluid model. Combining this simulation model with a genetic algorithm, plasma conditions with high plasma density and small variations in electron density were calculated. As a result, the optimum conditions were 175 to 210 MHz for the high-frequency generator and 0.5 to 4.0 MHz for the low-frequency generator.* (*Abstract*)

*Keywords—dual-frequency, etching, plasma simulation, RF frequency, genetic algorithm, optimization*

## I. INTRODUCTION

In the semiconductor etching process, dual-frequency plasma is used to precisely control the processing shape.[1][2] To increase the plasma density and improve the controllability, a dual-frequency plasma source has been proposed in which RF power supplies of different frequencies are connected to the upper and lower electrodes. [3][4] In dual frequency plasma, the plasma is generated by a high frequency power source and the bias is controlled by a low frequency power source. Such plasma optimization has hitherto been done experimentally, requiring a lot of time and cost.

Plasma simulation is effective for the development of plasma equipment and optimization of plasma conditions. By using plasma simulation, it is possible to visualize the distribution of electron density, positive ion density, and active species density. [5][6] In this study, we focused on the plasma simulation and the power frequency used.

In dual-frequency excited plasma, the plasma is excited by a high-frequency power supply and the bias Vdc is controlled by a low-frequency power supply. High electron density and low density fluctuation with respect to voltage fluctuation of the low-frequency power supply are expected. To optimize the power supply frequencies, plasma simulation and genetic algorithm were combined to optimize the frequencies of the upper and lower electrodes by calculation.

## II. PLASMA SIMULATION MODEL

The fluid model PHM (Plasma hybrid module) of Pegasus Software Inc. was used for the plasma simulation. PHM is a fluid model simulator, and the continuity equation of electrons is expressed by Eq. (1).

$$\frac{\partial n_e}{\partial t} + \nabla \cdot \boldsymbol{\Gamma}_e = R_e , \qquad (1)$$

where $n_e$ is the electron density, $\Gamma_e$ is the electron flux, and $R_e$ is the electron generation rate. Similarly, the ion formation rate is given by Eq. (2).

$$\frac{\partial n_i}{\partial t} + \nabla \cdot \boldsymbol{\Gamma}_i = R_i , \qquad (2)$$

where $n_i$ is the electron density, $\Gamma_i$ is the ion flux, and $R_i$ is the ion generation rate. The gas reaction model in the plasma simulation was constructed referring to Ref. 5.

A plasma simulation model is shown in Fig. 1. Plasma was generated at a gas pressure of 19.95 Pa in parallel plates 3 cm long and 0.5 cm wide. The standard conditions were a voltage ($V_{pp}$) of 50 V applied to the upper electrode and a frequency of 60 MHz, and a voltage ($V_{pp}$) of 50 V and a frequency of 2 MHz to the lower electrode. A fluid model plasma simulator PHM was used to calculate the electron density. [5]

A simulation result under this condition is shown in Fig. 2. The electron and Ar+ densities are almost the same, and the neutral state in which the negative and positive charges are

978-1-6654-7134-3/22 $31.00 © 2022 IEEE

equal, can be simulated. Both the electron density and Ar⁺ are high at the center of the electrode, and a low-density sheath region is formed near the electrode.

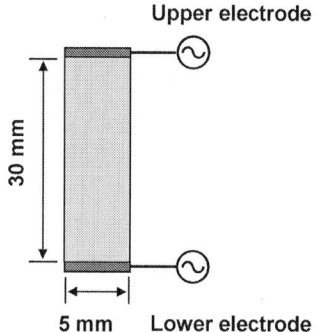

Fig.1. Simulation model.

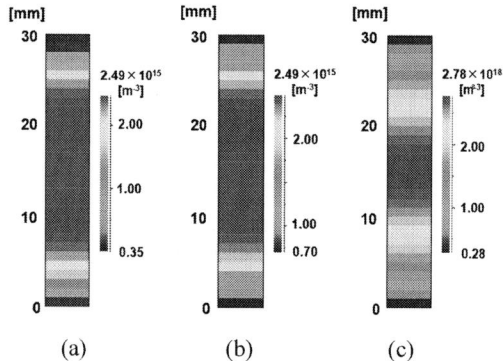

(a)      (b)      (c)

Fig. 2. Simulation results: (a)Electron density, (b) Ar⁺ density, and (c) Ar* density.

## III. DEPENDENCE OF ELECTRON DENSITY ON FREQUENCIES

We performed plasma simulations with a lower electrode frequency of 2 MHz and an upper electrode frequency from 20 to 300 MHz. The electron density at the central part of the plasma was extracted. Figure 3 shows the calculation results. The electron density increases with frequency, reaching a maximum at 175 MHz.

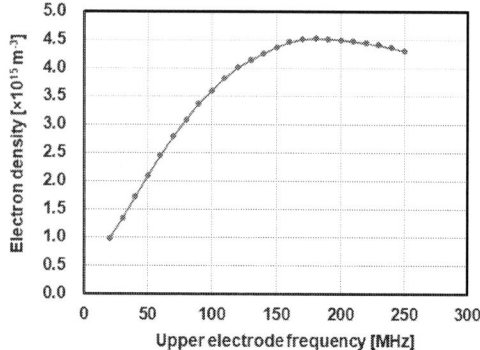

Fig. 3. Relationship between upper electrode frequency and electron density.

Next, we calculated the fluctuation rate of the electron density when the voltage was changed from 50 to 30 and 70 V by changing the frequency of the lower voltage. This calculation result is shown in Fig. 4. Fluctuations in the frequency range from 1.0 to 1.5 MHz are smaller than those at other frequencies.

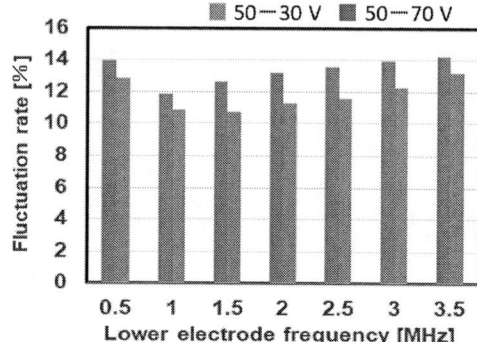

Fig. 4. Relationship between fluctuation rate and lower electrode frequency.

## IV. COUPLED CALCULATION OF PLASMA SIMULATION AND GENETIC ALGORITHM

In section III, the frequency of one electrode was fixed and the frequency of the other was optimized. As a next step, the frequencies of the upper and lower electrodes were optimized so that the electron density was high and the voltage fluctuation was small.

To optimize the plasma state, it is necessary to optimize the plasma conditions while changing both the frequencies of the upper electrode and the lower electrode. For this calculation, we developed a computational method combining plasma simulation and genetic algorithm. Figure. 5 shows the flow of optimization calculation. The electron density was calculated by combining the plasma simulation and the optimization software modeFRONTIER incorporated a genetic algorithm.

Genetic algorithm is a method of evolutionary computation to solve optimization problems. It is one of the multipoint search methods that can find the optimal solution at high speed. [7][8] This algorithm consists of three characteristic methods: natural selection, crossover, and mutation. Natural selection and crossover can efficiently generate better parameter combinations. There are two types of genetic algorithms: genetic algorithms [9] that handle a single objective function, and multi-objective genetic algorithms that handle multiple objective functions.

In this study, the objective variables of the optimization calculation were set to the electron density and the variation of the electron density with the voltage variation of the low-frequency power source. In the calculation, we used modeFRONTIER, which incorporates one of the genetic algorithms, NSGA-II. NSGA-II was proposed by Debi et al. and has excellent convergence and computational stability.[10]

We calculated the electron density for 300 combination cases of lower and upper electrode frequencies. Figure 6 shows the calculation results. The horizontal axis indicates the upper electrode frequency, and the vertical axis indicates the electron density. Also, the fluctuation in electron density accompanying the voltage fluctuation is displayed in color, and the variation in density increases as the color changes from blue to red. The results show that the electron density peaks at the upper electrode frequencies from 175 to 210 MHz, with a fluctuation of electron density in this region from 6 to 9%. In the region where the frequency of the upper electrode is 210 MHz or higher, the electron density fluctuates less, but the electron density decreases. These results indicate that the upper electrode frequency of 175–210 MHz and the lower electrode frequency of 0.5–4 MHz are the optimum conditions.

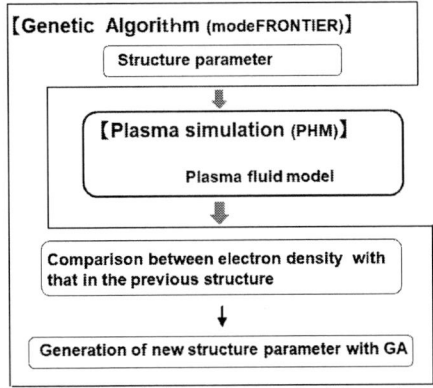

Fig. 5. Coupled calculation of plasma simulation and genetic algorism.

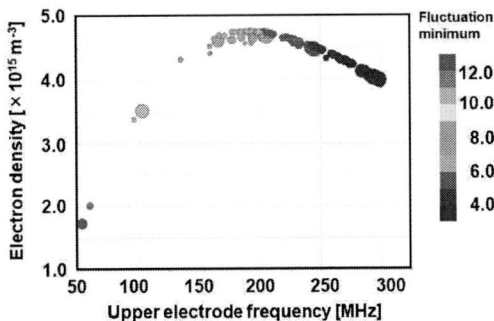

Fig.6. Optimization of electron density and fluctuation with upper electrode frequency.

## V. CONCLUSIONS

The frequencies of the upper and lower electrodes were optimized by combining two-frequency excitation plasma simulation and genetic algorithm. As a result, the optimum condition was that the upper electrode frequency ranged from 175 to 210 MHz and the lower electrode frequency range from 0.5 to 4 MHz.

## ACKNOWLEDGMENT

This work was carried out by the joint usage/research program of center for Low- temperature Plasma Sciences, Nagoya University.

## REFERENCES

[1] K. Ishikawa, K. Karahashi, T. Ichiki, J. P. Chang, S. M. George, W. M. M. Kessels, H. J. Lee, S. Tinck, J. H. Um, and K. Kinoshita, "Progress and prospects in nanoscale dry processes: How can we control atomic layer reactions? ", Jpn. J. Appl.Phys., vol. 56, 2017, pp. 06HA02-1–13.

[2] K. Ishikawa, K. Karahashi, T. Ishijima, S. Il Cho, S. Elliott, D. Hausmann, D. Mocuta, A. Wilson, and K. Kinoshita, "Progress in nanoscale dry processes for fabrication of high-aspect-ratio features: How can we control critical dimension uniformity at the bottom?', Jpn. J. Appl. Phys. vol. 56, 2018, pp. 06JA01–18.

[3] S. Rauf, "Dual radio frequency sources in a magnetized capacitively coupled plasma discharge", IEEE Trans. Plasma Sci. vol. 31, 2003, pp. 471– 478.

[4] C. O'Neill, J. Waskoenig and T. Gans, "Electron Heating in Dual-Radio-Frequency-Driven Atmospheric-Pressure Plasmas", IEEE Trans. Plasma Sci. vol. 39, 2011, pp. 2588–258.

[5] S. Takagi, S. Kawamura, and M. Sekine, "Plasma simulation for dual-frequency capacitively coupled plasma incorporating gas flow simulation", Jpn. J. Appl. Phys., vol. 60, 2021, pp. SAAB07-1–7.

[6] S. Takagi, S. Kawamura, and M. Sekine, "Ar/SF6 plasma simulation for dual-frequency capacitively coupled plasma incorporating gas flow simulation and secondary electron emission", Jpn. J. Appl. Phys., vol. 62, 2022, pp. SA1009-1–7.

[7] J. H. Holland, "Adaptation in natural and artificial systems: an introductory analysis with applications to biology, control, and artificial intelligence", University of Michigan Press, 1975.

[8] J. H. Holland, "Genetic Algorithms", Scientific American, vol. 267, No. 1, 1992, pp. 66–73.

[9] R.T. Marler and J.S. Arora, "Survey of multi-objective optimization methods for engineering", Struct. Multidisc. Optim., vol. 26, 2004, pp. 369–395.

[10] K. Deb, A. Pratap, S. Agarwal, and T. Meyaraivan, "A Fast and Elitist Multiobjective Genetic Algorithm: NSGA-II", IEEE Trans. Comput., vol.6, 2002, 182–197.

# Deposition rate dependence of the 5 nm-thick ferroelectric nondoped HfO₂ on MFSFET characteristics

Masakazu Tanuma
Electrical and Electronic Engineering
Tokyo Institute of Technology
Yokohama, Japan
tanuma.m.aa@m.titech.ac.jp

Joong-Won Shin
Electrical and Electronic Engineering
Tokyo Institute of Technology
Yokohama, Japan
shin.j.ag@m.titech.ac.jp

Shun-ichiro Ohmi
Electrical and Electronic Engineering
Tokyo Institute of Technology
Yokohama, Japan
ohmi@ee.e.titech.ac.jp

*Abstract*—**In this research, deposition rate dependence of 5 nm-thick ferroelectric nondoped HfO₂ (FeND-HfO₂) on the device characteristics was investigated. The equivalent oxide thickness (EOT) and leakage current were decreased by increasing deposition rate of HfO₂ from 5.0 nm/min to 6.0 nm/min. The subthreshold swing (SS) of 107 mV/dec. and saturation mobility ($\mu_{sat}$) of 150 cm²/(Vs) were obtained with deposition rate of 6.0 nm/min. Furthermore, the threshold voltage ($V_{TH}$) was controllable as the number of identical erase pulse of 4 V/1 μs was increased, which suggested the $V_{TH}$ control of approximately 10 mV.**

*Keywords*—*Ferroelectric nondoped HfO₂, RF magnetron sputtering, threshold voltage control.*

## I. INTRODUCTION

Metal-ferroelectric-semiconductor field-effect transistors (MFSFETs) are attracting much attention as next-generation nonvolatile memories. Conventional ferroelectric materials such as lead zirconate titanete (PZT) and strontium bismuth tantalate (SBT) were investigated for MFSFETs [1]. However, it was difficult to be realized because of interdiffusion issues of atoms such as Pb and Bi to Si substrate [2].

Amorphous HfO₂ is a high-k gate insulator material, while the crystallized HfO₂ was found to show ferroelectric property in 2011 [3]. Since ferroelectric HfO₂ has a high compatibility with Si CMOS process and its ferroelectricity is improved with thickness thinning, it is the most promising material to realize MFSFET. Among the crystal structure of HfO₂, the orthorhombic phase, which is a metastable phase, shows ferroelectricity. Because the formation of the orthorhombic phase is difficult due to the metastable phase, it is usually controlled by optimizing the annealing conditions and doping elements such as Zr and Si [4]. Annealing temperature should be increased to crystalize when the dopants are introduced in HfO₂. However, high temperature annealing led to the formation of SiO₂ interfacial layer (IL) between HfO₂ and Si substrate, which induces depolarization field [5]. It will degrade the retention characteristics as well as the operation voltage and speed of memory devices [6]. In this metal/ferroelectric/interlayer/Si (MFIS) structure, several studies such as high dielectric constant interlayer material and low area ratio of ferroelectric/interlayer capacitor have been reported to decrease the depolarization field [7].

Ferroelectric nondoped HfO₂ (FeND-HfO₂) has advantages compared to the doped HfO₂ such as the lower crystallization temperature and threshold voltage ($V_{TH}$) control to realize the MFSFET [8]. In previous work, it is assumed that the strained rhombohedral phase of HfO₂ was formed and it is origin of ferroelectricity [9]. Previously, we

have reported that the deposition rate was changed by changing the target-substrate distance and Ar/O₂ gas flow ratio during fabrication of MFS diode, and the formation of SiO₂ IL was suppressed [10]. We have considered that SiO₂ IL formation is dominated by the oxidation by O₂ plasma during HfO₂ deposition [10]. We have also reported that the variation of $V_{TH}$ of MFSFETs was decreased by direct deposition of HfO₂ on the Si substrate [11].

In this paper, we investigated the deposition rate dependence of 5 nm-thick FeND-HfO₂ on MFSFET characteristics.

## II. EXPERIMENTAL PROCEDURE

Figure 1 shows the experimental procedure and the schematic cross-sections of MFS diode and MFSFET used in this research. For MFS diode, p-Si(100) substrate was cleaned by SPM and DHF solutions. Then, 5 nm-thick HfO₂ and 20 nm-thick Pt were in-situ deposited by RF magnetron sputtering at room temperature (RT). The deposition rates of HfO₂ were 5.0 nm/min with the Ar/O₂ gas flow ratios of 2.0/0.2 sccm and 6.0 nm/min with the Ar/O₂ gas flow ratios of

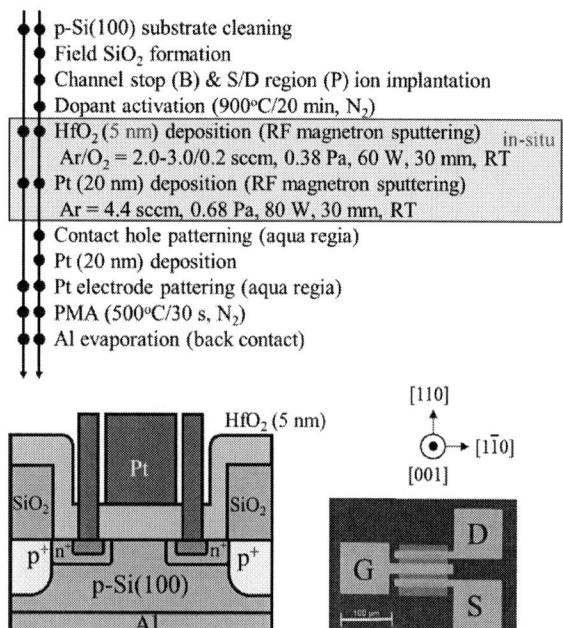

Fig. 1 Experimental procedure and schematics of fabricated device. A plane-view of the fabricated device is also shown.

3.0/0.2 sccm, respectively. After Pt patterning and etching by diluted aqua regia, the post-metallization-annealing (PMA)was carried out at 500°C for 30 s in $N_2$ ambient. Finally, Al back contact was formed by thermal evaporation.

For MFSFET fabrication, the gate-last process was utilized. Before the deposition, $p^+$ channel stop region ($B^+$, 100 keV, $1 \times 10^{14}$ cm$^{-2}$) and $n^+$ source/drain (S/D) region ($P^+$, 20 keV, $5 \times 10^{15}$ cm$^{-2}$) were formed with activation annealing at 900°C/20 min. For gate stacked structures, Pt/HfO$_2$(5 nm) were in-situ deposited by RF magnetron sputtering. The deposition rates of HfO$_2$ were 5.0 nm/min and 6.0 nm/min. PMA was carried out at 500°C for 30 s in $N_2$ ambient. The gate length (L) and width (W) were L/W = 10/90 µm. The fabricated devices were characterized by C-V, J-V, $I_D$-$V_G$, and $I_D$-$V_D$ measurements at RT in air.

## III. RESULTS AND DISCUSSIONS

Figure 2 shows the deposition rate dependence on C-V and J-V characteristics of MFS diodes with 5 nm-thick

FeND-HfO$_2$ [11]. The deposition rate of HfO$_2$ was changed from 5.0 nm/min to 6.0 nm/min by changing the Ar/O$_2$ gas flow ratio from 2.0/0.2 sccm to 3.0/0.2 sccm. It was found that the equivalent oxide thickness (EOT) was decreased from 3.2 nm to 2.8 nm by increasing the deposition rate from 5.0 nm/min to 6.0 nm/min. Furthermore, the decrease of leakage current was confirmed by increasing the deposition rate shown in Fig. 2(b). Utilizing the capacitance value of the C-V characteristics, the thickness of SiO$_2$ IL were estimated to be 2.7 nm and 2.1 nm with the deposition rate of 5.0 nm/min and 6.0 nm/min, respectively. It was considered that the deposition time was decreased by increasing the deposition rate, and the oxidation of Si substrate during HfO$_2$ deposition was suppressed.

Figure 3(a) shows the deposition rate dependence of $I_D$-$V_G$ characteristics of MFSFETs with 5 nm-thick FeND-HfO$_2$. FeND HfO$_2$ formed by the deposition rate of 5.0 nm/min showed negligible hysteresis characteristics. Although the

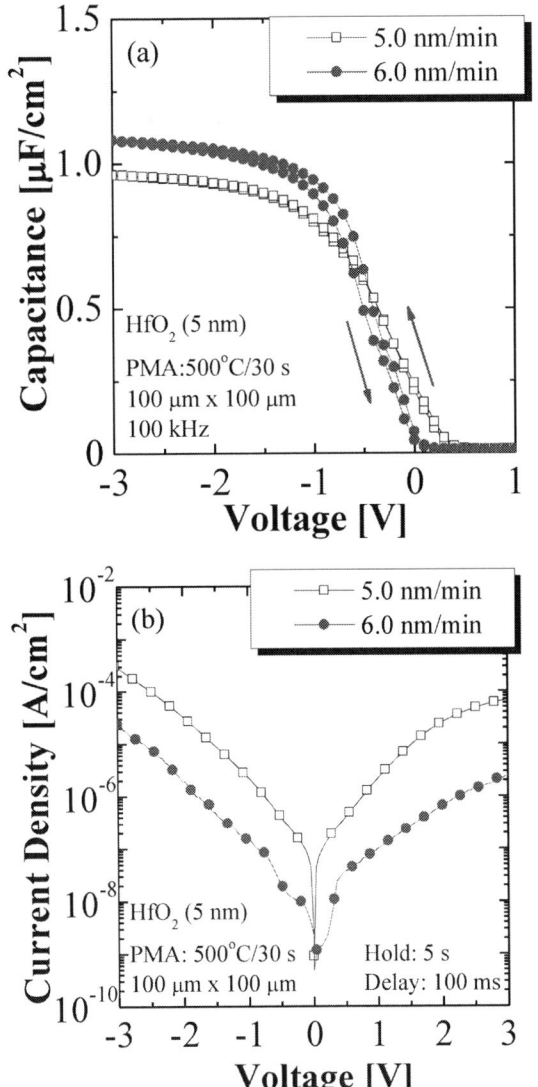

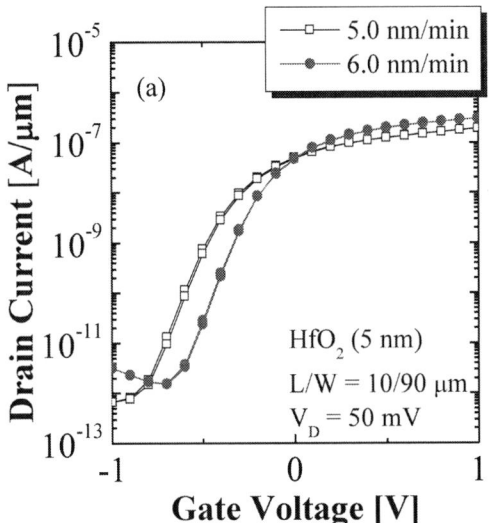

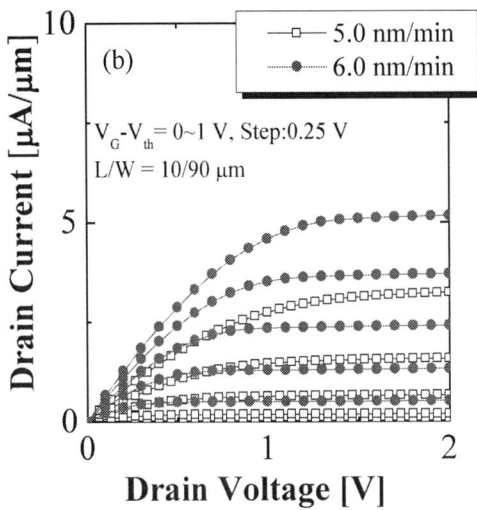

Fig. 2 Deposition rate dependence on (a) C-V characteristics and (b) J-V characteristics of MFS diodes.

Fig. 3 Deposition rate dependence of (a) $I_D$-$V_G$ characteristics and (b) $I_D$-$V_D$ characteristics of MFSFETs.

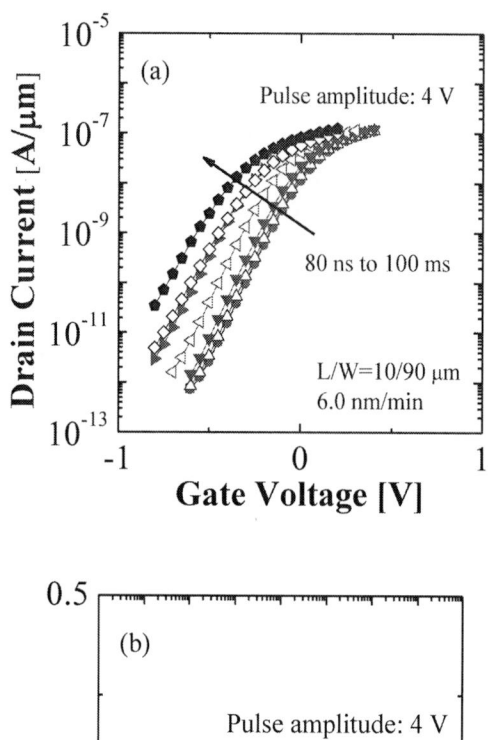

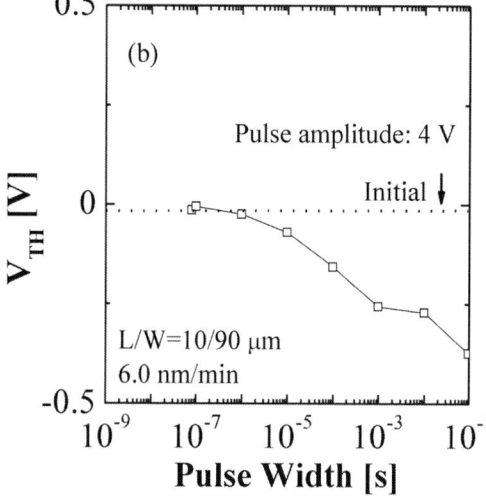

Fig. 4 Threshold voltage dependence on the input pulse width for erase operation. (a) $I_D$-$V_G$ characteristics and (b) Threshold voltage dependence of pulse width.

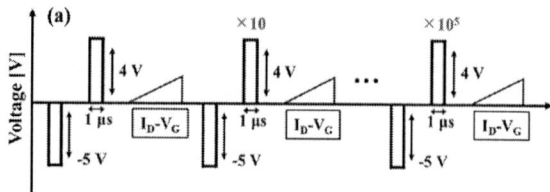

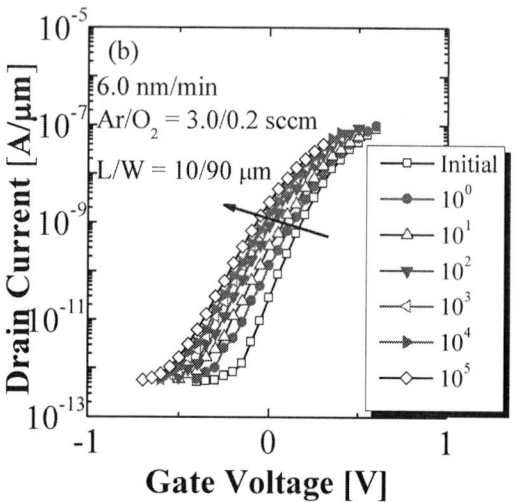

Fig. 5 Threshold voltage shift of MFSFET with the number of identical erase pulses. (a) Input pulse sequence and (b) $I_D$-$V_G$ characteristics.

the mobility was improved by increasing the deposition rate of $HfO_2$ and decreasing the $SiO_2$ IL thickness.

Figure 4 shows $V_{TH}$ dependence on the input pulse width for erase operation of MFSFET with the deposition rate of 6.0 nm/min. $V_{TH}$ was extracted at the drain current $I_D = 10^{-7} \times$ (W/L) [A]. Pulse amplitude was 4 V and pulse width was changed from 80 ns to 100 ms. The shift of $V_{TH}$ was confirmed when the pulse was 1 μs or larger shown in Fig. 4(b). The memory window (MW) of 0.41 V was obtained with the pulse width of 100 ms.

Figure 5 shows $V_{TH}$ shift of MFSFET with the number of identical erase pulses. Since the $V_{TH}$ shift was observed under the pulse width of 1 μs, the pulses of 4 V/1 μs were input from 1 to $10^5$ times. The $V_{TH}$ was shifted as the number of erase pulse was increased, and $V_{TH}$ shift of 0.1 V was obtained by the number of erase pulse as $10^1$, which suggested $V_{TH}$ control of approximately 10 mV.

## IV. CONCLUSIONS

In this paper, we investigated the deposition rate dependence of 5 nm-thick FeND-$HfO_2$ on MFSFET characteristics. It was found that EOT and leakage current were decreased by increasing deposition rate of $HfO_2$. The SS and $\mu_{sat}$ were also improved by increasing deposition rate. The $V_{TH}$ shift was confirmed by inputting erase pulse of 4 V/100 ms, and MW of 0.41 V was obtained. $V_{TH}$ was controllable as the number of identical erase pulse of 4 V/1 μs was increased, which suggested the $V_{TH}$ control of approximately 10 mV. Therefore, it is important to control the deposition rate to improve the device characteristics.

hysteresis of 90 mV was observed in case of 6.0 nm/min deposition for FeND-$HfO_2$, the decrease of EOT was confirmed. The applied voltage was swept from -1 V to 1 V and drain voltage was fixed at 50 mV. The subthreshold swing (SS) was improved from 113 mV/dec. to 107 mV/dec. by increasing the depositon rate of $HfO_2$ from 5.0 nm/min to 6.0 nm/min. The saturation mobility ($\mu_{sat}$) was extracted from $I_D$-$V_D$ characteristics of MFSFETs shown in Fig. 3(b). The applied voltage was swept from 0 V to 2 V and $V_G$-$V_{TH}$ was increased from 0 V to 1 V with 0.25 V step. In the $I_D$-$V_D$ characteristics, drain current was increased and the $\mu_{sat}$ was increased from 79 cm$^2$/(Vs) to 150 cm$^2$/(Vs) by increasing the deposition rate of $HfO_2$ from 5.0 nm/min to 6.0 nm/min. It is considered that the carriers are scattered by the fixed oxide charge in $SiO_2$ IL,

## ACKNOWLEDGMENT

The authors would like to thank Mr. D. Shoji of Tokyo Institute of Technology for his support and useful discussion for this research. This research was partially supported by JSPS KAKENHI Grant Number 19H00758, NEDO, JSW and CASIO Foundation.

## REFERENCES

[1] E. Tokumitsu, G. Fujii, and H. Ishiwara, "Nonvolatile ferroelectric-gate field-effect transistors using $SrBi_2Ta_2O_9/Pt/SrTa_2O_6/SiON/Si$ structures," Appl. Phys. Lett., vol. 75, no. 4, pp. 575-577, June 1999.

[2] J. Muller, P. Polakowski, S. Mueller, and T. Mikolajick, "Ferroelectric Hafnium Oxide Based Materials and Devices: Assessment of Current Status and Future Prospects," ECS J. Solid State Sci. Tech. vol. 4, no. 5, N30-N35, Feb. 2015.

[3] T. Boscke, J. Muller, D. Brauhaus, U. Schroder, and U. Bottger, "Ferroelectricity in hafnium oxide thin films," Appl. Phys. Lett. vol. 99, no. 10, 102903, Sept. 2011.

[4] K. Ni, P. Sharma, J. Zhang, M. Jerry, J. A. Smith, K. Tapily, R. Clark, S. Mahapatra, and S. Datta, "Critical role of interlayer in $Hf_{0.5}Zr_{0.5}O_2$ ferroelectric FET nonvolatile memory performance," IEEE Trans. Electron Dev., vol. 65, no. 6, pp. 2461-2469, Apr. 2018.

[5] Y. Zhang, J. Xu, D. Y. Zhou, H. H. Wang, W. Q. Lu and C. K. Choi, "Effects of Hf buffer layer at the Y-doped $HfO_2/Si$ interface on ferroelectric characteristics of Y-doped $HfO_2$ films formed by reactive sputtering," Ceram. Int., vol.44, no.11, pp.12841-12846, Aug. 2018.

[6] N. Gong and T. S. Ma, "Why Is FE-$HfO_2$ More Suitable Than PZT or SBT for Scaled Nonvolatile 1-T Memory Cell?," IEEE Electron Dev. Let., vol.37, no.9, pp.1123-1126, July 2016.

[7] C. H. Yeh, K. C. Chang, Y. H. Lin, T. C. Chang, Y. C. Chang, W. C. Chen, F. Y. Jin, F. M. Ciou, Y. S. Lin, W. C Ciou, Y. S. Lin, W. C. Hung, J. W. Huang, T. M. Tsai, and S. M. Sze, IEEE. Hung, J. W. Huang, T. M. Tsai, and S. M. Sze, "Analysis of Edge Effect Occurring in Non-Volatile Ferroelectric Transistors," IEEE Electron Dev. Lett., vol.42, no.3, pp.315-318, Jan. 2021.

[8] M. G. Kim and S. Ohmi, "Ferroelectric properties of undoped $HfO_2$ directly deposited on Si substrate by RF magnetron sputtering," Jpn. J. Appl. Phys., vol.57, no.11s, 11UF09, Sep. 2018.

[9] S. Ohmi, M. G. Kim, M. Kataoka, M. Hayashi, and R. M. D. Mailig, "Low-Voltage Operation of MFSFET with Ferroelectric Nondoped HfO2 Formed by Kr/O2-Plasma Sputtering," $78^{th}$ DRC Conf. Dig., vol. 96, June 2020.

[10] M. Tanuma, J. W. Shin, and S. Ohmi, "The effect of inter layers in the ferroelectric undoped $HfO_2$ formation," IEICE Trans. Electron., vol. E105-C, no. 10, Oct. 2022.

[11] M. Tanuma, J. W. Shin, and S. Ohmi, "Threshld voltage variation of MFSFET with 5 nm-thick ferroelectric nondoped $HfO_2$," 2022 Asia Pacific Workshop on Fundamentals and Applications of Advanced Semiconductor Devices, pp. 25-26, July 2022.

978-1-6654-7134-3/22 $31.00 © 2022 IEEE

# Ultra-fast Etching of Photoresist by Reactive Atmospheric-pressure Thermal Plasma Jet

Hibiki Kato
*Graduate School of Advanced Science and Engineering*
*Hiroshima University*
Hiroshima, Japan
m215781@hiroshima-u.ac.jp

Hiroaki Hanafusa
*Graduate School of Advanced Science and Engineering*
*Hiroshima University*
Hiroshima, Japan
hanafus@hiroshima-u.ac.jp

Takuma Sato
*Graduate School of Advanced Science and Engineering*
*Hiroshima University*
Hiroshima, Japan
takumas@hiroshima-u.ac.jp

Seiichiro Higashi
*Graduate School of Advanced Science and Engineering*
*Hiroshima University*
Hiroshima, Japan
sehiga@hiroshima-u.ac.jp

*Abstract*— **We have developed a new plasma source, reactive atmospheric-pressure micro-thermal-plasma-jet (R-μTPJ) for ultra-fast etching of photoresist. R-μTPJ was generated by DC arc discharge of $A_r$ and $O_2$ with input power of 260 W. local heating and simultaneous supply of reactive oxygen species has achieved an etching rate as high as 46.3 μm/s.**

*Keywords*—*dry process, RTA, style, atmospheric-pressure-thermal-plasma*

## I. INTRODUCTION

In recent years, the importance of high-speed etching of organic films has increased due to the frequent use of thick organic films. For instance, in the semiconductor manufacturing process, photoresists (PR) become thicker at the wafer edge after baking , so it is imperative to remove the PR before proceeding to the exposure process (Edge Bead Removal: EBR) [1]. Although chemical removal has been employed in the past, low etching speeds and handling issues make it extremely effective to introduce high-speed etching technology using dry processes such as ashing. Regarding the ashing of organic materials by plasma, the etching rate increases with increasing treatment temperature, and it has been reported that atmospheric pressure glow plasma enables uniform etching at a high rate of over 4 μm/min [2], and microwave-excited nonequilibrium plasma at a high rate of over 6 μm/s while heating the wafer to 598 K [3]. We assumed that local heating of the wafer edge and simultaneous ashing would enable high-speed PR removal. We attempted to further localize the heating and conducted experiments using reactive atmospheric-pressure micro-thermal plasma Jet(R-μTPJ) irradiation with $A_r$ and $O_2$.

## II. EXPERIMENT

After cleaning silicon (100) wafer, PR ( TOKYO OHKA KOGYO CO.,LTD.、 TSMR iP-3300 17cP) was spin coated at 4400 rpm and baked at 130 °C for 2 min to from 1 μm thick PR layer. The R-μTPJ was generated by DC arc discharge under atmospheric pressure with a supply current ($I$) = 20 A between a W cathode and a metal anode separated by $ES$ = 2.0 mm. Oxygen ($O_2$) gas flow rate ($f_{O2}$) was varied from 0.3 to

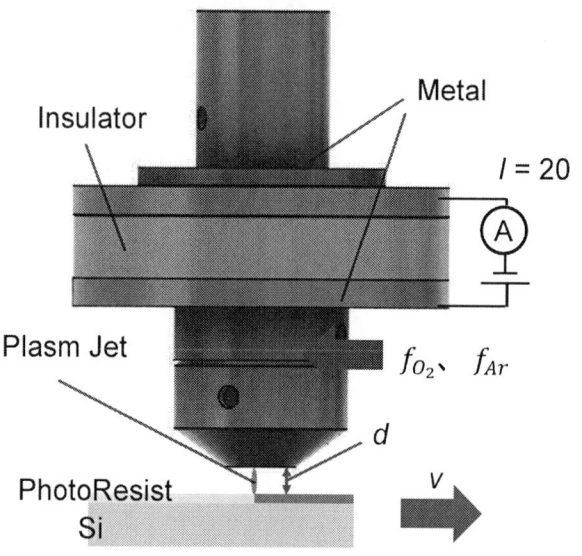

Fig. 1. Experimental set up for PR etching by R-μTPJ.

1.6 L/min, and argon ($A_r$) flow rate ($f_{Ar}$) was varied from 1.0 to 2.0 L/min. The samples were linearly moved by a motion stage in front of the R-μTPJ with a scanning speed ($v$) ranging from 20 to 45 mm/s. The distance between the plasma source and the substrate ($d$) was varied from 1.0 to 3.0 mm (Fig.1). The thickness decreases of the PR were observed by cross-sectional scanning electron microscope (SEM).

## III. RESILTS AND DISCUSSION

Fig. 2 shows the optical microscope images of PR before and after a R-μTPJ irradiation under different $f_{Ar}$ conditions. The R-μTPJ irradiation caused changes in the interference color as shown in Fig.2 (b) to (d). Samples center cross-section were observed by SEM with 20° tilt angle from the wafer surface, and the results are shown in Fig. 3 (a) to (d). The PR with a thickness of 0.87 μm before the treatment decreased their thicknesses after the treatment in conditions

---

Adaptable and Seamless Technology transfer Program through Target-driven R&D (A-STEP).

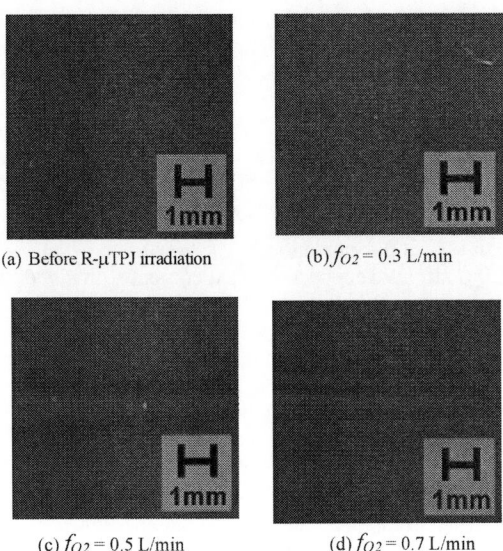

(a) Before R-μTPJ irradiation      (b) $f_{O2}$ = 0.3 L/min

(c) $f_{O2}$ = 0.5 L/min      (d) $f_{O2}$ = 0.7 L/min

Fig. 2.    Micrographs of photo resist (a) before and (b) to (d) after R-μTPJ irradiation

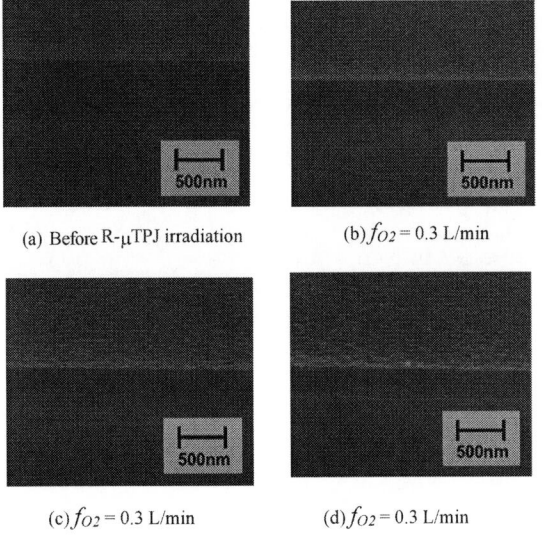

(a) Before R-μTPJ irradiation      (b) $f_{O2}$ = 0.3 L/min

(c) $f_{O2}$ = 0.3 L/min      (d) $f_{O2}$ = 0.3 L/min

Fig. 3.    SEM images of photo resist samples. (a) before and (b) to (d) after R-μTPJ irradiation. Samples were observed with 20° tilt angle from the wafer surface

(b) to (d). It is seen that the thickness decrement increases in response to the increasing oxygen flow rate.

We have confirmed increasing optical emission line from O* (777.4 nm) with increasing $f_{O2}$, which enhances etching of PR as confirmed in Fig. 4.

Fig. 5(a) and 5(b) show the thickness decreases of the PR after a R-μTPJ irradiation under different $f_{O2}$ and $d$ conditions. In all conditions, the change in film thickness reduction was found to be inversely proportional to the scan speed. This is thought to be due to the nearly constant etching speed of PR. The thickness decreases of the PR become larger as $d$ decreases and $f_{O2}$ increases.

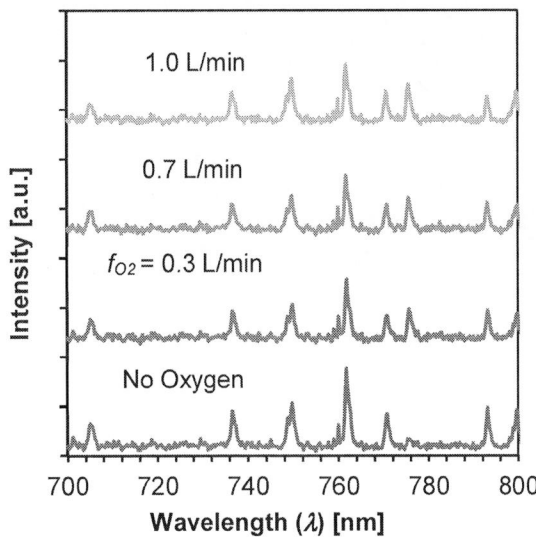

Fig. 4.    Emission Spectrometry of R-μTPJ

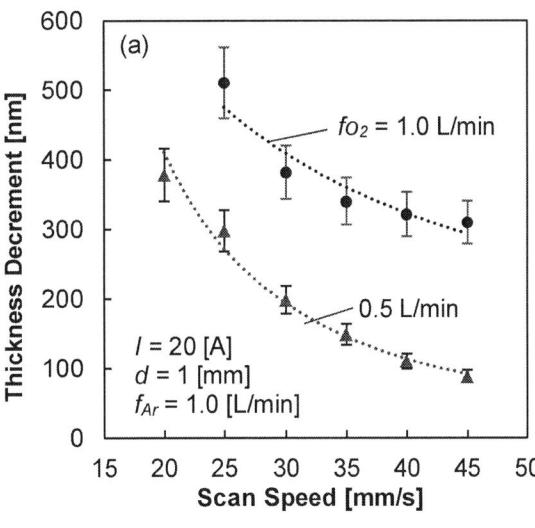

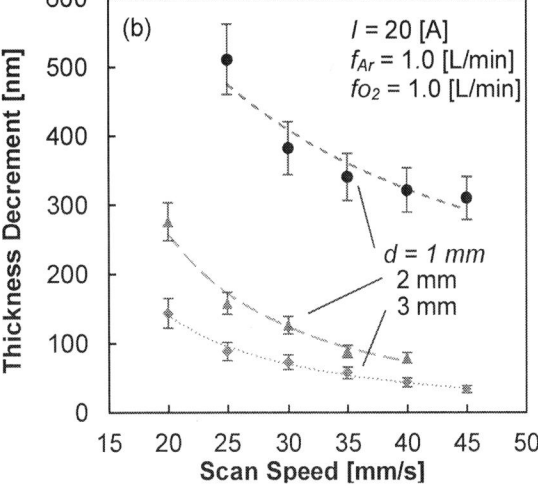

Fig. 5.    Thickness decrement of the PR with respect to scan speed. (a) $f_{O2}$ dependence and (b) working distance $d$ dependence are show.

PO-42

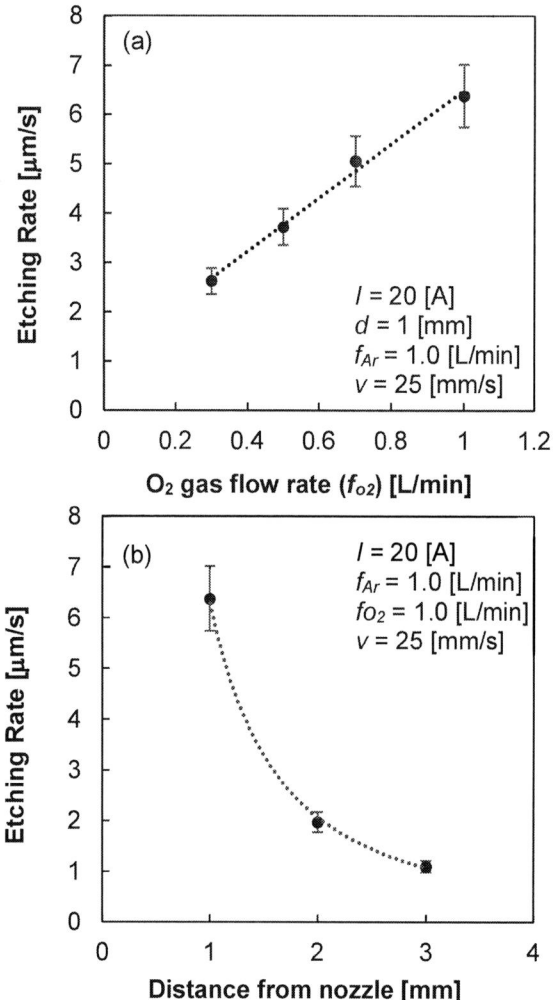

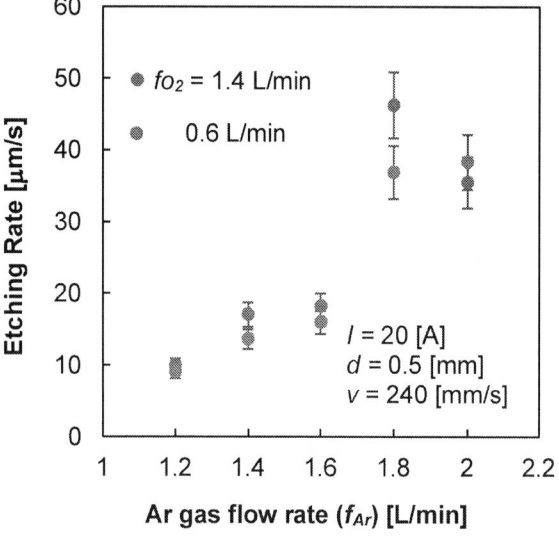

Fig. 6.    Etching rate of PR by R-μTPJ. (a) $f_{O_2}$ dependence and (b) working distance $d$ dependence are show.

Fig. 7.    Etching rate of PR with respect to $f_{Ar}$. With an addition of $O_2$, a high etching rate of 45.3 μm/s was obtained.

Fig. 6(a) and 6(b)shows the etching rate given that the 0.87-μm-thick PR is etched by R-μTPJ with $f_{O_2}$ varied from 0.3 to 1.0 L/min and $d$ varied from 1.0 to 3.0 mm. The etching rate of $f_{O_2}$ = 1.0 L/min is estimated to be 6.38 μm/s. The etching rate with $d$ decreasing 3.0 to 1.0 mm become rapidly larger. This result suggests that R-μTPJ not only generates reactive oxygen species, but rapidly heats the PR surface, which enhances etching reaction markedly.

Fig. 7 shows the etching rate with respect to vried from 1.0 to 2.0 L/min. We conducted an experiment with higher $A_r$ and $O_2$ flow rates. By increasing $f_{Ar}$, the heat transport to the PR surface is remarkably increased and at the same time, addition of $O_2$ enhances surface reaction. The etching rate of $f_{O_2}$ = 1.8 L/min is estimated to be 46.3 μm/s, which is more than 9 times higher compared to the conventional reports. [3]

## IV. CONCLUSION

We have developed a new rapid heat treatment source, R-μTPJ. R-μTPJ irradiation with $A_r$ and $O_2$ under different $f_{O_2}$, d or $f_{Ar}$ conditions showed the possibility of RP removal. The etching rate as high as 46.3 μm/s was obtained. This technique is quite hopeful to apply to EBR by dry process.

## ACKNOWLEDGMENT

A part of this work is supported by Adaptable and Seamless Technology transfer Program through Target-driven R&D (A-STEP) from Japan Science and Technology Agency (JST) Grant Number JPMJTR20RS.

## REFERENCES

[1]    I. Jekauc, M. Watt, T. Hornsmith, and J Tiffany :,Proc. SPIE 5376, 1255-1263 (2004).

[2]    K. Taniguchi, K. Tanaka, T. Inomata and M. Kogoma: J. Photopolym. Sci. Technol. 10 (1997).

[3]    Yamakawa, Koji, Hori, Masaru, Goto, Toshio, Den, Shoji, Katagiri, Toshirou, Kano, Hiroyuki:, Journal of Applied Physics 98, 043311 (2005).

# Obtaining of carbon nanowalls with a specified morphology

Yerassyl Yerlanuly
*Kazakh-British Technical University*
*al-Farabi Kazakh National University*
Almaty, Kazakhstan
yerlanuly@physics.kz

Maratbek T. Gabdullin
*Kazakh-British Technical University*
*al-Farabi Kazakh National University*
Almaty, Kazakhstan
gabdullin@physics.kz

Renata R. Nemkayeva
*Kazakh-British Technical University*
*al-Farabi Kazakh National University*
Almaty, Kazakhstan
quasisensus@mail.ru

Rakhymzhan Zhumadilov
*Kazakh-British Technical University*
*al-Farabi Kazakh National University*
Almaty, Kazakhstan
rakimzhan@gmail.com

Balaussa Ye. Alpysbayeva
*Kazakh-British Technical University*
*al-Farabi Kazakh National University*
Almaty, Kazakhstan
balau@list.ru

Tlekkabul S. Ramazanov
*al-Farabi Kazakh National University*
Almaty, Kazakhstan line 5: email
ramazan@physics.kz

*Abstract*— **This work presents experimental results on the synthesis of carbon nanowalls (CNWs) with predefined morphology on the surface of nanoporous alumina membrane using method of radio-frequency plasma-enhanced chemical vapor deposition. Obtained samples were characterized by the methods of scanning electron microscopy and Raman spectroscopy. From the microstructure analyses of CNWs, it has been observed that there is a time dependence of the reproducibility of membrane morphology by CNWs. At the early stage of CNWs growth, CNWs grow preferably around the edges of nanopores and continue to grow vertically with time. Nanopores size begin to shrink drastically and pores are completely covered by secondary flake-like CNWs after 25 minutes of growth.**

*Keywords*— *carbon nanowalls, radio frequency plasma enhanced chemical vapor deposition, capacitively coupled plasma, nanoporous alumina membrane.*

## I. INTRODUCTION

Carbon Nanowalls (CNWs) are three-dimensional networks of vertically oriented graphene layers [1], [2]. Compared to the classical graphene, the carbon nanowalls usually have a larger specific surface area. At the same time, due to the vertical orientation of the small graphene sheets, there is always a rather high level of defectiveness introduced by the crystallite edges [3], [4]. Due to their unique physico-chemical properties [5]–[7], CNWs have found practical applications in electrochemical devices (supercapacitors) [8], electron field emitters [2], temperature control of electronic devices [9], materials for gas storage [10], substrate for catalyst [2], deformation sensors [11], fuel cells [12], solar cells [13], etc. In addition, since CNWs are mainly composed of graphene sheets, they are expected to exhibit high mobility and significant sustained current density, accordingly, CNWs can be used in the field of nanoscale electronic devices, including various bio-, photo- and gas-sensitive sensors [14]–[17].

Despite the wide range of possible applications, control of the synthesis process and final morphology of CNW films is still a complicated task, particularly obtaining of CNWs with the required morphology and properties [18]–[20]. There are already several papers on the control of CNW morphology. For example, the growth and shape of carbon nanowalls can be controlled through the adjustment of carbon atom concentration [5], [21], [22]. Authors of work [4] reported that it is possible to control the average distance between the walls by changing the parameters of plasma discharge. Thus, they managed to control the walls separation distance using the system of radical injection RI-PECVD simultaneously changing the input voltage from 90V to 150V to change the electrostatic power of impulses. It was revealed that the higher voltages lead to a wider distance between the walls.

This work is dedicated to the synthesis of CNWs on nanoporous alumina membranes as substrate with predefined morphology using plasma enhanced chemical vapor deposition (PECVD). The nanoporous alumina membranes with different morphology and thickness were obtained by the method of two-step electrochemical anodization. Synthesized carbon nanostructures were studied using Raman spectroscopy and scanning electron microscopy. The dependence of morphology and the height of carbon nanowalls on the synthesis time is revealed.

## II. EXPERIMENTAL SECTION

Synthesis of CNWs on the surface of nanoporous alumina membrane was conducted using RF (radio frequency) – PECVD method, detailed description of the experimental setup and the synthesis process is reported in papers [23], [24]. Synthesis parameters in this method are the following: power - 11 W, heater temperature - 500 °C (substrate temperature - 460 °C), duration of Ar plasma processing - 10 min at the flow rate 7 sccm, synthesis time 25-35 min in the flow of gas mixture Ar/methane – 7/0.8 sccm, respectively. As substrates we used nanoporous membranes of aluminium oxide obtained by two-step method of electrochemical anodic oxidation [20] with pores diameter ~150 nm and membrane thickness of 10.5 micrometers.

## III. RESULTS AND DISCUSSIONS

Fig. 1 shows SEM images of CNWs samples synthesized on the surface of nanoporous alumina membrane using RF-PECVD method. Before the deposition, the membrane demonstrates the pore diameter of ~150 nm (figure 1a) and thickness (or pore depth) of 10 micrometers. After 25 minutes of the synthesis, one can see that CNWs replicate (reproduce) the membrane's morphology (Fig. 1b). At the following increase of synthesis time to 30 min, the density of CNWs on the membrane surface increases (Fig. 1c) and at 35 min pores of membrane are almost completely covered by CNWs (Fig. 1d), since long-term synthesis leads to increasing density and height of the resulting CNWs [23], [25].

978-1-6654-7134-3/22 $31.00 © 2022 IEEE

For complete study and comprehension of the synthesized CNWs' features, Raman spectroscopy measurements were performed. Fig.2 shows the results of structural analysis of CNW films obtained by Raman spectroscopy. The Raman spectra of the samples presented in Fig. 2a) show a typical CNWs spectrum with clear characteristic graphite peaks D, G, D', G' (2D), and G + D [22]. Analysis of Raman spectra (Fig. 2 b) shows the ratios of peaks' intensities - $I_D/I_G$, $I_G/I_{2D}$, $I_G/I_{D'}$. One can see that with increasing synthesis time, there is a decline in the intensity of peaks D and D' that are induced by defects in graphite structure. Moreover, the decrease in $I_D/I_G$ ratio indicates the increase in few-layer graphene lateral sizes with increasing synthesis time. This is also proved by SEM images, as white lines that correspond to the vertically oriented carbon nanowalls getting longer and of higher contrast. A clear downward trend in $I_G/I_{2D}$ value for longer synthesis duration can be induced by increasing levels of graphitization and long-range order.

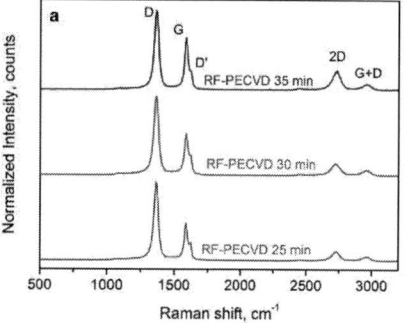

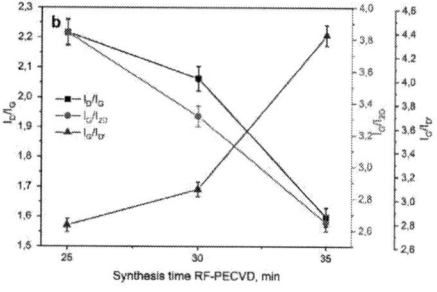

Fig. 2. Raman spectroscopy analysis of the CNW films synthesized by RF-PECVD method. a) Spectra of the synthesized CNW films. b) Dependences of the ratio of the $I_D$ and $I_G$ peaks, $I_G$ and $I_{2D}$ peaks, $I_G$ and $I_{D'}$ peaks on synthesis time

## IV. CONCLUSION

In conclusion, we have demonstrated that CNWs of well-defined morphology was successfully synthesized on the alumina membrane using RF-PECVD method. The obtained Raman spectra of the samples are typical for CNWs, there is a definite tendency of improving the quality of CNW with increasing synthesis time. Therefore, the obtained experimental results can be used to control the morphology of CNWs, which is of great importance for further practical

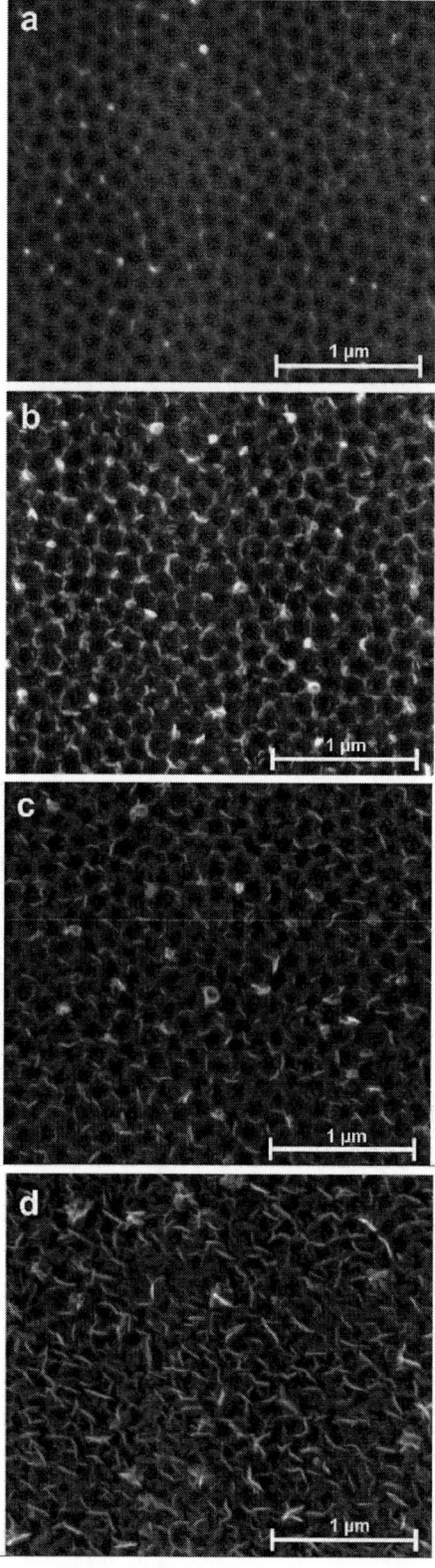

Fig. 1. SEM images of CNWs synthesized on the surface of nanoporous alumina membrane by RF-PECVD method: a) SEM image of nanoporous alumina membrane before the synthesis process; b) c) and d) SEM images of CNWs synthesized by RF-PECVD at 25, 30 and 35 min, respectively.

applications of carbon nanostructures for creating various sensors (gas, pressure, light, etc.) and supercapacitors.

### ACKNOWLEDGMENT

This research has been funded by the Science Committee of the Ministry of Education and Science of the Republic of Kazakhstan (Grant No. AP08856684).

### REFERENCES

[1] Y. Wu, P. Qiao, T. Chong, and Z. Shen, "Carbon Nanowalls Grown by Microwave Plasma Enhanced Chemical Vapor Deposition," *Advanced Materials*, vol. 14, no. 1, pp. 64–67, Jan. 2002, doi: 10.1002/1521-4095(20020104)14:1<64::AID-ADMA64>3.0.CO;2-G.

[2] M. Hiramatsu and M. Hori, *Carbon Nanowalls*. Vienna: Springer Vienna, 2010. doi: 10.1007/978-3-211-99718-5.

[3] S. Kurita *et al.*, "Raman spectra of carbon nanowalls grown by plasma-enhanced chemical vapor deposition," *Journal of Applied Physics*, vol. 97, no. 10, p. 104320, May 2005, doi: 10.1063/1.1900297.

[4] T. Ichikawa, N. Shimizu, K. Ishikawa, M. Hiramatsu, and M. Hori, "Synthesis of isolated carbon nanowalls via high-voltage nanosecond pulses in conjunction with CH4/H2 plasma enhanced chemical vapor deposition," *Carbon*, vol. 161, pp. 403–412, May 2020, doi: 10.1016/j.carbon.2020.01.064.

[5] H. J. Cho, H. Kondo, K. Ishikawa, M. Sekine, M. Hiramatsu, and M. Hori, "Density control of carbon nanowalls grown by CH4/H2 plasma and their electrical properties," *Carbon*, vol. 68, pp. 380–388, Mar. 2014, doi: 10.1016/j.carbon.2013.11.014.

[6] Vesel, Zaplotnik, Primc, and Mozetič, "Synthesis of Vertically Oriented Graphene Sheets or Carbon Nanowalls—Review and Challenges," *Materials*, vol. 12, no. 18, p. 2968, Sep. 2019, doi: 10.3390/ma12182968.

[7] Y. Yerlanuly *et al.*, "Physical properties of carbon nanowalls synthesized by the ICP-PECVD method vs. the growth time," *Scientific Reports*, vol. 11, no. 1, p. 19287, Dec. 2021, doi: 10.1038/s41598-021-97997-8.

[8] E. Ghoniem, S. Mori, and A. Abdel-Moniem, "An efficient strategy for transferring carbon nanowalls film to flexible substrate for supercapacitor application," *Journal of Power Sources*, vol. 493, p. 229684, May 2021, doi: 10.1016/j.jpowsour.2021.229684.

[9] Z. Wang *et al.*, "Near Room-Temperature Synthesis of Vertical Graphene Nanowalls on Dielectrics," *ACS Applied Materials & Interfaces*, vol. 14, no. 18, pp. 21348–21355, May 2022, doi: 10.1021/acsami.2c02381.

[10] S. Chul Shin *et al.*, "Carbon nanowalls as platinum support for fuel cells," *Journal of Applied Physics*, vol. 110, no. 10, p. 104308, Nov. 2011, doi: 10.1063/1.3662142.

[11] P. Slobodian *et al.*, "Transparent elongation and compressive strain sensors based on aligned carbon nanowalls embedded in polyurethane," *Sensors and Actuators A: Physical*, vol. 306, p. 111946, May 2020, doi: 10.1016/j.sna.2020.111946.

[12] T. Ohta, H. Iwata, M. Hiramatsu, H. Kondo, and M. Hori, "Power Generation Characteristics of Polymer Electrolyte Fuel Cells Using Carbon Nanowalls as Catalyst Support Material," *C*, vol. 8, no. 3, p. 44, Aug. 2022, doi: 10.3390/c8030044.

[13] W. Wei, K. Sun, and Y. H. Hu, "Synthesis of Mesochannel Carbon Nanowall Material from $CO_2$ and Its Excellent Performance for Perovskite Solar Cells," *Industrial & Engineering Chemistry Research*, vol. 56, no. 7, pp. 1803–1809, Feb. 2017, doi: 10.1021/acs.iecr.6b04768.

[14] A. Palla-Papavlu, S. Vizireanu, M. Filipescu, and T. Lippert, "High-Sensitivity Ammonia Sensors with Carbon Nanowall Active Material via Laser-Induced Transfer," *Nanomaterials*, vol. 12, no. 16, p. 2830, Aug. 2022, doi: 10.3390/nano12162830.

[15] P. K. Roy *et al.*, "Ultrasensitive Gas Sensors Based on Vertical Graphene Nanowalls/SiC/Si Heterostructure," *ACS Sensors*, vol. 4, no. 2, pp. 406–412, Feb. 2019, doi: 10.1021/acssensors.8b01312.

[16] J. Cong *et al.*, "Direct Growth of Graphene Nanowalls on Silicon Using Plasma-Enhanced Atomic Layer Deposition for High-Performance Si-Based Infrared Photodetectors," *ACS Applied Electronic Materials*, vol. 3, no. 11, pp. 5048–5058, Nov. 2021, doi: 10.1021/acsaelm.1c00807.

[17] T. Ichikawa, K. Ishikawa, H. Tanaka, N. Shimizu, and M. Hori, "Scaffolds with isolated carbon nanowalls promote osteogenic differentiation through Runt-related transcription factor 2 and osteocalcin gene expression of osteoblast-like cells," *AIP Advances*, vol. 12, no. 2, p. 025216, Feb. 2022, doi: 10.1063/5.0075530.

[18] J. Fang, I. Levchenko, T. van der Laan, S. Kumar, and K. (Ken) Ostrikov, "Multipurpose nanoporous alumina–carbon nanowall bi-dimensional nano-hybrid platform via catalyzed and catalyst-free plasma CVD," *Carbon*, vol. 78, pp. 627–632, Nov. 2014, doi: 10.1016/j.carbon.2014.07.053.

[19] N. Jiang, H. X. Wang, H. Zhang, H. Sasaoka, and K. Nishimura, "Characterization and surface modification of carbon nanowalls," *Journal of Materials Chemistry*, vol. 20, no. 24, p. 5070, 2010, doi: 10.1039/c0jm00446d.

[20] Y. Yerlanuly *et al.*, "Synthesis of carbon nanowalls on the surface of nanoporous alumina membranes by RI-PECVD method," *Applied Surface Science*, vol. 523, p. 146533, Sep. 2020, doi: 10.1016/j.apsusc.2020.146533.

[21] N. Santhosh *et al.*, "Oriented Carbon Nanostructures by Plasma Processing: Recent Advances and Future Challenges," *Micromachines*, vol. 9, no. 11, p. 565, Nov. 2018, doi: 10.3390/mi9110565.

[22] W. Takeuchi, H. Sasaki, S. Kato, S. Takashima, M. Hiramatsu, and M. Hori, "Development of measurement technique for carbon atoms employing vacuum ultraviolet absorption spectroscopy with a microdischarge hollow-cathode lamp and its application to diagnostics of nanographene sheet material formation plasmas," *Journal of Applied Physics*, vol. 105, no. 11, p. 113305, Jun. 2009, doi: 10.1063/1.3091279.

[23] D. Batryshev, Y. Yerlanuly, B. Alpysbaeva, R. Nemkaeva, T. Ramazanov, and M. Gabdullin, "Obtaining of carbon nanowalls in the plasma of radio-frequency discharge," *Applied Surface Science*, vol. 503, p. 144119, Feb. 2020, doi: 10.1016/j.apsusc.2019.144119.

[24] D. G. Batryshev, Y. Yerlanuly, T. S. Ramazanov, M. K. Dosbolayev, and M. T. Gabdullin, "Elaboration of carbon nanowalls using radio frequency plasma enhanced chemical vapor deposition," *Materials Today: Proceedings*, vol. 5, no. 11, pp. 22764–22769, 2018, doi: 10.1016/j.matpr.2018.07.088.

[25] S. Kondo *et al.*, "Initial growth process of carbon nanowalls synthesized by radical injection plasma-enhanced chemical vapor deposition," *Journal of Applied Physics*, vol. 106, no. 9, p. 094302, Nov. 2009, doi: 10.1063/1.3253734.

**Yerassyl Yerlanuly** received the bachelor's degree in electrical power engineering and the master's degree in nanomaterials and nanotechnology from Al-Farabi Kazakh National University, Almaty, Kazakhstan, in 2016, 2018, respectively, where he continued his studies and research activities. His areas of interests include physics of plasmas, materials science, carbon nanomaterials and its practical application.

**Maratbek T. Gabdullin** has made a significant contribution to the theoretical and experimental study of physical processes of nanomaterials synthesis, proceeding in a plasma environment, obtaining new composite nanomaterials, hydrogen storage in nanomaterials, and practical aspects of nanomaterials application in industry. He maintains close creative contacts with leading scientists from various scientific centers worldwide. Recipient of the RK State Scholarship for talented young scientists (2011-2012), Best Lecturer 2014, Laureate of Al-Farabi State Prize in Science and Technology in 2015, According to Web of Science Awards (Clarivate Analytics) "Leader of Science"

(2017, 2019). In 2018 he was awarded "Person of the Year - Altyn Adam" in the category of "Science Worker". Prof. Gabdullin is the author of more than 15 patents of the RK, more than 120 publications in journals with high impact factor, included in the Web of Science and Scopus bases, and more than 300 publications in journals from the list of CQASES of the MES RK. He has repeatedly made scientific presentations at prestigious international conferences (UK, Japan, Korea, Germany, France, Italy, Portugal, etc.).

**Renata R. Nemkayeva** in 2009, she received a bachelor's degree in physics, 2012 - a master of natural sciences in physics. Since 2010, she has been working in the scientific field, her activities cover nanomaterials, characterization of materials, carbon nanostructures, two-dimensional materials.

**Rakhymzhan Zhumadilov** is a PhD doctoral student of the 3rd course. Works mainly in the field of synthesis of carbon nanomaterials and analysis of the obtained samples.

**Balaussa Ye. Alpysbayeva** has a PhD in the specialty "6D071000-Materials Science and Technology of New Materials". Highly qualified specialist in scanning probe microscopy (SPM). Scientific direction of Alpysbayeva B.Ye. covers nanotechnology, research of the properties and obtaining of various nanomaterials.

**Tlekkabul Ramazanov** received the bachelor's degree from the Faculty of Physics, Kirov Kazakh State University, Almaty, Kazakhstan, in 1983, and the degree of Doctor of Physical and Mathematical Sciences from Al-Farabi Kazakh National University in 1996. Since 1987, he has been an Engineer, a Junior Researcher, an Assistant, a Senior Lecturer, an Associate Professor, and a Professor with Al-Farabi Kazakh National University, Almaty. He is currently the Leader of the research group at the Research Institute of Experimental and Theoretical Physics (IETP), Almaty, where he is involved in theory, computer simulation, and experimental study of the properties of non-ideal and gas discharge plasmas. His current research interests include theory, computer modeling, and experimental study of the properties of non-ideal plasma.

PO-55

2022 International Symposium on Semiconductor Manufacturing (ISSM).
December 12-13, 2022

# Process Optimizations for Ge-on-Si Depletion Mode Transistors using Mesa Architecture

Sumit Choudhary
*School of Computing and Electrical Engineering.*
*Indian Institute of Technology, Mandi*
Mandi, India
0000-0002-3127-4691

Daniel Schwarz
*Institute of Semiconductor Engineering*
*University of Stuttgart*
Stuttgart, Germany
0000-0003-2702-4697

Hannes S. Funk
*Institute of Semiconductor Engineering*
*University of Stuttgart*
Stuttgart, Germany
0000-0001-8485-2400

Kumar Palit Sharma
*School of Computing and Electrical Engineering*
Indian Institute of Technology, Mandi
Mandi, India
0000-0002-9711-6506

Satinder K. Sharma
*School of Computing and Electrical Engineering*
*Indian Institute of Technology, Mandi*
Mandi, India
0000-0001-9313-5550

Jörg Schulze
*Institute of Semiconductor Engineering*
*University of Stuttgart*
Stuttgart, Germany
0000-0003-3621-7888

*Abstract*— The p-Ge layers are epitaxially grown by MBE over the n-Ge and strain-free Ge buffer layers on the Si substrate. The drain-source & channel mesa is patterned in the p-Ge layer to create the raised active channel. Post-plasma oxidation was carried out to improve the interface properties of Ge channel. The proposed process doesn't involve source-drain implants, ease channel patterning using MAPDST, a -ve tone resist with high etch resistance and selectivity w.r.t. Ge. The process flow scheme will utilize the "beyond Si" channel materials over Si substrates, concurrently exploiting the standard well, established state-of-art Si CMOS fabrication technology.

*Keywords—FET fabrication, Germanium, Ge-FETs, Electron beam lithography, MAPDST, Molecular beam epitaxy, optimization, Process flow, resist, Reactive ion etching, SiGe technology.*

## I. Introduction

The Si CMOS feature size miniaturization approaching 3nm technology node at the chip integration level [1]. Apart planar FET scaling eventually halted, since the distance between the nearby atoms has their physical limit. Therefore, different methods must be utilized to enhance the FET performance. Furthermore, it is asserted that electron transport in FETs is ballistic or nearly ballistic in the current technology node, meaning that a physical factor relevant to determining the driving current is the injection velocity [2], [3]. Besides this, materials with higher carrier mobility, can be integrated into the FET active channel regions, as carrier mobility is a crucial parameter for boosting the injection velocity. Alternatively, germanium (Ge) is energy-efficient material and has substantially higher mobilities than silicon, desirable for high-speed and low-voltage applications [4][5]. The bottleneck concern for the chip industry is an integration of high-mobility materials at chip-level; on the other hand, industrial scale production of single crystalline silicon wafers is dominating in the industry. One of the solution to that is Ge can only be used as the active channel region and standard state-of-art Si wafer will be used as substrate [4]. Another significant concern about the stability of the GeO₂, which is still a serious challenge, however significant improvement has been achieved using the high pressure $O_2$ (HPO) oxidation, low temperature annealing (LOA) [6][7] and Plasma post oxidation processes to form the interface $GeO_x$ required to keep the interface trap density ($D_{it}$) under control [8].

Unfortunately, Ge-on-insulator (GOIs) production is neither cost-effective nor process-compatible when compared to the production of SOI with Smart-Cut technology. The molecular beam epitaxy (MBE) enables ultrathin epitaxial growth of Ge over the Si substrate, with intermediate buffer layer to minimize the strain related current leakage and active channel can be patterned using high etch resistance electron beam lithography (EBL) resist (4-(methacryloyloxy) phenyl) dimethylsulfoniumtriflate (MAPDST). The patterned mesa permits harnessing the qualified Ge properties as an active channel and simultaneously exploiting the standard state-of-the-art Si platform. Furthermore, the simplified and robust process flow for realizing the Ge channel over the Si substrate is the need of the hour for extracting the Ge channel FET performance characteristics, and projecting the Ge as "beyond Si" channel material.

In this article, we have proposed the optimized MBE growth of epi-Ge active layers with a buffer layer over the standard Si wafer. The simplified process flow is designed to harness and study the Ge FETS properties. Mesa structures are patterned using the -ve tone resist by electron beam lithography, and used as the mask during reactive ion etching. Apart from optimized recipes, process concerns have been discussed and proposed for the realization of Ge FET$_S$ on Si substrates.

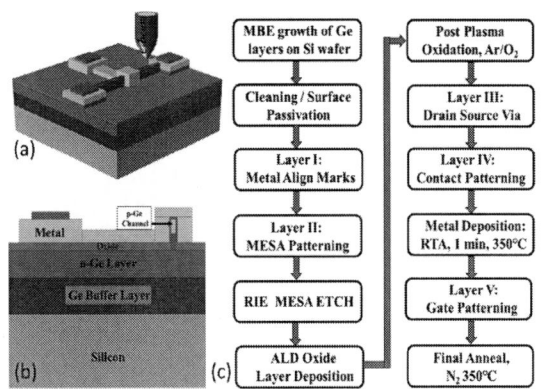

Fig. 1(a) Bird eye view of the proposed device architecture. Fig. 1(b) Cross-sectional view. Fig 1(c). Proposed device fabrication process flow.

978-1-6654-7134-3/22 $31.00 © 2022 IEEE

## II. Ge-on-Si Device Fabrication

### A. Device Concept

The epitaxially grown narrow p-Ge channel is patterned over the bulk n-Ge layer. The n-Ge layer has grown over the strain-free Ge buffer layer (to minimize the strain-related leakage), and further buffer layer has epitaxially grown over the standard state-of-art silicon wafer, forming the stack of p-Ge/n-Ge/Ge(buffer)/Si. The doping concentration is constant throughout in the patterned p-Ge channel, source, and drain regions. Typically, the channel concentration is kept high i.e., $1E18$ cm$^{-3}$ whereas beneath the n-Ge layer has a doping concentration of $1E16$ cm$^{-3}$. The device can be turned OFF through the gate electric field by inducing the full depletion of the channel carriers. Apart from nicely coupled p-channel can be further depleted by the low work function of the gate material and bottom p-Ge/n-Ge layers junction depletion. The thickness and channel widths can be precisely controlled by epitaxial growth and advanced lithography process, respectively. The substrate leakage current has been eliminated by the use of p-n junction between the p-channel and beneath the n-Ge layer. The proposed device design and cross-sectional view are shown in Fig. 1(a,b), respectively.

### B. MBE Epitaxial Growth

Initially 6-inch Si (100) wafer with Boron (B) doping, with specific resistance of 10 $\Omega$cm $\leq \rho \leq$ 20 $\Omega$cm selected as the substrate. Ge, B and Sb were evaporated from thermal effusion cells using solid source MBE (SS-MBE). A Si buffer layer with thickness d = 50 nm was deposited. On the Si buffer layer, a virtual Ge substrate (Ge-VS) with d = 100 nm was grown and annealed at 850 °C to minimize the lattice mismatch [9]. Over the Ge-VS, an n-Ge buffer layer with d = 400 nm was grown. The n-Ge buffer layer was grown with Sb doping at a nominal concentration $N_D$ = 1E16 cm$^{-3}$ at a growth temperature T = 250 °C. The n-Ge buffer layer further reduces the threading dislocation density. On top of the n-Ge buffer layer, the p-Ge channel was grown. The selected channel thickness is d = ~75 nm with B doping at a nominal concentration $N_A$ = 1E18 cm$^{-3}$ [10].

### C. Electron Beam Patterning

The key 5-layer fabrication process flow is explained in Fig. 1(c). The GDSII mask file was prepared using layout software for patterning the raised channel and drain source (D/S) regions (mesa) structures, together with aligned mask layers for example electrode contact, via and gate, etc. While designing the mask files concerns about the write field alignments (WFA) to avoid the overlay and laser interferometry stage movements cause patterning errors must be considered during mask preparation. To expose fine patterns, sub 50nm highly collimated beam having high acceleration of 20 to 30kV, write field corrections to avoid the pattern overlays, beam current 20 to 30pA, aperture size of 10μm, and a low working distance of 6 mm implemented using Raith nanofabrication e-line plus EBL tool. Firstly, the metal alignment marks (Layer I), was patterned using a positive tone of the (polymethyl methacrylate) PMMA and MMA resist bilayer process to create clean undercut profiles and able to fabricate clean residue-free devices after lift-off due to difference in molecular weight of both resists. For PMMA exposures the optimized dose of 70μC/cm$^2$ was used. After the exposure the samples were developed in the MIBK:

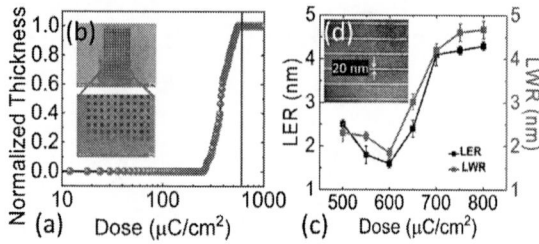

Fig. 2 MAPDST resist optimization. Fig. 2(a) NRT curve over a dose matrix of 10 to 1000 μC/cm$^2$. Inset Fig. 2(b) shows the exposed image of dose matrix. Fig. 2(c) After exposure the resist line edge roughness (LER) and line width roughness (LWR) were extracted by SUMMIT$^{TM}$ software over L/10S exposures.

IPA (1:3) solution for 30 secs. Next to that for the MBE-grown Ge samples the descumming has been done under weak O$_2$ plasma. Thereafter 20 nm of aluminium metal was deposited using thermal evaporation then metal lift-off has been done in warm acetone solution (50°C). After optical microscope inspection, weak plasma exposure can be done if resist residues are still present. Similarly, the exposure and patterning of layer IV and V masks can be done with different metals Pt/Ni and Ti/TaN for drain source and gate patterning, respectively.

### D. Resist Optimization

The active p-Ge layer patterning (Layer II) to create a MESA structure is the critical and very crucial part. Altogether the above-mentioned prerequisite, there is a requirement for the highly stable negative tone resist, which should have the following properties like, sensitivity, line edge roughness (LER), line width roughness (LWR), no residues, stable dose, high etch resistance, selectivity to reactive ion etching (RIE) processing, ease of post stripping. The mesa was patterned using EBL and with in-lab formulated, MAPDST negative tone resist, the detailed synthesis process is discussed elsewhere [11]. After the resist formulation, the resist was spin-coated over the Si samples. The resist dose sensitivity was measured by exposing the dose matrix having a dose variation from 10μC/cm$^2$ to 2000μC/cm$^2$. Furthermore, after the exposure, post-exposure baking at 115°C was done, thereafter the exposed samples were developed in the developer tetramethyl ammonium hydroxide (TMAH) solution prepared in DI water at PH of 11.5 at room temperature for 30 sec and then samples were dried by N$_2$ gas purging. The exposed and developed dose matrix was scanned by atomic force microscopy (AFM) to extract the normalized remaining thickness (NRT) curve for the optimum resist exposure dose estimation, Fig. 2 (a, b). Henceforth, the LER and LWR were extracted by SUMMIT$^{TM}$ software over L/10S exposures, over the varied exposure dose range. At 600μC/cm$^2$ dose the LER and LWR both are extracted minimum, Fig. 2 (c, d).

### E. MESA paterning

After the resist dose optimization, mesa patterning (layer II) has been exposed by EBL. The formulated MAPDST resist was in powder form, to make a solution it was dissolved in 3% by weight in the methanol. The MAPDST resist was spin-coated over the samples at 2000 RPM for 1 minute. Thereafter, pre-exposure bake (PEB) was done at 100°C for 120 sec. The EBL exposure was carried out at a dose of

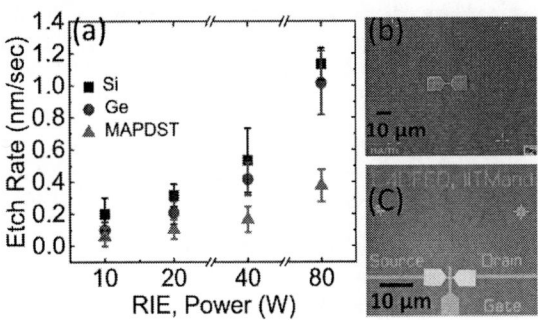

Fig. 3 MAPDST resist etch optimization w.r.t. Ge and Si in reactive ion etching under $Sf_6$ and $O_2$ plasma. Fig. 3(b) depicts the etched mesa. Fig. 3(c) Gate alignment patterning w.r.t. drain source mesa structure alignment.

$600\mu C/cm^2$. Other parameters for EBL and PEB were kept constant as discussed section C & D, respectively. After the development the samples were exposed to UV light having a wavelength of 365nm for 5 minutes each to improve the stability of patterned edges. Subsequently, the samples were exposed to a customized RIE tool for the creation of raised channel and D/S pads (mesa structure). The etch resistance is optimized by controlling the parameters like power, $Sf_6$ flow rate, chamber process vacuum, and percentage of $O_2$. Fig 3(a) shows the variation of the power while keeping the other parameters kept constant like $Sf_6/O_2$ (4:1), process pressure was set to 1 mtorr. The anisotropic profile and optimum etch rate were achieved at 20W RIE power. Fig. 3(b) illustrates the etched mesa.

### F. Dielectric Deposition & Drain/Source Contacts

After executing the Layer II processing steps, to meet the desired gate-to-channel capacitance the hafnium oxide ($HfO_2$) high-κ dielectric was deposited using the atomic layer deposition (ALD) to maintain the conformality of deposition over the raised structures [5]. Ultratech ALD tool was utilized for ALD depositions. The tetrakis-dimethylamide hafnium (TDMA-Hf) for $HfO_2$ and DI water were used as precursors. The $HfO_2$ precursor was heated to 75°C and chamber temperature was maintained at 150°C during the deposition. The pulse time for $HfO_2$ and $H_2O$ cycles was used 0.15 and 0.015 sec respectively, with an intermediate gap of 20 sec between two successive pulses. The growth rate achieved approx. 1 Å /cycle on germanium. To improve the Ge channel and oxide interface properties leading to low interface trap density ($D_{it}$), thin atomic layer of $GeO_x$ is formed by the plasma post oxidation (PPO), beneath the $HfO_2$ oxide layer by exposing the samples to $O_2$ plasma for 1 min [8], the $GeO_x$ formation at interface to helps to keep the $D_{it}$ under control, $HfO_2$ layer act as barrier for oxygen diffusion, as GeO is volatile in nature [12]. After the oxide deposition over the full substrate, the drain/source contact needs to etch through the oxide to form the Ni metal and Ge contact. To etch the via over the drain/source (D/S) pads is accomplished by EBL alignment patterning and exposure (Layer III) by spin coating the multiple (3 to 4) layers of PMMA resist each at 2000 RPM. After exposure and development of resist, the open windows of the samples were exposed in the RIE tool for etching the D/S pads under $Ar/Sf_6$ plasma. Etched via were inspected under the scanning electron microscope, the complete removal of oxides was further cross checked by fabricating the Ni/Ge contact on the control samples and measuring the contact

resistivity. Ni metal is chosen for ohmic contact formation and NiGe (germanide) formation at low temperature [13], [14]. Thereafter, the contact metal deposition (Layer IV) patterning was accomplished similarly, discussed in previous sections. The platinum/nickel (Pt/Ni) metal (20/30nm) was deposited for contact pads by sputtering technique, using DC power of 8 watts, process pressure was kept 6E-3 mbar, under Ar ambient, the lift-off process removed further undesired Pt/Ni. To achieve the desired ohmic contacts post-metal rapid thermal annealing (RTA) was performed at 350°C, under $N_2$ environment for 1 min.

### G. Gate Patterning

The gate patterning step (Layer V) was performed by EBL keeping the concerns and followed the critical steps/process parameters discussed in section C. The aligned patterned gate is described in Fig. 3(c). The gate metal was chosen TaN due to low work function of TaN close to 4.3 eV, helps in the depletion of channel carriers, further it as capped by Ti metal. Finally, the whole sample was subjected to final annealing at 350°C under $N_2$ environment for 10 mins.

## III. ELECTRICAL CHARACTERIZATION OF PROTOTYPES

The proposed junctionless Ge-on-Si p-FETs operated in the accumulation to depletion region. Beneath n-Ge and Ge-VS layers are utilized for the reduction of substrate leakage optimum biasing keeps the pn junction layers in reverse bias. To ensure high $I_{ON}$ the D/S contact resistance must be optimized beforehand as it limits the $I_{ON}$ drain current due to this device's performance is compromised. Fig. 4 depicts the ohmic behaviour of Ni/Ge devices measured by fabricating the CTLM structures with spacing 4 to 48μm using the fabrication procedures discussed earlier, then the samples were subject to rapid thermal annealling at 350°C for 1 min for NiGe (germanide) formation. The D/S contact resistivity is computed $\sim 7.49 \times 10^{-7}$ $\Omega cm^2$ through CTLM measurements [15]. Fig. 5(a, b) Shows the transfer and output characteristics, respectively. The $I_{ON} = \sim 10\mu A/\mu m$ and $I_{ON}/I_{OFF} \sim 10^3$ is attributed to contact resistance optimisation, $Ge/GeO_x/HfO_2$ interface improvement by plasma post oxidation [8],

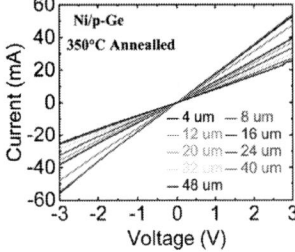

Fig. 4 Ni/p-Ge ohmic contact optimization by fabricating the CTLM structures with spacing ranging from 4 to 48 μm.

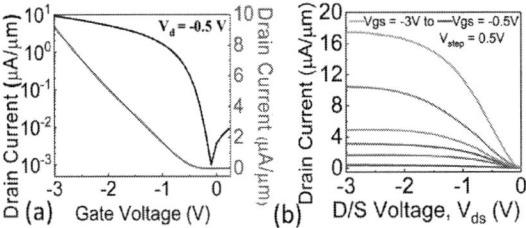

Fig. 5(a) Transfer Characteristics, Fig.5(b) Output characteristics of p-Ge-on-Si devices.

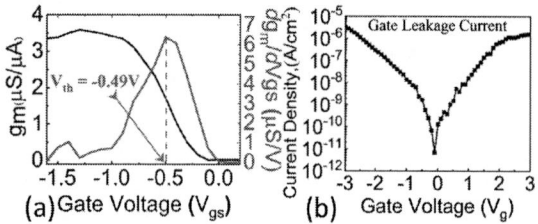

Fig. 6(a) Depicts the transconductance and dgm/dVgs vs gate voltage characteristics on a double y-axis plot. Fig. 6(b) The gate leakage current vs gate voltage characteristics of Ti/TaN/HfO$_2$/p-Ge-on-Si FETs.

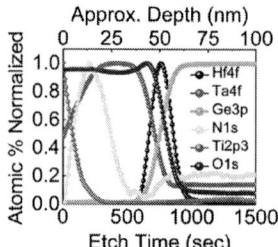

Fig. 7(b) XPS depth profile for Ti/TaN/HfO$_2$/p-Ge over the gate stack.

suppression of leakage current to the substrate, MBE layers growth engineering. At low drain voltage up to 1 V, linear behaviour of $I_{ds}$ versus $V_{ds}$ is observed ascribed to ohmic D/S contacts. By varying the $V_{gs}$, the $I_{ds}$ have modulated due to accumulation, and the depletion of charge carriers demonstrating strong channel coupling. The $V_{th}$ of the devices further modulated by body contact at Si substrate (not included here). Fig. 6(a) depicts transconductance $d_{gm}/dV_{gs}$ vs $V_{gs}$ characteristics, gm = ~3.5µS/µm. Fig. 6(b) depicts the low gate leakage current density ~1E-8 A/cm$^2$ at $V_{gs}$= 1V. Fig. 7 represents the XPS depth profile for Ti/TaN/HfO$_2$/p-Ge over the gate stack demonstrates the atomic percentage of individual elements in the stack at respective depths.

## IV. CONCLUSION

The process flow for first prototypes of Ge-on-Si FETs using strain-free Ge buffer layers and mesa architecture has been implemented. Mesa formation minimizes the active area interaction with the substrate to minimize the leakage current. The critical process step is processing a -ve tone resist which could help high resolution, low LER and LWR patterning, ease of stripping after processing. The plasma post oxidation exhibits the low interface traps would eventually help enhance the devices' performance. The process flow and optimization projections would be utilized to integrate "beyond Si" channel materials, simultaneously exploiting the legacy of state-of-art Si CMOS fabrication technology.

## ACKNOWLEDGMENT

The authors would like to thank the DST-DAAD collaboration between (University of Stuttgart, Stuttgart, Germany and the Indian Institute of Technology, (IIT), Mandi, India) for supporting this work. Institute of Semiconductor Engineering (IHT), Stuttgart, Germany, for

**Author Bio: Sumit Choudhary** (Member, IEEE) received the M.Tech. degree in Microelectronics and VLSI Design from the Department of Electronic Science, Kurukshetra University, Kurukshetra, India. From 2012 to 2017, he was a faculty with the Department of Electronic Science, Kurukshetra University, India. Since 2018, he has been doing doctoral

optimized layered growth of Ge over Si substrates with MBE. The center for design and fabrication of electronic devices (C4DFED), IIT Mandi, class 100 clean room facility for device process flow optimizations and advance material research center (AMRC), IIT Mandi for material characterizations. Mr. Sumit Choudhary is financially supported by Visvesvaraya, Ph.D. fellowship scheme, (MeitY), Government of India.

## REFERENCES

[1] M. LAPEDUS, "Transistors Reach Tipping Point At 3nm," 2022. Accessed:Nov.19,2022Available:https://semiengineering.com/transistors-reach-tipping-point-at-3nm/.

[2] K. Natori, "Ballistic metal-oxide-semiconductor field effect transistor," *J. Appl. Phys.*, vol. 76, no. 8, pp. 4879–4890, Oct. 1994, doi: 10.1063/1.357263.

[3] M. Lundstrom and Z. Ren, "Essential physics of carrier transport in nanoscale MOSFETs," *IEEE Trans. Electron Devices*, vol. 49, no. 1, pp. 133–141, 2002, doi: 10.1109/16.974760.

[4] A. Toriumi and T. Nishimura, "Germanium CMOS potential from material and process perspectives: Be more positive about germanium," *Jpn. J. Appl. Phys.*, vol. 57, no. 1, 2018, doi: 10.7567/JJAP.57.010101.

[5] S. Choudhary, *et al.*, "Impact of Charge Trapping on Epitaxial p-Ge-on-p-Si and HfO$_2$ Based Al/HfO2/p-Ge-on-p-Si/Al Structures Using Kelvin Probe Force Microscopy and Constant Voltage Stress," *IEEE Trans. Nanotechnol.*, vol. 20, pp. 346–355, 2021, doi: 10.1109/TNANO.2021.3069820.

[6] C. H. Lee, T. Tabata, T. Nishimura, K. Nagashio, K. Kita, and A. Toriumi, "Ge/GeO2 interface control with high-pressure oxidation for improving electrical characteristics," *Appl. Phys. Express*, vol. 2, no. 7, pp. 1–4, 2009, doi: 10.1143/APEX.2.071404.

[7] C. H. Lee, T. Nishimura, K. Nagashio, K. Kita, and A. Toriumi, "High-electron-mobility Ge/GeO2 n-MOSFETs with two-step oxidation," *IEEE Trans. Electron Devices*, vol. 58, no. 5, pp. 1295–1301, 2011, doi: 10.1109/TED.2011.2111373.

[8] R. Zhang, T. Iwasaki, N. Taoka, M. Takenaka, and S. Takagi, "High-mobility Ge pMOSFET With 1-nm EOT Al 2O 3GeO xGe gate stack fabricated by plasma post oxidation," *IEEE Trans. Electron Devices*, vol. 59, no. 2, pp. 335–341, 2012, doi: 10.1109/TED.2011.2176495.

[9] M. Oehme, J. Werner, M. Jutzi, G. Wöhl, E. Kasper, and M. Berroth, "High-speed germanium photodiodes monolithically integrated on silicon with MBE," *Thin Solid Films*, vol. 508, no. 1–2, pp. 393–395, 2006, doi: 10.1016/j.tsf.2005.06.106.

[10] D. Schwarz, *et al.*, "Alloy Stability of Ge1−xSnx with Sn Concentrations up to 17% Utilizing Low-Temperature Molecular Beam Epitaxy," *J. Electron. Mater.*, vol. 49, no. 9, pp. 5154–5160, 2020, doi: 10.1007/s11664-020-08188-6.

[11] V. S. V. Satyanarayana *et al.*, "Radiation-sensitive novel polymeric resist materials: Iterative synthesis and their EUV fragmentation studies," *ACS Appl. Mater. Interfaces*, vol. 6, no. 6, pp. 4223–4232, Mar. 2014, doi: 10.1021/am405905p.

[12] S. K. Wang, K. Kita, T. Nishimura, K. Nagashio, and A. Toriumi, "Isotope tracing study of GeO desorption mechanism from GeO2/Ge stack using 73Ge and 18O," *Jpn. J. Appl. Phys.*, vol. 50, no. 4 PART 2, p. 04DA01, 2011, doi: 10.1143/JJAP.50.04DA01.

[13] K. Gallacher, P. Velha, D. J. Paul, I. MacLaren, M. Myronov, and D. R. Leadley, "Ohmic contacts to n-type germanium with low specific contact resistivity," *Appl. Phys. Lett.*, vol. 100, no. 2, p. 022113, Jan. 2012, doi: 10.1063/1.3676667.

[14] S. Choudhary *et al.*, "A Steep Slope MBE-Grown Thin p-Ge Channel FETs on Bulk Ge-on-Si Using HZO Internal Voltage Amplification," *IEEE Trans. Electron Devices*, vol. 69, no. 5, pp. 2725–2731, 2022, doi: 10.1109/TED.2022.3161857.

[15] H. Yu *et al.*, "A simplified method for (Circular) transmission line model simulation and ultralow contact resistivity extraction," *IEEE Electron Device Lett.*, vol. 35, no. 9, pp. 957–959, 2014, doi: 10.1109/LED.2014.2340821.

research (Ph.D) with the School of Computing and Electrical Engineering (SCEE), Indian Institute of Technology Mandi, Mandi, India. His research interests include Ge FET channel integration on Si substrates, FinFET devices nanofabrication, Advanced lithography, CMOS micro-nanoelectronics process flow optimization and characterization.

978-1-6654-7134-3/22 $31.00 © 2022 IEEE

# Experimentally study on the effect of RIE etching power on etching rate of β-Ga₂O₃ thin film

line 1: 1st Wang Xu
line2: *College of Big Data and Information Engineering, Guizhou University*
line3: *Guizhou Provincial Key Laboratory of Micro and nano Electronics and Software Technology*
line 4: *Guiyang City, Guizhou Province*
line 5: *1712066349@qq.com*

line 1: 2nd Ran Jing yang
line 2: *College of Big Data and Information Engineering, Guizhou University*
line3: *Guizhou Provincial Key Laboratory of Micro and nano Electronics and Software Technology*
line 4: *Guiyang City, Guizhou Province*
line 5: *1669077509@qq.com*

line 1: 3rd Yang Lai
line 2: *College of Big Data and Information Engineering, Guizhou University*
line 3: *Guizhou Provincial Key Laboratory of Micro and nano Electronics and Software Technology*
line 4: *Guiyang City, Guizhou Province*
line 5: *1305612762@qq.com*

line 1: 4th Yang Fa shun
line2: *College of Big Data and Information Engineering, Guizhou University*
line3: *Guizhou Provincial Key Laboratory of Micro and nano Electronics and Software Technology*
line 4: *Guiyang City, Guizhou Province*
line 5: *fashun@126.com*

line 1: 5th Ma Kui
line 2: *College of Big Data and Information Engineering, Guizhou University*
line3: *Guizhou Provincial Key Laboratory of Micro and nano Electronics and Software Technology*
line 4: *Guiyang City, Guizhou Province*
line5: *kma@gzu.edu.cn*
*(Corresponding Author )*

*Abstract*—As a member of ultra-wide band gap semiconductor materials, β-Ga₂O₃ materials have attracted wide attention from researchers in the semiconductor field in recent years. Etching process is crucial to realize semiconductor devices and integrated circuits based on β-Ga₂O₃ materials. Based on the reaction ion etching process commonly used in silicon-based semiconductor technology, the etching experiment research of β-Ga₂O₃ thin film is carried out. The β-Ga₂O₃film is etched with SF6, based on the induction coupled reaction ion etching. The effect of RIE etching power, excitation power and bias power, on etching rate of β-Ga₂O₃ thin film has been studied. SEM characterization results show that the etching rate is the highest at 600W excitation power. The etching rate increases with the increase of bias power. The etching rate at 200W bias power is slightly higher than that at 150W bias power. However, the photoresist used as the etch mask will be damaged at 200W bias power.

*Keywords—ultra-wide band gap semiconductor material, β-Ga₂O₃ thin film, etching power, etching rate*

## I. INTRODUCTION

As one of the wide band gap semiconductors, gallium oxide has good application prospects in high temperature, high pressure, high frequency and so on. Studying gallium nitride films since 1970s, Ichinose N.[1] disclosed a method for growing β-Ga₂O₃ single crystals, a method for preparing high-quality thin film single crystals, and β-Ga₂O₃ light emitting devices which are capable of emitting in ultraviolet regions and their manufacturing methods. Jeliazova Y. et al. [2] grown on a nickel (100) substrate yielded an ultra-thin β-Ga₂O₃ membrane. H Zhuang [3] et al. successfully prepared gallium nitride films on a silicon (111) substrate by electrophoretic deposition, and characterized the structure and composition of the formed film by using Fourier transform infrared transmission spectrum, X-ray diffraction and X-ray photoelectron spectroscopy. The results showed that the resulting film was a polycrystalline structure.

In the 20th century, technology changes rapidly, with each passing day, and chips are applied in various extreme conditions. Therefore, the semiconductor materials used to make chips are very high requirements, so the research on wide-band semiconductor materials has become more popular.

The wet etching of β-Ga₂O₃ film was mainly realized by various acids such as HF[4], HCl[5], H₂SO₄[6], and HNO₃[7], etc. Shigeo Ohira et al. studied HCl, H₂SO₄, HF, KOH, NaOH, HNO₃, H₂O₂ mixed H₂SO₄ etching Ga₂O₃, they found that under the experimental conditions of 60 ℃, the effect of NaOH etching is very weak, HCl can't etch, HNO₃ heating time corrosion speed can reach 86.6nm/h, HF etching rate can reach 58.7nm/h, and Ga₂O₃ corrosion effect is very good. No considerable temperature dependence on etch rate has been shown by Shah et al. for BCl₃/Ar or Cl₂/Ar with ICP RIE up to 200 °C [8].

A significant reduction in breakdown voltages of dry-etched β-Ga₂O₃ Schottky diode has been reported, and the breakdown voltage further reduces as the dry-etched power increases [9]. Yang et al. measured the I–V characteristics of Schottky diode fabricated with bulk and ICP-etched β-Ga₂O₃ [10]. The highest etch rate achieved was ~ 1300 Å min−1 using 800 W ICP source power and 200 W chuck power (13.56 MHz) with either Cl₂/Ar or BCl₃/Ar.[11]

Until now, various gallium oxide-based semiconductor devices have been used. Professor Ting Hao Chang[12]'s team studied the electrical properties of amorphous indium gallium oxide thin-film transistors as deep UV photoelectric transistors. Zhe Li et al. obtained a β-Ga₂O₃ film on a flexible polyimide substrate and prepared a flexible β-Ga₂O₃ solar-blind UV photoelectric detector with high response rate and high light / dark current ratio. Shinji Nakagomi, Tsubasa Sai, and Yoshihiro Kokubun [13] have successfully prepared field-effect hydrogen sensors with a self-temperature compensation function based on β-Ga₂O₃ materials. A team of Professor Shigeyuki Imura [14] at the

University of Tokyo has developed a stacked CMOS image sensor covered with a thin-film gallium oxide ($Ga_2O_3$) / crystal selenium (c-se) hetero junction photodiode that successfully controls the size of the polycrystalline particles to make it much smaller than the pixel size of the image sensor. In 2016, Andrew J. Green et al.[15] grew a single-crystal-doped $\beta$-$Ga_2O_3$ epitaxial layer on a single-crystal-doped semi-insulating (100) $\beta$-$Ga2O3$ substrate by using a metal-organic gas-phase epitaxial layer. Metal oxide semiconductor field-effect transistors with 2μm gate length, 3.4μm source spacing, and 0.6 μm gate spacing were prepared. In 2018, Zhanbo Xia et al.[16] reported a silicon-doped delta $\beta$-$Ga_2O_3$ Metal semiconductor field effect transistor, characterized by n-type $Ga_2O_3$ The source leakage is formed in the ohmic contact, with a peak leakage current of 140 m A/mm and a breakdown voltage of 170V. In 2019, Dr Uttam Singisetti's team[17] reported a $\beta$-$Ga_2O_3$ of 5 microns thick Base MOSFET, with a breakdown voltage of up to 1850V。

From the above literature, gallium oxide can be applied in Daily-blind UV-photoelectric detector, sensors, MOSFET and other devices, while Etching is an essential step in the device manufacturing process. If we can study the etching process of gallium oxide film, then it will lay a good foundation for the practical application of gallium oxide film, the RIE etching method is used to study the gallium oxide film.

## II. DESIGN OF THE EXPRIMENTAL PROTOCOL

Firstly, we use magnetic sputtering equipment to prepare gallium oxide film with a thickness of about 400nm. The process parameters are shown in Table 1.

Table 1 Magnetic sputter process parameters

| Process parameters | Value |
|---|---|
| Target material | Gallium oxide ceramic target (99.99%) |
| Substrate | Sapphire (0001 crystal) |
| Background vacuum(Pa) | 9.0E-4 |
| Target-substrate distance (cm) | 6 |
| Splutter Power (W) | 150 |
| Working Pressure (Pa) | 1.0 |
| Oxygen flow rate (sccm) | 2.3 |
| Argon gas flow rate (sccm) | 46.2 |
| Spluttering time (min) | 90 |
| Substrate heating temperature (℃) | 500 |
| Annealing temperature (℃) | 900 |
| Annealing time (min) | 90 |
| The annealing atmosphere | nitrogen |

Nextly, etching methods are investigated. RIE dry etching mainly chlorine and fluorine gas etching [18,19], Etching graphics have small deviation and high accuracy. Combined with the laboratory conditions, the $\beta$-$Ga_2O_3$ film was etched by ICP of fluorine-based gas.

It is known that in the ICP dry etching technology, the main factors affecting the film etching rate are excitation power, bias power, reaction pressure and gas flow rate. Among them, the influence of excitation power and bias power is the most obvious. On this basis, experiments are carried out to study the influence of excitation power and bias power on the etching rate of gallium oxide thin film.

The experiment first fixed a bias power of 150W,

reaction pressure of 1.5Pa, SF6 flow of 18sccm and Ar gas flow of 2sccm, and changed the excitation power from 500 to 700 W to etch the β-Ga2O3 film. Specific experimental parameters are shown in Table 2. After determining the excitation power as the maximum of the 600W moment eclipse rate, Changed the bias power to etch the β-Ga2O3 film and the specific experimental parameters are shown in Table 3.

Table.2 Etching conditions with varied exciting power and fixed other parameters

| Etching conditions | value |
|---|---|
| This low vacuum（Pa） | 1.0E-3 |
| SF6（sccm） | 18 |
| Ar（sccm） | 2 |
| Excitation powe（W） | 500、550、600、650、700 |
| Bias power（W） | 150 |
| Working pressur（Pa） | 1.5 |
| Etching time（min） | 5 |
| Flushing time（min） | 10 |

Table.3 Etching conditions with varied bias power and fixed other parameters

| Etching conditions | value |
|---|---|
| This low vacuum（Pa） | 1.0E-3 |
| SF6（sccm） | 18 |
| Ar（sccm） | 2 |
| Excitation power（W） | 600 |
| Bias power（W） | 50、100、150、200 |
| Working pressure（Pa） | 1.5 |
| Etching time（min） | 5 |
| Flushing time（min） | 10 |

## III. INTERPRETATION OF RESULT

### A. Influence of excitation power on thin film etching

After etching the β-Ga2O3 film with changing the excitation power, the etched samples were detected and measured using SEM to obtain the SEM images as shown in Figure 1

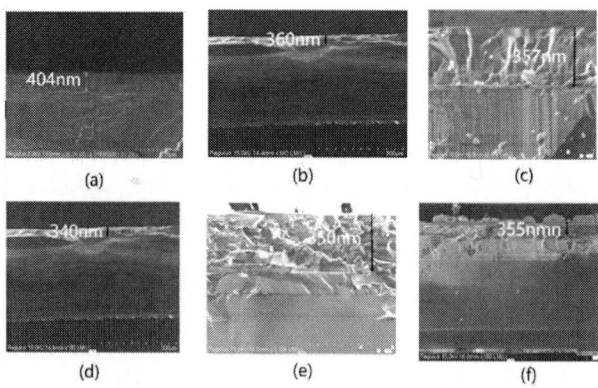

Fig.1 SEM images: (a) no n etched β-Ga2O3thin film, (b) to (f) corresponding to β-Ga2O3thin film samples etched at excitation power of 500W to 700W in steps of 50W.

The etching thickness is converted to the etching rate to obtain the line plot as shown in Fig. 2

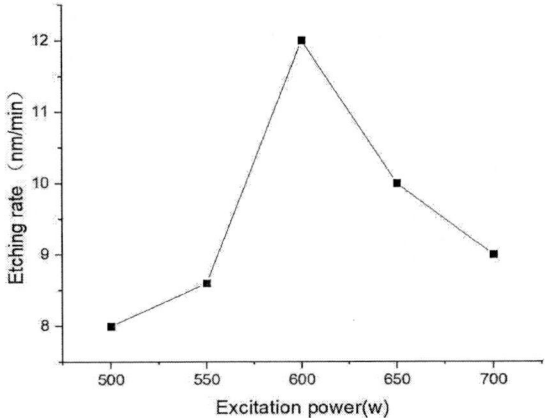

Fig.2 Curve of etching rate versus excitation power

According to the above picture is not difficult to see, with the increase of power, thin film etching rate also gradually increased, but when the power reached 600W power found the etching rate began to decrease, cause the above reason may be when the power is less than 600W, with the increase of power, ionization of plasma density also increases, plasma energy also increases. The chemical reaction accelerates, and the physical bombardment of the plasma also increases, so the etching rate increases. When the excitation power continues to increase, although the plasma density and energy increase, while the same density increases, the mutual collision between ions also increases, and the average free path between ions decreases. As a result, the energy of ion bombardment on the surface of the film decreases, so the etching rate of the film also decreases.

### B. Influence of bias power on film etching

After etching the $\beta$-Ga$_2$O$_3$ film by changing the bias power, the etched samples were detected and measured using SEM to obtain the SEM image as shown in Fig. 3

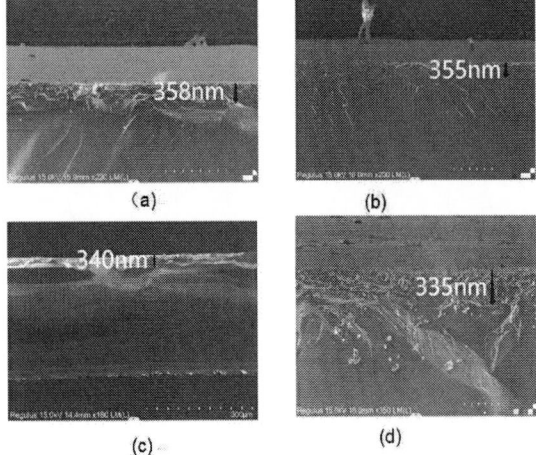

Fig 3.SEM images: (a) to (d) corresponding to $\beta$-Ga$_2$O$_3$thin film samples etched at bias power of 50W to 200W in steps of 50W.

The etching thickness is converted to the etching rate to obtain the line plot as shown in Fig. 4

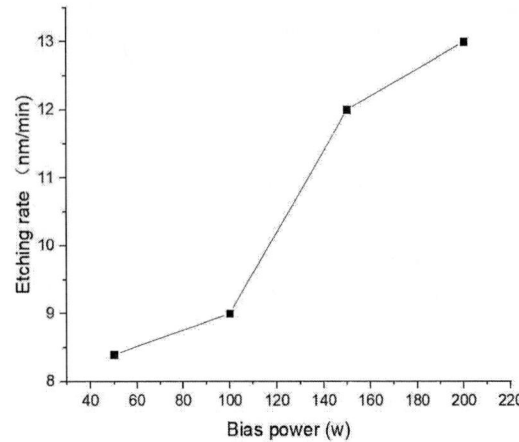

Fig.4 Curve of etching rate versus bias power

As can be seen from Figure 4, as the bias frequency power increases, the etching rate increases positively proportional to the bias power, because the ion bombardment energy increases and the physical etching increases. However, the photoresist used as an etching mask can be damaged at 200W bias power, so the best bias power is 150W.

## IV. SUMMARIZE

We constantly adjust multiple process parameters to etch the $\beta$-Ga$_2$O$_3$ films. The thickness of the post etched film is measured by the SEM after the etching ends, and then the etched film is converted into the etching rate. From the above experimental data, it is clearly found that when the excitation power reaches 600W and the RF bias reaches 150W. The etching rate of the film as well as the quality of the film is optimal.

## V. REFERENCES

[1] Ichinose N.$\beta$-Ga$_2$O$_3$single crystal growing method, thin-film single crystal growing method, Ga$_2$O$_3$light-emitting device, and its manufacturing method[D].EP, 2004.

[2] Jeliazova Y, Franchy R.Vibrational properties of ultra thin Ga$_2$O$_3$film grown on Ni(100)[J].Acta Horticulturae, 2001, 889(889):131-136.

[3] Zhuang H, Yang L I, Xue C, et al.Formation of GaN Film by Ammoniating Ga$_2$O$_3$Deposited on Si Substrate with Electrophoresis [J].International Journal of Modern Physics B, 2002.

[4] Lee Y, Johnson N R, George S M.Thermal Atomic Layer Etching of Gallium Oxide Using Sequential Exposures of HF and Various Metal Precursors[J].Chemistry of Materials, 2020.

[5] S.Yu.Nekrasov and Art.A.Migdisov and A.E.Williams-Jones and A.Yu.Bychkov.An experimental study of the solubility of Gallium(III) oxide in HCl-bearing water vapour[J].Geochimica et Cosmochimica Acta, 2013.

[6] Ohira S, Arai N.Wet chemical etching behavior of $\beta$-Ga$_2$O$_3$single crystal[J].physica status solidi (c), 2008, 5(9):3116-3118.

[7] Xia Li, Wang Guisu. Method for synthesizing Gd$_3$Ga$_5$O$_{12}$(GGG) transparent ceramic nano crystals:, 2009.

[8] J. Yang, F. Ren, R. Khanna, K. Bevlin, D. Geerpuram, L.-C. Tung, J. Lin, H. Jiang, J. Lee, E. Flitsiyan, L. Chernyak, S.J. Pearton, A. Kuramata, Annealing of dry etch damage in metallized and bare (-201) Ga2O3. J. Vac. Sci. Technol. B 35(5), 051201 (2017).

[9] J. Yang, S. Ahn, F. Ren, R. Khanna, K. Bevlin, D. Geerpuram, S.J. Pearton, A. Kuramata, Inductively coupled plasma etch damage in (-201) Ga2O3 Schottky diodes. Appl. Phys. Lett. 110(14), 1 – 5 (2017).

[10] A.P. Shah, A. Bhattacharya, Inductively coupled plasma reactive-ion etching of β-Ga2O3: comprehensive investigation of plasma chemistry and temperature. J. Vac. Sci. Technol. 35(4), 041301

[11] J. Yang, S. Ahn, F. Ren, S. Pearton, R. Khanna, K. Bevlin, D. Geerpuram, A. Kuramata, Inductively coupled plasma etching of bulk, single-crystal Ga2O3. J. Vac. Sci. Technol. B 35(3), 031205

[12] Li, Z, et al.Flexible Solar-Blind $Ga_2O_3$Ultraviolet Photodetectors with High Responsivity and Photo-to-Dark Current Ratio.IEEE Photonics Journal, 2019. 11(6):1-9.

[13] Nakagomi S, Sai T, Kokubun Y.Hydrogen gas sensor with self temperature compensation based on $\beta$-$Ga_2O_3$thin film[J].Sensors & Actuators B Chemical, 2013, 187(oct.):413-419.

[14] Imura, S, et al.High-Sensitivity Image Sensors Overlaid with Thin-Film Gallium Oxide/Crystalline Selenium Heterojunction Photodiodes.IEEE Transactions on Electron Devices, 2016. 63(1): p.86-91.

[15] Chabak K D, Moser N, Green A J, et al.Enhancement-mode $Ga_2O_3$wrap-gate fin field-effect transistors on native (100)

$\beta$-$Ga_2O_3$substrate with high breakdown voltage[J].Applied Physics Letters, 2016, 109(21):213501.

[16] Xia, Z, et al.Delta Doped $\beta$-$Ga_2O_3$Field Effect Transistors with Regrown Ohmic Contacts.IEEE Electron Device Letters, 2018. 39(4): p.568-571.

[17] Chatterjee B, Zeng K, Nordquist C D, et al.Device-Level Thermal Management of Gallium Oxide Field-Effect Transistors[J].IEEE Transactions on Components, Packaging, and Manufacturing Technology, 2019, pp(99):1-1.

[18] Ahmadi, Elaheh, Oshima, et al.Chlorine-based dry etching of $\beta$-$Ga_2O_3$[J].Semiconductor Science and Technology, 2016, 31(6):65006.1.

[19] Liang H, Chen Y, Xia X, et al.A preliminary study of SF6 based inductively coupled plasma etching techniques for beta gallium trioxide thin film[J].Materials Science in Semiconductor Processing, 2015, 39:582-586.

PO-63

2022 International Symposium on Semiconductor Manufacturing (ISSM).
December 12-13, 2022

# Nanoimprint Lithography with $CO_2$ ambient

Toshiki ITO
*Semiconductor Production Equipment Group*
*Canon Inc.*
Tochigi, Japan
ito.toshiki@mail.canon

Yuto ITO
*R&D Headquaters*
*Canon Inc.*
Tokyo, Japan
ito.yuto@mail.canon

Isao KAWATA
*R&D Headquaters*
*Canon Inc.*
Tokyo, Japan
kawata.isao@mail.canon

Ken-ichi UEYAMA
*R&D Headquaters*
*Canon Inc.*
Tokyo, Japan
ueyama.ken-ichi@mail.canon

Kouhei NAGANE
*R&D Headquaters*
*Canon Inc.*
Tokyo, Japan
nagane.kohei@mail.canon

Weijun LIU
*Materials Div.*
*Canon Nanotechnologies, Inc.*
Austin, USA
wliu@cnt.canon.com

Timothy STACHOWIAK
*Materials Div.*
*Canon Nanotechnologies, Inc.*
Austin, USA
TStachowiak@cnt.canon.com

Wei ZHANG
*Applications Div.*
*Canon Nanotechnologies, Inc.*
Austin, USA
wzhang@cnt.canon.com

Teresa ESTRADA
*Applications Div.*
*Canon Nanotechnologies, Inc.*
Austin, USA
testrada@cnt.canon.com

*Abstract*—**In Jet and Flash Imprint Lithography (JFIL), ambient gas is trapped between the resist, the substrate and the mold. The volume of the trapped ambient gas is estimated about 9.7 ~ 21.5% of the resist volume. It takes time for the bubbles to disappear in the closed space. In case that carbon dioxide ambient is applied in JFIL, it was theoretically and experimentally demonstrated that the trapped carbon dioxide gas dissolved rapidly into organic liquid or organic solid layer in imprint stack. The trapped carbon dioxide gas bubble disappeared more rapidly than that of helium gas, which resulted in higher throughput and fewer defect number.**

*Keywords—Nanoimprint, Lithography, Photoresist, Throughput, Helium, Carbon dioxide*

## 1. Introduction

The JFIL imprint process [1] is depicted in Fig. 1.

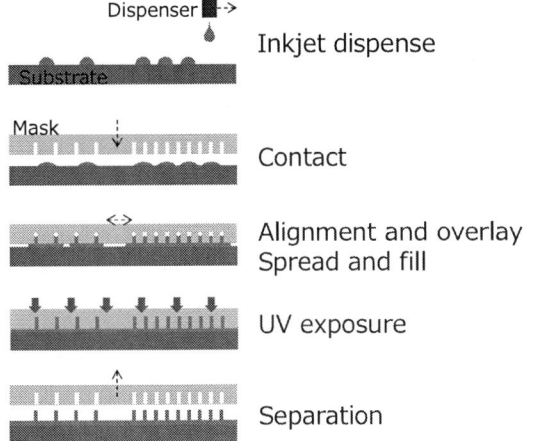

Fig. 1. J-FIL process

A resist liquid is discretely dispensed into the pattern forming region on the substrate. The drops of resist dispensed on the patterning region spread over the substrate.

This phenomenon is called pre-spreading. Next, a mold having a pattern is contacted to the resist drops on the substrate. Thereby, the drops of the resist spreads to the whole area of the gap between the substrate and the mold by the capillary force. This phenomenon is called spreading. The spreading process is classified into dynamic spreading (DS), which occurs after the mold surface contacts with the resist droplets until the droplets of the resist combine with each other, and static spreading (SS), which occurs after the bonding and the trapped gas disappears. The resist is also filled in the recesses forming the pattern of the mold by capillary force. This filling phenomenon is called filling. The time it takes to spread and fill is called the filling time. After the filling of the resist is completed, the resist is irradiated with UV light through the mold to cure the resist. Thereafter, the mold is pulled away from the cured resist. By performing these steps, the pattern of the mold is transferred to the resist on the substrate to form a pattern of the resist.

When JFIL is applied to semiconductor device manufacturing, high throughput is required to improve productivity. In JFIL, the volume of the atmospheric gas trapped between the resist, the substrate and the mold is 9.7% or more by the resist volume ratio when the arrangement of the resist droplets is a diamond-like array, and 21.5% or more when the arrangement is a square array. The time required for SS accounts for more than 50% of the total process time of JFIL because it takes time to make the gas disappear in the closed space [2].

In JFIL using spin-on-carbon (SOC) as an intermediate transfer layer on a device substrate, it is known that the use of air as atmospheric gases results in shorter SS time than helium on a certain SOC [3]. The helium bubble is absorbed into the quartz of the mold by diffusion of helium atoms. The main mechanism of air bubbles is that oxygen and nitrogen molecules are dissolved and occluded in SOC.

The authors found that the thinner the SOC layer, the longer the SS time. The smaller the circuit pattern to be

978-1-6654-7134-3/22 $31.00 © 2022 IEEE 105

formed, the thinner the intermediate transfer layer must be. Therefore, the authors proposed carbon dioxide as an atmospheric gas that is expected to have a shorter SS time due to its high solubility to SOC than helium or air.

## 2. Calculation

### 2.1. Gas disappearance time

In order to theoretically calculate the time required for DS and SS, a simplified model of the JFIL process was developed as shown in Fig. 3.

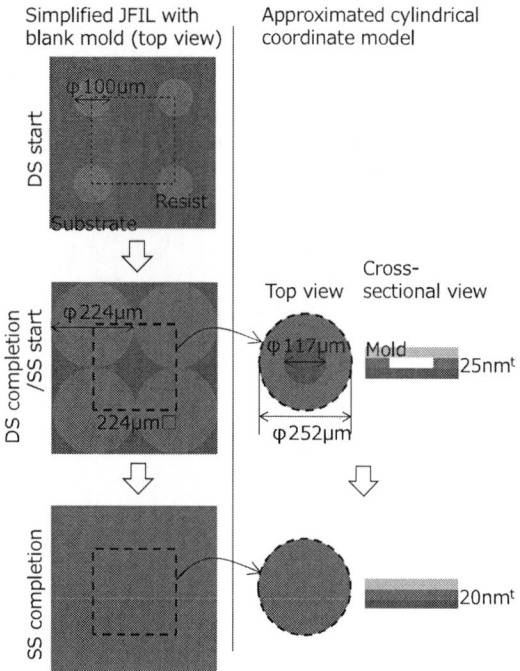

Fig. 3. Calculation model

The mold assumes a blank mold with no recess pattern formed. The drops of resist dispensed onto the substrate in a square arrangement are independent of each other before the mold is contacted. When the mold is contacted, the mold surface is brought into contact with the resist, and dynamic spread (DS) is started. The drop of the resist is assumed to have a cylindrical shape and to spread while maintaining a cylindrical shape. It is defined that DS is completed and SS starts at the moment when the drops of the resist contact each other. In SS, the ambient gas is trapped by the resist, the mold and the substrate.

The unit cell of the square array in the SS process is approximated to a cylindrical coordinate system model as shown in Fig. 3. The same volume of gas and resist as the unit cell of the Cartesian coordinate system is sandwiched between the disk-shaped substrate and the mold in the cylindrical coordinate system. The outer boundary of the disk is a closed boundary. Atmospheric gas disappears by diffusion and dissolution with Laplace pressure as driving force. If the resist is assumed to be an incompressible fluid, the mold descends to compensate for the volume of the missing gas. When the volume of gas becomes zero, SS completion is defined. The various numerical values

illustrated in Fig. 3 are calculated assuming a droplet volume of 1 pL, a drop diameter of 100 μm at the start of DS, and a resist liquid film thickness of 20 nm at the end of SS.

The DS time was calculated by coupling the lubrication equation for a cylindrical resist droplet with the mold equation of motion. The reason for adopting the lubrication equation is that the spacing of the applied droplets is about 100 μm and the height of the droplet distribution is on the order of several nm to several μm, and the pressure gradient in the height direction is negligible because of the high aspect. Since the Reynolds number can be considered sufficiently small, it was treated as an incompressible fluid. The above coupled equation has an analytical solution and the DS time $T_{DS}$ is given below:

$$T_{DS} = \frac{\pi(R^4 - R_0^4)}{4V} \frac{3\mu\pi}{4\sigma(\cos\theta_u + \cos\theta_d)} \text{ (eq.1)}$$

Here, $V$ is the volume of one droplet, $R$ is half the distance between the droplets, $R_0$ is the radius of the cylinder of the droplet of the resist, $\mu$ is the viscosity of the resist, $\sigma$ is the surface tension, $\theta_u$ is the contact angle between the resist and the mold, and $\theta_d$ is the contact angle between the resist and the underlayer.

The calculation method of SS time in the cylindrical coordinate system model of Fig. 3 is described in detail below. The bubbles trapped by the collision of the resist drops are approximated by a cylinder of radius $r_g$ having the same volume, and the calculation region is also approximated by a cylinder of radius $r_c$ having the same volume. The equation of state of the ideal gas, the equation of mass of the gas, the equation of diffusion of the gas into the mold, the equation of diffusion of the gas into the SOC, the equation of diffusion of the gas into the resist, the equation of mass conservation of the resist, the equation of lubrication of the resist region, and the equation of motion of the mold are coupled to obtain the equation of time evolution of the bubble volume, the pressure $p_g$, and the height $h$ of the mold. The equation for the ideal gas is given below.

$$p_g = \rho_g RT \text{ (eq.2)}$$

where $p_g$ is the pressure of the gas, $\rho_g$ is the density of the gas, $R$ is the gas constant, and $T$ is the temperature. The gas mass equation is given below.

$$\pi r_g^2 h \rho_g = M_g \text{ (eq.3)}$$

where $M_g$ is the mass of the gas.

The diffusion equation of the gas is given below, for example in the case of a mold.

$$\frac{dc_g}{dt} = D_g \nabla^2 c_g \text{ (eq.4)}$$

Here, $C_g$ is the gas concentration of the mold, and $D_g$ is the diffusion coefficient of the gas into the mold. The boundary condition is that saturated dissolution according to Henry's law occurs only near the interface where the mold and the gas are in contact. The diffusion equation of the gas into the resist and the substrate was the same as the diffusion equation of the gas into the mold, and the same was applied to the boundary conditions.

The equation for the mass conservation of the resist is given below.

$$\rho_r \pi(r_c^2 - r_g^2)h = \rho_r V_d = M_r \text{ (eq.5)}$$

where $\rho_r$ is the density of the resist, $V_d$ is the volume of the droplet of the resist, and $M_r$ is the total mass of the resist.

The lubrication equation for the region of the resist is given by.

$$\frac{h^3}{12\mu}\frac{1}{r}\frac{\partial}{\partial r}r\frac{\partial p_r}{\partial r} = \frac{dh}{dt} \text{ (eq.6)}$$

Here, $\mu$ is the viscosity of the resist as described above. The equation of motion of the mold is given below.

$$\pi r_c^2 \rho_m h_m \frac{d^2 h}{dt^2} = \pi r_g^2 p_g + \int_{\theta=0}^{2\pi}\int_{r=r_g}^{r_c} p_r(r) r\,dr\,d\theta$$
(eq.7)

The target film thickness is defined as $RLT$ (residual layer thickness), and the resist thickness is defined as $h$. The completion condition in the SS calculation was set as described below:

$$\frac{(h-RLT)}{RLT} < 10^{-5} \text{ (eq.8)}$$

### 2.2. Molecular dynamics calculation

The diffusion coefficient of the gas in the resist was calculated by using molecular dynamics calculation for a molecular assembly containing 10 gas molecules per 500 molecules of isobornyl acrylate that can be used as a component of the JFIL resist. GROMACS -2016.4 (Copyright © 2001 -2017, The GROMACS development team at Uppsala University, Stockholm University and the Royal Institute of Technology, Sweden.) was used for molecular dynamics calculations.

The sampling of the equilibrium state in the molecular dynamics calculation was obtained by placing the object molecule in the unit cell which imposed the periodic boundary condition, calculating the force acting between atoms included in each molecule for each time, and calculating the locus of all atoms for the time evolution.

The diffusion coefficients were calculated from the mean square displacements of gas molecules from the history of molecular motion obtained by Production Run.

### 3. Experimental

#### 3.1. Measurement of gas disappearance time

FNIS-031A developed by FUJIFILM Corporation was used as the JFIL resist. ODL -301 provided by Shin-Etsu Chemical Industry Co., Ltd. was used as SOC.

Drops of 3.25 pL of resist in a square array of 140 μm per side were uniformly dropped in a range of 26 × 33 mm on a silicon substrate or on a silicon substrate coated with SOC having a thickness of 200 nm, and a quartz blank mold was brought into contact with the droplets.

The spreading behavior of the drops of the JFIL resist through the mold, that is, the shrinking and disappearance behavior of the gas bubble trapped in the three-phase interface of the mold, the substrate, and the resist were photographed by an optical microscope with magnification of 5 times. The time from the contact of the resist and the mold to the disappearance of the area occupied by the void down to 1% or less was measured as the gas disappearance time. An example of the measurement of gas disappearance time is shown in Fig. 4.

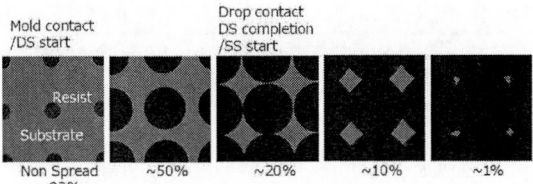

Fig.4. Measurement of gas disappearance time

#### 3.2. Imprinting

The filling time in the imprint equipment NZ2 developed by Canon Inc. was measured by the following experiment. As the JFIL resist, FNIS-031A developed by FUJIFILM Corporation was used. ODL -301 provided by Shin-Etsu Chemical Industry Co., Ltd. was used as the SOC material. 0.6 pL drops of resist was dispensed into a 26 × 33 mm area on a silicon substrate coated with a 200 nm-thick ODL -301 at a uniform density so as to have an average liquid film thickness of 32 nm, carbon dioxide or helium was used as an atmospheric gas, and a quartz blank mold was brought into contact with the droplet. After a predetermined waiting time elapses after contact, ultraviolet light is irradiated through the mold to cure the resist, and the mold is separated to obtain a cured film of the resist. The number of non-fill defects originating from residual gas in the cured film was measured. The waiting time when the number of non-fill defects is less than 10/cm² was defined as the filling time. Although the filling time does not always coincide with the gas disappearance time, it is expected that the shorter the gas disappearance time, the shorter the filling time.

### 4. Results and discussion

The gas disappearance time measured in a helium atmosphere was 2.7 seconds when the SOC layer was 0 nm, and 0.7 seconds when the SOC layer was 200 nm thick. In an air atmosphere, the gas did not disappear after 24 seconds when the SOC layer was 0 nm, and 0.6 seconds when the SOC layer was 200 nm thick. In a carbon dioxide atmosphere, when the SOC layer was 0 nm, the gas disappearance time was 2.67 times as long as that in a helium atmosphere. In the carbon dioxide atmosphere, the gas disappearance time was not measured when the SOC layer was 200 nm thick.

#### 4.1. Coefficients of solubility and diffusion

The diffusion and solubility coefficients of the resist, SOC, and gas (helium, nitrogen, oxygen, and carbon dioxide) in the mold necessary for the theoretical calculation of the SS time are summarized as shown in Table 1.

Table 1. Coefficient of solubility and diffusion

|        | He | N₂ | O₂ | CO₂ |
|--------|--------|--------|--------|--------|
| $Ms$ | 0.0027 | 0 | 0 | 0 |
| $Md$ | $1.2 \times 10^{-12}$ | 0 | 0 | 0 |
| $Rs$ | 0.004 | 0.11 | 0.13 | 1.8 |
| $Rd$ | $5 \times 10^{-9}$ | $2.5 \times 10^{-10}$ | $2.5 \times 10^{-10}$ | $5 \times 10^{-11}$ |
| $Cs$ | 0.0018 | 0.18 | 0.21 | 2.8 |
| $Cd$ | $1.4 \times 10^{-9}$ | $2.0 \times 10^{-10}$ | $1.75 \times 10^{-10}$ | $1.30 \times 10^{-10}$ |
| $Ss$ | 0 | 0 | 0 | 0 |

| $Sd$ | 0 | 0 | 0 | 0 |
| --- | --- | --- | --- | --- |

$Ms$: Solubility coefficient to mold (kg/m³*atm)
$Md$: Diffusion coefficient to mold (m²/s)
$Rs$: Solubility coefficient to resist (kg/m³*atm)
$Rd$: Diffusion coefficient to resist (m²/s)
$Cs$: Solubility coefficient to SOC (kg/m³*atm)
$Cd$: Diffusion coefficient to SOC (m²/s)
$Ss$: Solubility coefficient to silicon (kg/m³*atm)
$Sd$: Diffusion coefficient to silicon (m²/s)

The method how to determine the coefficients described in Table 1 is explained below.

### 4.1.1. Si substrate

Solubility coefficients and diffusion coefficients of helium, nitrogen, oxygen, and carbon dioxide to the silicon substrate were all set to zero. This is based on the known fact that the lattice spacing of silicon crystals is smaller than the molecular diameters of these gases and cannot be permeated [4].

### 4.1.2. Quartz mold

The diffusion coefficient and solubility coefficient of helium for quartz, which is the material of the mold, are known values [5]. For carbon dioxide, nitrogen and oxygen, the solubility coefficient and diffusion coefficient for quartz were both zero. This is based on the fact that the lattice spacing of quartz is smaller than the molecular diameters of carbon dioxide, nitrogen, and oxygen and is impermeable [5].

### 4.1.3. Resist

The diffusion coefficients of helium, nitrogen, oxygen, and carbon dioxide to the JFIL resist were determined using the molecular dynamics calculation described above.

The permeability coefficient ($P$) is the product of the diffusion coefficient ($D$) and the solubility coefficient ($S$).

$$P = S \times D \text{ (eq.9)}$$

The permeability coefficients to the mold and the silicon substrate are known in either gas. Therefore, the permeability coefficient which gives the same calculation result as the measured value of the gas disappearance time at zero nm of the SOC layer can be calculated. A solubility coefficient in the resist was calculated based on eq. 9 from the obtained permeability coefficient.

### 4.1.4. SOC

The permeability coefficient of helium to the SOC was calculated by fitting to give the same calculated value as the measured gas disappearance time at the SOC of 200 nm. The solubility coefficient of helium for SOC was assumed to be equivalent to that for organic polymers such as natural rubber [6]. The diffusion coefficient of helium for SOC was calculated based on eq.9.

The permeability coefficients of nitrogen and oxygen to SOC were calculated by fitting to give the same calculated values as the measured values of the gas disappearance time when SOC was 200 nm. The diffusion coefficients of nitrogen and oxygen to SOC were calculated from the

molecular weight ratio to helium, assuming that the diffusion coefficient of gas molecules was inversely proportional to the molecular weight of gas molecules. The solubility coefficients of nitrogen and oxygen for SOC were calculated based on eq.9.

The diffusion coefficient of carbon dioxide relative to SOC was calculated from the molecular weight ratio of helium and gas molecules assuming that the diffusion coefficient of gas molecules in SOC is inversely proportional to the molecular weight of gas molecules.

As for the solubility coefficient of carbon dioxide for SOC, the value obtained by assuming that the ratio of the solubility coefficient of carbon dioxide for SOC to the solubility coefficient of nitrogen for SOC coincides with the ratio of the solubility coefficient of carbon dioxide for resist to the solubility coefficient of nitrogen for resist was used.

### 4.1.5. Schematic image of diffusion and dissolution

Diffusion and dissolution coefficients of the ambient gasses are schematically summarized in Fig.5.

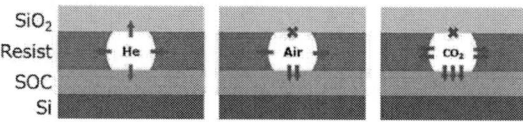

Fig.5. Diffusion and dissolution of gasses

Helium is the only gas that can diffuse into $SiO_2$ layer. Solubility of helium to resist and SOC is not so fast.

Oxygen, nitrogen and carbon dioxide cannot diffuse into $SiO_2$ layer because of their large molecules. However, oxygen, nitrogen and especially carbon dioxide have some solubility into organic layers.

### 4.2. Theoretical calculation of gas disappearance time

In the SOC thickness range of 0 ~ 200 nm, the gas disappearance time when the ambient gas (gas in the gap) is helium, nitrogen, or carbon dioxide was calculated by inputting various coefficients described in Table 1. The volume of the droplet of the resist was set to 3.5 pL, the film thickness after filling was set to 28 nm, and the diameter of the droplet at the start was set to 100 μm. The mold thickness was 1 mm. The calculation results are shown in Fig. 5.

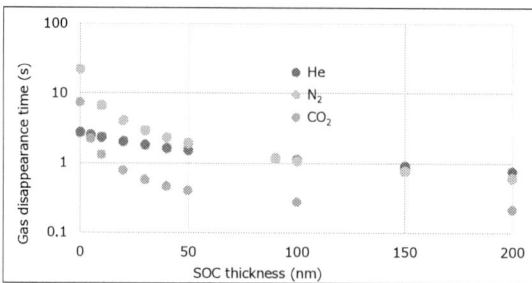

Fig.6. SOC thickness and gas disappearance time

In Fig. 6, when the SOC layer is zero nm, the value is based on the experimentally measured value, the helium atmosphere is the fastest. It is believed that helium diffused into the quartz mold. The thicker the SOC layer, the shorter the gas disappearance time for both gases. When the SOC layer is 5 nm or thicker, carbon dioxide is the fastest. This is considered due to the high solubility of carbon dioxide in SOC. Nitrogen is faster than helium when the SOC layer is larger than 100 nm.

### 4.3. Experiment

Imprinting experiment was carried out on a Si substrate coated with the SOC of 200 nm thick, with the JFIL resist of averaged thickness of 40 nm. Fig. 7 shows the measurement results of the waiting time dependence of the number of non-fill defect density on the Si substrate with SOC layer.

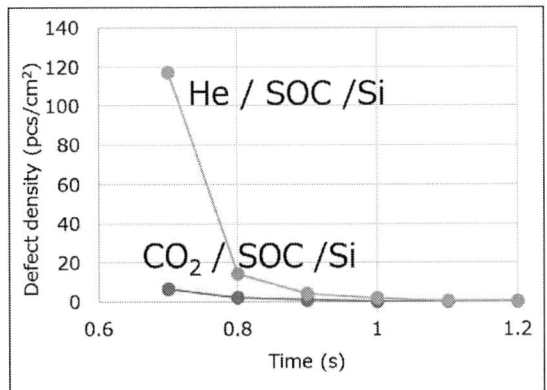

Fig.7. Waiting time and defect density on SOC substrate

The longer the waiting time, the smaller the defect density was. In this defect inspection experiment, filling time is defined as a time required to get defect density less than 10 pcs/cm$^2$. The filling times were less than 0.7 seconds for carbon dioxide and 0.9 seconds for helium. It is considered that the ambient carbon dioxide dissolved rapidly into SOC faster than diffusion of the ambient helium.

Another imprinting experiment was carried out on a bare-Si substrate with the JFIL resist of averaged thickness of 500 nm. Fig. 8 shows the measurement results of the waiting time dependence of the number of non-fill defects on the Si substrate without SOC layer.

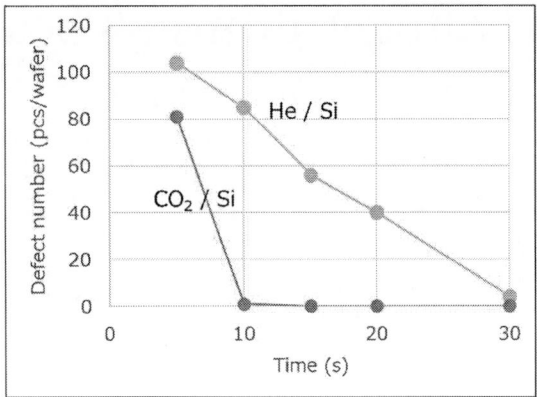

Fig.8. Waiting time and defect number on bare-Si substrate, with resist of 500 nm thick

Despite that SOC was not coated on Si wafer, defect number in carbon dioxide ambient was less than that in helium ambient in any time. This is considered due to the high solubility of carbon dioxide in the resist.

## 5. Conclusion

In imprint lithography with ambient gas of carbon dioxide, the trapped ambient gas dissolves rapidly into organic liquid layer and/or organic solid layer, which results in higher throughput and fewer defect number.

## References

[1] M. Colburn, S. C. Johnson, M. D. Stewart, S. Damle, T. C. Bailey, B. Choi, M. Wedlake, T. B. Michaelson, S. V. Sreenivasan, J. G. Ekerdt, C. G. Willson, *Proc. SPIE*, **3676** (1999) 379.

[2] Wei Zhang, Brian Fletcher, Ecron Thompson, Weijun Liu, Tim Stachowiak, Niyaz Khusnatdinov, J. W. Irving, Whitney Longsine, Matthew Traub, Van Truskett, Dwayne LaBrake, Zhengmao Ye, *Proc. SPIE*, **9777** (2016), 97770A

[3] K. Okabe, T. Higuchi, M. Komori, T. Kono, *J. Micro/Nanopatterning, Materials, and Metrology,* **21**(2022)011006

[4] P. Jung, *Nuclear Instruments and Methods in Physics Research Section B: Beam Interactions with Materials and Atoms*, **91**(1994) 362

[5] J. F. Shackelford, *Procedia Materials Science*, 7(2014)278

[6] M. Khawaja, A. P. Sutton, A. A. Mostofi, *J. Phys. Chem. B*, **121**(2017) 287

# Hydrogen diffusion behavior in CH₄N-molecular-ion-implanted wafers for three-dimensional stacked CMOS image sensors

Ryosuke Okuyama
Advanced Evaluation & Technology
Development Department
SUMCO Coroporation
Imari, Japan
rokuyama@sumcosi.com

Takeshi Kadono
Advanced Evaluation & Technology
Development Department
SUMCO Coroporation
Imari, Japan
tkadono@sumcosi.com

Ayumi Masada
Advanced Evaluation & Technology
Development Department
SUMCO Coroporation
Imari, Japan
aonaka@sumcosi.com

Akihiro Suzuki
Advanced Evaluation & Technology
Development Department
SUMCO Coroporation
Imari, Japan
asuzuki1@sumcosi.com

Koji Kobayashi
Advanced Evaluation & Technology
Development Department
SUMCO Coroporation
Imari, Japan
kkobayas@sumcosi.com

Satoshi Shigematsu
Customer Engineer Department
SUMCO Coroporation
Imari, Japan
sshigema@sumcosi.com

Ryo Hirose
Advanced Evaluation & Technology
Development Department
SUMCO Coroporation
Imari, Japan
rhirose@sumcosi.com

Yoshihiro Koga
Advanced Evaluation & Technology
Development Department
SUMCO Coroporation
Imari, Japan
ykoga4@sumcosi.com

Kazunari Kurita
Advanced Evaluation & Technology
Development Department
SUMCO Coroporation
Imari, Japan
k-kurita@sumcosi.com

*Abstract*— In this study, the diffusion behavior of hydrogen in a CH₄N-molecular-ion-implanted epitaxial wafer was invested by reaction kinetic analysis. Two hydrogen-trapping sites, carbon aggregate and end of range (EOR) defects, were formed in the CH₄N-implanted region. The C–H₂ binding state was formed in the carbon aggregate region. On the other hand, the N–H binding state was formed in the EOR defect region. This result indicates that CH₄N-molecular-ion-implanted epitaxial wafers contribute to the reduction in $D_{it}$ at the SiO₂/Si interface due to hydrogen desorption from the CH₄N-implanted region during heat treatment in the device process.

*Keywords—silicon, molecular ion implantation, hydrogen*

## I. Introduction

In three-dimensional stacked CMOS image sensors (3D-CISs), the SiO₂/Si interface state density ($D_{it}$) increases owing to the adoption of deep trench isolation (DTI). [1–3] It is an important technology issue to reduce the noise generated by the increase in $D_{it}$. [4–6] We previously reported that epitaxial silicon wafers implanted with hydrocarbon molecular ions have a $D_{it}$ reduction effect (hydrogen termination effect) owing to the desorption of hydrogen trapped in the region implanted with hydrocarbon molecular ions. [7–12] However, as mentioned above, it is important to improve the hydrogen termination effect in 3D-CISs. Thus, we have developed CH₄N-molecular-ion-implanted epitaxial silicon wafers in which nitrogen is added to hydrocarbon molecular ions. [12] It has been reported that the CH₄N-implanted region has end-of-range (EOR) defects, which are not observed in the hydrocarbon-molecule-ion-implanted region, and shows an improved gettering capability. [12] However, the diffusion behavior of hydrogen trapped in the CH₄N-implanted region

has not been reported. The purpose of this study is to clarify the behavior of hydrogen diffusing from the CH₄N-molecular-ion-implanted region of epitaxial silicon wafers.

## II. Experimental Procedure

A wafer was prepared by CH₄N implantation into a p-type Si (100) substrate with a carbon dose of $1.0 \times 10^{15}$ atoms/cm², followed by the growth of a 5.0-μm-thick silicon epitaxial layer. Then, the wafer was isothermally heat-treated in 100 % nitrogen atmosphere at 700, 900, 1000 and 1100 °C for 10, 30, 60 and 120 min. Hydrogen concentration was evaluated by secondary ion mass spectrometry (SIMS) analysis.

## III. Results and Discussion

### A. Hydrogen diffusion behavior in CH₄N-implanted region

Fig. 1 shows the depth profile of hydrogen after epitaxial growth. By focusing on the shape of the hydrogen concentration peak, we observed two peaks and performed peak separation using the Lorentzian function. Fig. 2 shows the results of peak separation after (a) epitaxial growth and (b) isothermal heat treatment at 700 °C for 30 min. The results indicate that for the hydrogen peak concentration after the heat treatment, peak 2 (red line), which is a steep peak, decreased from peak 1 (black line), which is a shallow peak on the surface, and that the amount of hydrogen diffusing from the peak 2 region was larger than that from the peak 1 region. A previous study has shown that carbon aggregates are formed in the shallow-peak region and EOR defects are formed in the steep-peak region. [12]

978-1-6654-7134-3/22 $31.00 © 2022 IEEE

2022 International Symposium on Semiconductor Manufacturing (ISSM).
December 12–13, 2022

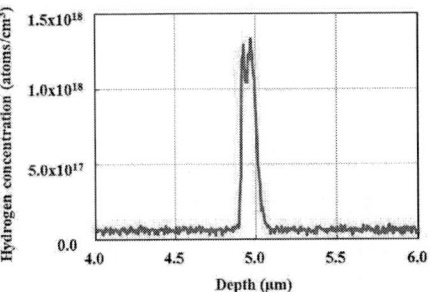

Fig. 1.   Depth profile of hydrogen concentration in CH$_4$N after epitaxial growth.

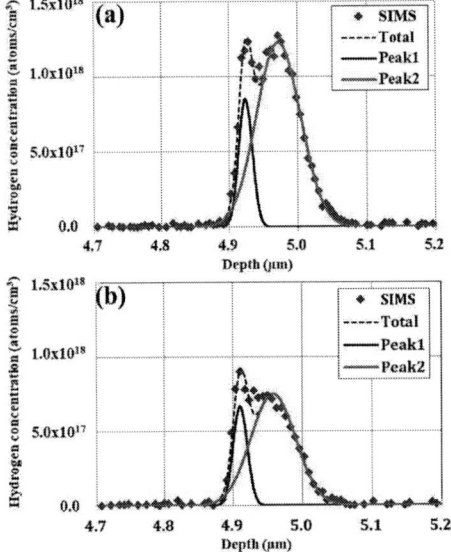

Fig. 2.   Depth profiles of hydrogen concentration of SIMS (blue), total (dot), peak 1 (black) and peak 2 (red) after (a) epitaxial growth and (b) heat treatment at 700 ℃ for 30 min.

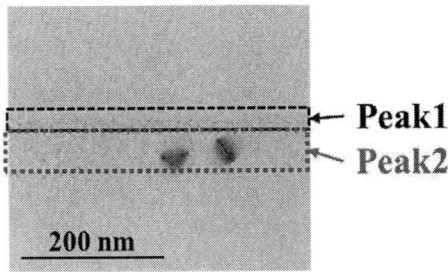

Fig. 3.   Cross-sectional transmission electron microscopy (TEM) image of CH$_4$N-implanted region after epitaxial growth.

Fig. 3 shows the cross-sectional transmission electron microscopy (TEM) image of the CH$_4$N-implanted region. We observed carbon aggregates in the peak 1 region and EOR defects in the peak 2 region. As shown in Fig. 3, it is inferred that peak 1 corresponds to hydrogen trapped in the carbon aggregate region and peak 2 corresponds to hydrogen trapped in the EOR defect region. The amount of hydrogen that diffused from the CH$_4$N-implanted region after 700 ℃ heat treatment was larger in the peak 2 region than in the peak 1 region. The previously reported hydrocarbon-molecular-ion-implanted epitaxial wafers have only one peak equivalent to

peak 1 corresponding to hydrogen trapped in the carbon aggregate region. [10] Therefore, we consider that the diffusion behavior of hydrogen trapped in the peak 1 region (carbon aggregates) is similar to that in the hydrocarbon-molecular-ion-implanted region. However, the diffusion behavior of hydrogen trapped in the peak 2 region (EOR defects) is not clear. In addition, our previous study has clarified the diffusion behavior of hydrogen in oxygen-added hydrocarbon-molecular-ion (CH$_3$O)-implanted regions with EOR defects similarly to that in CH$_4$N-implanted regions. [13] In the case of the diffusion behavior of hydrogen in the CH$_3$O-implanted region, the amounts of hydrogen diffusing from the peak 1 and peak 2 regions were almost the same. In the case of CH$_4$N, the peak 2 region showed a larger amount of hydrogen diffusing than the peak 1 region. It is considered that the diffusion behavior of hydrogen in EOR defects is different between CH$_3$O and CH$_4$N. Thus, in order to clarify the hydrogen diffusion behavior in the CH$_4$N-implanted region, we attempted to derive the desorption activation energy of hydrogen in the CH$_4$N-implanted region by reaction kinetics. We performed the reaction kinetic analysis of hydrogen in the CH$_4$N-implanted region. We then utilized a reaction model to analyze the desorption activation energy of hydrogen trapped in the CH$_4$N-implanted region. A simple reversible reaction model was considered, since hydrogen adsorption and desorption would both occur in the CH$_4$N-implanted region. Equation (1) shows the reversible reaction of hydrogen in the CH$_4$N-implanted region. H$_D$ is hydrogen in the CH$_4$N-implanted region. H is hydrogen out-diffusing from the CH$_4$N-implanted region after heat treatment. The reaction rate constants of the forward (desorbed) and reverse (adsorbed) reactions are $k_d$ and $k_a$, respectively.

$$H_D \underset{k_a}{\overset{k_d}{\rightleftarrows}} H \quad (1).$$

Then, the desorbed hydrogen concentration can be expressed as

$$\frac{dC_{HD}}{dt} = -k_d C_{HD} + k_a C_H, \quad (2)$$

where $C_{HD}$ is the concentration of $H_D$, $C_H$ the concentration of desorbed hydrogen after heat treatment and $t$ the heat treatment time. The concentrations $C_{HD}$ and $C_H$ before heat treatment are considered to be $C_0$ and zero as the initial conditions, respectively. Therefore, $C_H$ during heat treatment can be expressed by the difference between $C_0$ and $C_{HD}$. Consequently, (3) can be derived from (2).

$$C_{HD} = \frac{k_a + k_d \, exp\left[-\left(k_d + k_a\right)t\right]}{k_d + k_a} C_0 \quad (3)$$

Fig. 4 shows the plot of ratios of $C_0$ to $C_{HD}$ of peaks (a) 1 and (b) 2 versus heat treatment time, and dotted lines indicating curves fitted by (3). $C_{HD}$ is the integrated value of the hydrogen concentration in the CH$_4$N-implanted region. Both peaks 1 and 2 can be fitted according to the considered reversible reaction model. The amount of hydrogen diffusing for peak 2 was larger than that for peak 1 at 700 ℃ heat treatment. Consequently, the desorption activation energies for peaks 1 and 2 seem to be different. Fig. 5 shows Arrhenius plots of $k_d$ and $k_a$ obtained by fitting the results in Fig. 4. The desorption ($E_D$) and adsorption ($E_A$) energies for peak 1 were derived to be 1.02. and 0.31 eV, respectively. On the other

978-1-6654-7134-3/22 $31.00 © 2022 IEEE

hand, $E_D$ and $E_A$ for peak 2 were derived to be 0.53 and 0.11 eV, respectively. In the next section, we will discuss these activation energies.

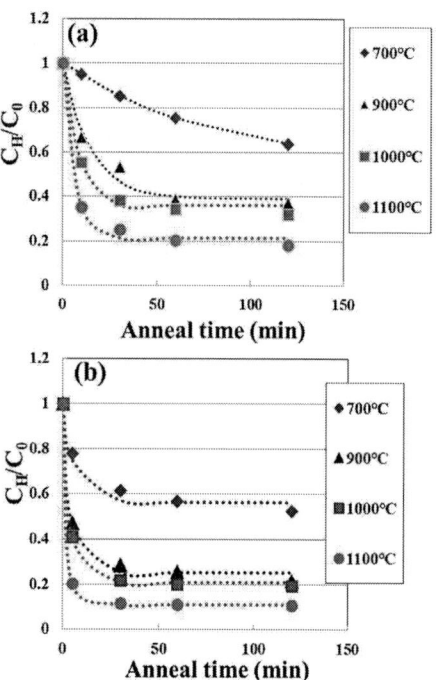

Fig. 4. Plot of ratios of $C_0$ to $C_{HD}$ of peaks (a) 1 and (b) 2 versus heat treatment time, and dotted lines indicating curves fitted by (3).

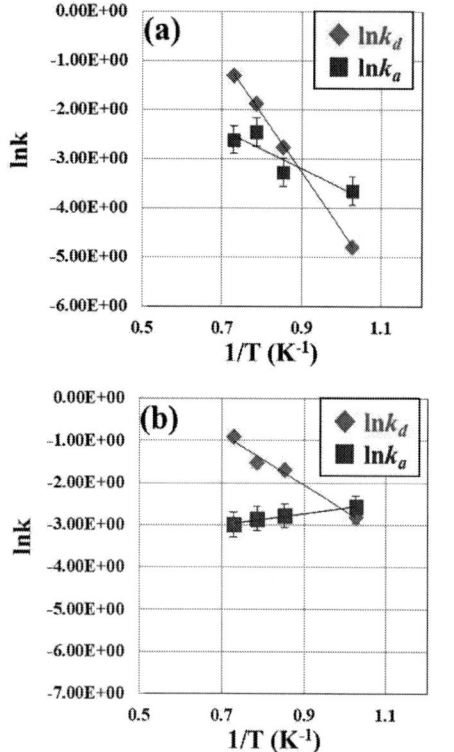

Fig. 5. Arrehnius plots of $k_d$ and $k_a$ of peaks (a) 1 and (b) 2 obtained by fitting the results in Fig. 4.

## B. Model of hydrogen diffusion behavior from CH$_4$N-implanted region

First, we discuss the activation energies of adsorption and desorption obtained from the analytical results of peak 1. In the case of the activation energies in the reversible reaction, the difference between the activation energies during adsorption and desorption can be interpreted as the binding energy. Therefore, the binding energy for peak 1 can be estimated as 1.02 - 0.31 = 0.71 eV. In a previous study, 0.76 eV was obtained as the hydrogen desorption activation energy in the hydrocarbon-molecular-ion-implanted region. [10] Regarding the hydrogen binding state in the hydrocarbon-molecular-ion-implanted region, the activation energy is close to a value in the range from 0.70 to 0.80 eV, which is the C–H$_2$ binding energy in silicon wafers. [14] It has been reported that hydrogen in the hydrocarbon-molecular-ion-implanted region forms the C–H$_2$ binding state with carbon aggregates. Moreover, as shown in Fig. 3, peak 1 in the CH$_4$N-implanted region is the depth position of carbon aggregates. Peak 1 is considered to be formed by the C–H$_2$ binding state due to hydrogen trapped in carbon aggregates. Therefore, the hydrogen diffusion from the peak 1 region is considered to be the desorption of hydrogen molecules (H$_2$) from the C–H$_2$ binding state.

Then, regarding peak 2, Fig. 4 shows that peak 2 is the depth position where EOR defects are formed. However, it has been reported that the EOR defects formed in the CH$_4$N-implanted region are {111} stacking faults [12]. Hydrogen trapping and desorption behaviors in similar defects have been reported for {111} platelet defects [15–17]. However, these studies showed that hydrogen molecules are desorbed from platelet defects during low-temperature heat treatment below 700 °C. It has also been reported that {111} platelet defects are formed at depths smaller than the hydrogen implantation range (Rp) by monomer hydrogen ion implantation [18]. Since the {111} stacking faults, which are the EOR defects in the CH$_4$N-implanted region, are formed at a depth larger than the Rp of CH$_4$N implantation, they are considered to have a morphology different from that of conventional {111} platelet defects. Carbon is also distributed near EOR defects. However, the desorption and adsorption activation energies obtained by peak 2 analysis were 0.53 eV and 0.11 eV, respectively. The binding state is different from the C–H$_2$ binding energy. Therefore, it is unlikely that carbon is the hydrogen-trapping site in peak 2. Considering the possibility that nitrogen bonds to hydrogen in the CH$_4$N-implanted region, the depth profile of nitrogen was confirmed by SIMS analysis. Fig. 6 shows the depth profiles of nitrogen and hydrogen in the CH$_4$N-implanted region obtained by SIMS. It was found that nitrogen also formed a peak in the CH$_4$N-implanted region. In addition, it can be seen that nitrogen is distributed in both regions of peaks 1 and 2. Here, the desorption activation energy of 0.53 eV obtained for hydrogen in peak 2 is discussed. In a previous study, an activation energy of 0.58 ± 0.05 eV was obtained for the desorption of hydrogen from N–H bonds in SiN films [19]. Therefore, it is considered that the hydrogen in the peak 2 region forms a N–H binding state. Nitrogen is distributed in both regions of peaks 1 and 2. However, it is inferred that hydrogen bonds to nitrogen in the peak 2 region, that is, near the EOR defects. Our previous studies showed that nitrogen atoms bind to the EOR defects in the CH$_4$N-implanted region [12, 20]. Therefore, the hydrogen in the peak 2 region forms a N–H binding state with nitrogen on the EOR defects. Furthermore, we consider that hydrogen desorbs and diffuses from the N–H binding state as atomic hydrogen (H).

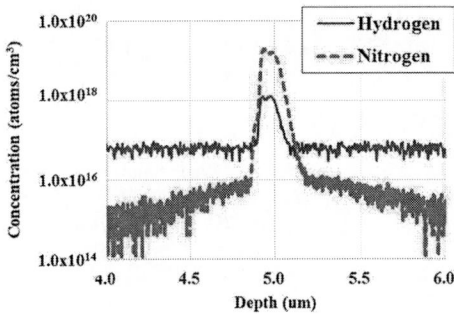

Fig. 6. Depth profiles of nitrogen and hydrogen in CH₄N-implanted region obtained by SIMS.

In the CH₄N-implanted epitaxial wafer, it was clarified that there are two types of hydrogen, namely these that, desorb from the C–H₂ binding state and these that desorb from the N–H binding state.

## IV. SUMMARY

In this study, the diffusion behavior of hydrogen from the CH₄N-implanted region was analyzed by reaction kinetics. It was found that two hydrogen-trapping sites, carbon aggregates and EOR defects, were formed corresponding to a peak in the CH₄N-implanted region. In the carbon aggregate region, we observed that the C–H₂ binding state is formed in the CH₄N-implanted region. On the other hand, in the EOR defect region, a desorption activation energy of 0.53 eV was derived, which is close to that of hydrogen from the N–H bond in the SiN film. We consider that the N–H binding state is formed in the CH₄N-implanted region. In addition, the desorption activation energy of hydrogen from the N–H binding state is lower than that from the C–H₂ binding state. Therefore, this result indicates that CH₄N-molecular-ion-implanted epitaxial wafers contribute to the reduction in $D_{it}$ at the $SiO_2/Si$ interface due to hydrogen desorption from the CH₄N-implanted region even during low-temperature heat treatment in the device process.

## ACKNOWLEDGMENT

The authors would like to thank Dr. Hisashi Furuya and Mr. Naoki Ikeda of the Technology Division, SUMCO Corporation for their support and helpful advice.

## REFERENCES

[1] F. Russo, G. Nardone, M. L. Polignano, A. D'Ercole, F. Pennella, M. D. Felice, A. D. Monte, A. Matarazzo, G. Moccia, G. Polsinelli, A. D'Angelo, M. Liverani, and F. Irrerac, ECS J. Solid State Sci. Technol. 6, 217 (2017).

[2] J. L. Regolini, D. Benoit, and P. Morin, Microelectron. Reliab. 47, 739 (2007).

[3] F. Russo, G. Moccia, G. Nardone, R. Alfonsetti, G. Polsinelli, A. D'Angelo, A. Patacchiola, M. Liverani, P. Pianezza, T. Lippa, M. Carlini, M. L. Polignano, I. Mica, E. Cazzini, M. Ceresoli, and D. Codegoni, Solid-State Electron. 91, 91 (2014).

[4] J. Jung, D.-W. Kwon, and J. Kim, Jpn. J. Appl. Phys. 45(1), 3466 (2006).

[5] J. Jung, D.-W. Kwon, and J. Kim, Jpn. J. Appl. Phys. 47(1), 139 (2008).

[6] B. Park, J. Jung, C.-R. Moon, S. Hwang, Y. Lee, D. Kim, K. Paik, J. Yoo, D. Lee, and K. Kim, Jpn. J. Appl. Phys. 46, 2454 (2007).

[7] K. Kurita, T. Kadono, R. Okuyama, R. Hirose, A. Onaka-Masada, Y. Koga, and H. Okuda, Jpn. J. Appl. Phys. 55, 121301 (2016).

[8] K. Kurita, T. Kadono, R. Okuyama, S. Shigematsu, R. Hirose, A. Onaka-Masada, Y. Koga, and H. Okuda, Phys. Status Solidi A 214, 1700216 (2017).

[9] K. Kurita, T. Kadono, S. Shigematsu, R. Hirose, R. Okuyama, A. Onaka-Masada, H. Okuda, and Y. Koga, Sensors 19, 2073 (2019).

[10] R. Okuyama, A. Masada, T. Kadono, R. Hirose, Y. Koga, H. Okuda, and K. Kurita, Jpn. J. Appl. Phys. 56, 025601 (2017).

[11] R. Okuyama, T. Kadono, A. Onaka-Masada, A. Suzuki, K. Kobayashi, S. Shigematsu, R. Hirose, Y. Koga, and K. Kurita, Jpn. J. Appl. Phys. 2020, 59, 125502.

[12] A. Suzuki, T. Kadono, R. Hirose, R. Okuyama, A. Masada, S. Shigematsu, K. Kobayashi, Y. Koga, and K. Kurita, Phys. Status Solidi A 2019, 216, 1900172.

[13] R. Okuyama, T. Kadono, A. Onaka-Masada, A. Suzuki, K. Kobayashi, S. Shigematsu, R. Hirose, Y. Koga, and K. Kurita, Jpn. J. Appl. Phys. 57, 011301 (2018).

[14] B. Hourahine, R. Jones, S. Oberg, P. R. Briddon, V. P. Markevich, R. C. Newman, J. Hermansson, M. Kleverman, J. L. Lindstrom, L. I. Murin, N. Fukata, and M. Suezawa, Physica B 308, 197 (2001).

[15] N. Fukata, S. Sasaki, K. Murakami, K. Ishioka, K. G. Nakamura, M. Kitajima, S. Fujimura, J. Kikuchi, and H. Haneda, Phys. Rev. B 56, 6642 (1997).

[16] N. Fukata, S. Sasaki, S. Fujimura, H. Haneda, and K. Murakami, Jpn. J. Appl. Phys. 35, 3937 (1996).

[17] N. Fukata, S. Sato, H. Morihiro, K. Murakami, K. Ishioka, M. Kitajima, and S. Hishita, J. Appl. Phys. 101, 046107 (2007).

[18] G. F. Cerofolini, F. Corni, S. Frabboni, C. Nobili, G. Ottaviani, and R. Tonini, Mater. Sci. Eng. B 27, 1 (2000).

[19] C. Boehme, J.Appl. Phys., 88(10), 15 (2000).

[20] A. Suzuki, T. Kadono, R. Hirose, R. Okuyama, A. Masada, S. Shigematsu, K. Kobayashi, Y. Koga, and K. Kurita, J. Electrochem. Soc. 169, 047521 (2022).

# Preparation of Uniform SiO₂ Insulating Layer on the Inner Wall of TSV by Thermal Oxidation

line 1: 1st Guo Fengjie
line2: *College of Big Data and Information Engineering, Guizhou University*
line3: *Guizhou Provincial Key Laboratory of Micro and nano Electronics and Software Technology*
line 4: *Guiyang City, Guizhou Province*
line 5: *2307691725@qq.com*

line 1: 2nd Ran Jing yang
line 2: *College of Big Data and Information Engineering, Guizhou University*
line3: *Guizhou Provincial Key Laboratory of Micro and nano Electronics and Software Technology*
line 4: *Guiyang City, Guizhou Province*
line 5: *1669077509@qq.com*

line 1: 3rd Wang Shuo
line 2: *College of Big Data and Information Engineering, Guizhou University*
line 3: *Guizhou Provincial Key Laboratory of Micro and nano Electronics and Software Technology*
line 4: *Guiyang City, Guizhou Province*
line 5: *1712066349@qq.com*

line 1: 4th Ma Kui
line 2: *College of Big Data and Information Engineering, Guizhou University*
line3: *Guizhou Provincial Key Laboratory of Micro and nano Electronics and Software Technology*
line 4: *Guiyang City, Guizhou Province*
line 5: *kma@gzu.edu.cn*

line 1: 5th Yang Fa shun
line2: *College of Big Data and Information Engineering, Guizhou University*
line3: *Guizhou Provincial Key Laboratory of Micro and nano Electronics and Software Technology*
line 4: *Guiyang City, Guizhou Province*
line 5: *fashun@126.com*
(Corresponding Author)

*Abstract*—SiO₂ insulating layer is an indispensable part of a TSV. In the current process, the SiO₂ insulating layer is commonly deposited on the inner wall of the TSV based on deep trench sputtering method. The thickness at different position (neck, middle, bottom) of the SiO₂ insulating layer, deposited by deep trench sputtering, is non-uniform. In this paper, the thickness uniformity of SiO₂ insulating layer prepared on the inner wall of TSV based on CVD&PVD process and thermal oxidation method is comparatively studied. The experimental results show that, based on the CVD&PVD process, the average thickness of the SiO₂ insulating layer at middle and bottom position of the TSV has changed by −54.02% and −58.30% compared with that at the top position, respectively. Based on the thermal oxidation method, the average thickness of the SiO₂ insulating layer at middle and bottom position of the TSV has changed by 1.17% and 0.26% compared with that at the top position, respectively. The thermal oxidation method can realize the SiO₂ insulating layer with uniform thickness on the inner wall of TSV.

*Keywords*—TSV, SiO₂ insulating layer, thermal oxidation, uniform

## I. INTRODUCTION

With the development of integrated circuits, electronic devices as a whole are moving in the direction of small size and light weight[1]. Three-dimensional stacking achieves this by reducing the size of the device and increasing the level of integration[2]-[3]. TSV, as a very important part of three-dimensional systems, enables multi-layer chip stacks, which can be interconnected vertically to reduce interconnect lengths, reduce footprints and improve integration[4]-[8].

As one of the key technologies of TSV process, preparation of insulation layer has been attracting much attention[9]-[11]. As a functional layer between the silicon and the conductor, the insulating layer is used to prevent the formation of conductive channels between the metallic material and the silicon substrate, improving the electrical reliability of the chip[12]. For the preparation of insulation layer, there are various equipment and processes to meet the needs of mass production. CVD, PVD, thermal oxidation and other processes have been widely used in various applications[13]-[14]. Duan Haoze used CVD to prepare insulating layer films. The thickness of the top of the film is 11.82 μm, the thickness of the middle is 8.94 μm, and the thickness of the bottom is 5.82 μm [15]. Zheng Shuai used the wet process to prepare the insulating layer film. The thickness of the insulating layer is 4 μm at the top and 850 nm at the bottom [16].

In this paper, through the optimization of thermal oxidation, the method of dry O2 + wet O2 + dry O2 is used to prepare the insulating layer thin film. By comparing with the insulating layer film prepared by CVD&PVD, we explore the process of preparing insulating layer film with more uniform thickness.

## II. EXPERIMENTAL PROTOCOL DESIGN

This experiment is based on the combination of dry oxygen oxidation and water vapor oxygen oxidation, that is, the film obtained by wet oxygen oxidation method. Its reaction formula is as follows:

Dry Oxygen Oxidation:

$$Si(g) + O_2(g) \longrightarrow SiO_2(s) \qquad (1)$$

Water vapor oxidation:

$$Si(g) + 2H_2O(g) \longrightarrow 2SiO_2(s) + 2H_2(g) \qquad (2)$$

Put the silicon wafer flat into the oxidation furnace, and raise the temperature to 1150°C. A flow rate of 1L/min O₂ is introduced to keep the silicon wafer in a dry oxygen environment. After 10 minutes, switch to a humid oxygen environment and perform constant temperature heating for 2 hours. Finally, the O₂ flow rate of 1L/min is still maintained for dry oxygen oxidation.

## III. INTERPRETATION OF RESULT

In this experiment, a cylindrical TSV with a diameter of 10 μm and a depth of 80 μm was obtained by Bosch etching. The SEM images of the $SiO_2$ insulating layer at different positions deposited by CVD&PVD process are shown in Fig 1.

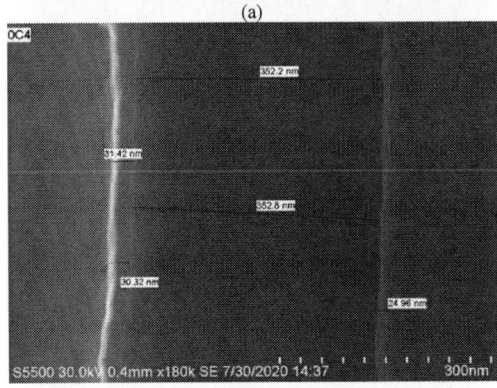

(a)

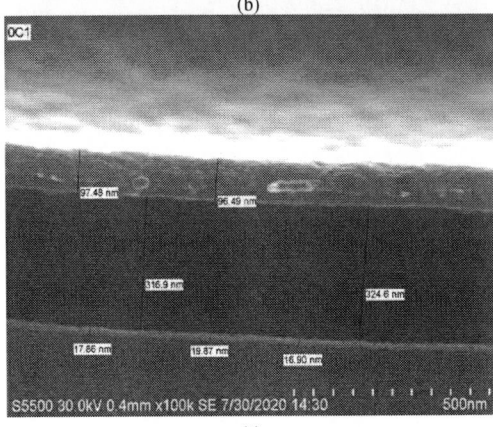

(b)

(c)

Fig.1 SEM images of $SiO_2$ insulating layer at different position deposited by CVD&PVD process: (a) neck, (b) middle, and (c) bottom

It can be seen that the thickness of the insulating layer at the top of the TSV is the largest, reaching 767.3nm, and the thickness of the insulating layer in the middle is 352.8nm, which is almost the same as the thickness of the

insulating layer at the bottom of 320nm. Generally speaking, the thickness of the insulating layer obtained by the CVD&PVD process presents a trend of widening at the top and narrowing at the bottom. The difference between the maximum and minimum thickness is 447.3nm. This difference is very large. In the subsequent filling material, the top of the TSV will be closed first, and the bottom will not be completely filled, forming a large cavity, which will have a very large negative impact on the experiment.

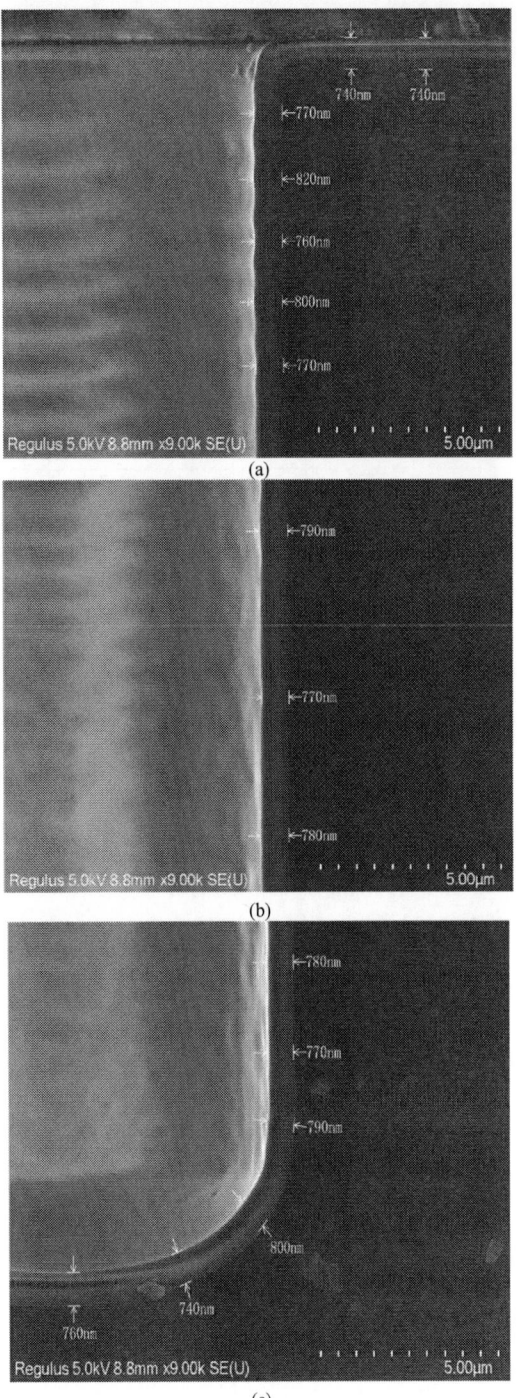

(a)

(b)

(c)

Fig.2 SEM images of $SiO_2$ insulating layer at different position prepared by thermal oxidation method: (a) neck, (b) middle, and (c) bottom

The SEM image of the $SiO_2$ insulating layer obtained by thermal oxidation is shown in Fig 2. The average thickness of the top of the insulating layer is 771 nm, the average thickness of the middle is 780 nm, and the average thickness of the bottom is 773 nm. On the whole, the thickness of the insulating layer is relatively consistent and uniform, which can achieve better performance.

Table.1 Thickness uniformity of $SiO_2$ insulating layer prepared by different methods

| Position / Samples | Top | Middle | | Bottom | |
|---|---|---|---|---|---|
| | T (nm) | T (nm) | DTNT(%) | T (nm) | DTNT(%) |
| prepared by CAD&PVD | 767.3 | 352.8 | -54.02 | 320 | -58.30 |
| Ref. [1] | 1370 | 1120 | -18.25 | 903.4 | -34.06 |
| Ref. [2] | 432.7 | 396.6 | -8.34 | 324.5 | -25.01 |
| this paper | 771 | 780 | ~1.17 | 773 | ~0.26 |

PS. T is the average thickness of the $SiO_2$ layer.
DTNT is the Deviation from thickness near the top

The thickness uniformity of the insulating layer obtained by thermal oxidation and other methods is shown in Table 1. For the insulating layer obtained by CVD&PVD method, the DDNT values are all negative, and the values are all relatively large. However, for the insulating layer obtained by the thermal oxidation method, DDNT is positive and the value is small. It can be seen from the comparison that the thermal oxidation method can effectively change the film thickness of the insulating layer, make the overall thickness tend to be consistent, achieve better performance, and improve the disadvantages of uneven film thickness prepared by traditional processes.

## IV. SUMMARIZE

The uniform thickness of the insulating layer film is conducive to improving the performance of the through-silicon hole and making it easier to fill the material. In this paper, the insulating layer film is prepared by thermal oxidation method. Compared with CVD&PVD, it is found that the thickness of the insulating layer film prepared by thermal oxidation is more uniform and achieves better performance.

## V. REFERENCES

[1]. Yuan Ye, DING Bingrui. Application Status and Development Trend of Integrated Circuit Technology [J]. Digital Communications World,2019(08):176.

[2]. Irene, E. A. The Effects of Trace Amounts of Water on the Thermal Oxidation of Silicon in Oxygen[J]. Journal of the Electrochemical Society, 1974, 121(12):1613.

[3]. Tong Zhiyi, ZHAO Zhang. Vertical Integration: An Effective Way to Extend Moore's Law [J]. Special Equipment for Electronics Industry,2012,41(01):1-7+32.

[4]. Yang Xueyan. Advantages and Applications of Integrated Circuits [J]. Science & Technology Innovation and Application,2018(29):185-186.

[5]. Yuan Ye, Ding Bingrui. Application status and development trend of integrated circuit technology [J]. Digital Communications World, 2019, 000(008):176-176.

[6]. Salah K . More than moore and beyond CMOS: New interconnects schemes and new circuits architectures[C]// 2017 IEEE 19th Electronics Packaging Technology Conference (EPTC). IEEE, 2017.

[7]. Segura J . Integrated microelectromechanical systems in the More than Moore era[C]// International Conference on Design & Technology of Integrated Systems in Nanoscale Era. IEEE, 2017.

[8]. Wu Xiang-dong. Research Progress of TSV Interconnect Technology in 3D Integrated Packaging [J]. Electronics and Packaging, 2012, 12(9):6.

[9]. Lu Hao,Dong Yaqiang,Liu Xincai,Liu Zhonghao,Wu Yue,Zhang Haijie,He Aina,Li Jiawei,Wang Xinmin. Enhanced Magnetic Properties of FeSiAl Soft Magnetic Composites Prepared by Utilizing PSA as Resin Insulating Layer[J]. Polymers,2021,13(9).

[10]. LI Jing. Application of PECVD in TSV Field [J]. Special Equipment for Electronics Industry,2014,43(07):6-8+12.

[11]. Wen Jiechao, Kong Quancun, Liu Guili, Niu Xianli, Tian Yuanbo. Micro electrochemical machining in preparation of hollow electrode wall insulation technology research [J]. Journal of mechanical science and technology, 2020, 33 (3) 6:411-418. The DOI: 10.13433 / j.carol carroll nki. 1003-8728.20190150.

[12]. Huang Z G. Wet Preparation Technology of $SiO\_2$ Insulation Layer for TSV Interconnection [D]. Shanghai jiaotong university, 2016. DOI: 10.27307 /, dc nki. Gsjtu. 2016.006147.

[13]. Huang Cui. Research on High-performance Through-hole (TSV) Three-dimensional Interconnect [D]. Tsinghua University,2015.

[14]. Lehmann, V. Formation Mechanism and Properties of Electrochemically Etched Trenches in n-Type Silicon[J]. J.electrochem.soc, 1990, 137(2):653-659.

[15]. Duan Haoze. $SiO\_2$ electrochemical wet insulation method and application [D]. Shanghai jiaotong university, 2020. The DOI: 10.27307 /, dc nki. Gsjtu. 2020.003615.

[16]. Zheng Shuai. Wet preparation technology of $SiO\_2$ insulating layer based on porous silicon and its application [D]. Shanghai jiaotong university, 2017. DOI: 10.27307 /, dc nki. Gsjtu. 2017.000616.

# Yield prediction with Machine Learning and parameter limits in semiconductor production

Rebecca Busch
University of Siegen
Siegen, Germany
rebecca.busch@uni-siegen.de

Michael Wahl
University of Siegen
Siegen, Germany
michael.wahl@uni-siegen.de

Peter Czerner
Elmos Semiconductor SE
Dortmund, Germany
peter.czerner@elmos.de

Bhaskar Choubey
University of Siegen
Siegen, Germany
bhaskar.choubey@uni-siegen.de

*Abstract*—Yield is an important cost factor in wafer production. Therefore, continuous data-driven yield monitoring and optimization provides opportunities to reduce production costs. Predicting yield during production would reveal its relationships with production parameters enabling dynamic optimization with a preventive and active increase in yield. In our investigations, we will first predict the yield based on one yield critical process step and later on with the data of four process steps. We will use different machine learning methods for this. Furthermore, we will look at whether the classification into good and bad yield values with these methods provides better results for the prediction. Another point of our investigations are the parameter limits of the individual methods. We show that these can be controlled by a simple method and optimised, if necessary.

*Keywords*—semiconductor manufacturing, yield prediction, machine learning, parameter limits

## I. INTRODUCTION

The demands and challenges in the production of semiconductor manufacturing are becoming ever greater. Increasing functionality and high quality are expected preferably at reduced prices. However, this can only be achieved if the costs are kept low. Cost can be reduced by increasing the yield in production. One would force to produce as few defective parts as possible. Yield defines this ability. In semiconductor production, a wafer must go through several hundred different process steps. Depending on its complexity, a few or several hundred microchips are produced on a wafer. These process steps can be doping, deposition, lithography, etching, and shaping [1]. Most of these may be repeated multiple times until the chip is ready. Each of these steps has the potential to introduce errors. Hence, it is important to monitor each step closely and to control these parameters. Within each production step in semiconductor manufacturing, inline measurement values such as layer thickness or etching rate are collected. Additionally, hundreds of sensors typically measure data such as pressure, plasma parameters or temperature, which are collected as time series. In general, this huge amount of data is too large for direct evaluation. Therefore, a well-defined filtering and statistical pre-processing performed by Fault Detection and Classification (FDC) methods [2] is required. FDC provides statistical process parameters which are often a result of a targeted compression of the time series, and which are used for semi-automatic monitoring collection. The data are typically collected for every wafer. This means that for each wafer and process flow, we have several hundred different parameters. These parameters have been used for the prediction of the

yield in combination with suitable mathematical techniques. A measure to evaluate the production quality is this wafer yield [3], which is defined as the number of working chips, based on the total amount of chips that are produced on a wafer. To increase productivity and decrease the cost per chip, enhancing the wafer yield is one of the core challenges in semiconductor manufacturing. In literature, several yield prediction models intended to improve wafer yield have been presented so far. According to the conceptual paper of Lee and co-authors [4], there are two approaches for controlling the yield in semiconductor production. In the first approach, an attempt is made to find the optimal process parameters to increase the yield. Some ideas are explained for example by Zhang and co-authors [5]. We intend to identify better parameter limits as used so far in production.

In the second approach of Lee and co-authors [4], an attempt is made to identify errors by analysing the process and predicting the yield by scrutinizing the process parameters. We will predict the yield, first using one process step and then adding further process steps. For this, we apply different machine learning (ML) methods to the data and compare the results. This paper is structured as follows: In section II we explain our methodology. In the next section, section III, we present our results. In section IV, we look at the parameter limits evaluated so far and narrow them down. In the last section, we briefly summarize our results and provide a glace on our further work.

## II. METHODOLOGY

For our investigations, we used the database of a leading foundry. At this, an attempt was made to draw conclusions about the yield by observing the test parameters of the wafers already produced by Random Forest investigations [6]. Test parameters are the results of various in-line tests of the wafers. They allow to estimate certain parameters, e.g., the layer thickness of a deposited material. In these tests, wafers that significantly deviate from the default values are sorted out directly. The results, however, still show strong deviations between prediction and real values. In our study, we concentrate on the FDC parameters (Fault Detection and Classification), which are acquired during production. Thus, we can predict the yield in an earlier stage of production. The FDC data reflect, for example, the average temperature, the maximum material application, or the minimum duration of exposure, etc. They show in summary which settings on the machine prevail in this process during production. Data from 35,336 wafers was extracted from the production database over a time slot of six months. A significant problem with real data is that some data are not always present. This is due to

978-1-6654-7134-3/22 $31.00 © 2022 IEEE

various reasons. Not all wafers are measured for reasons of cost or time. This is called structural measurement errors. On the other hand, the measurements could have failed at various points in time for a member of reasons. Therefore, values would be simply missing. In addition, not every machine delivers the same sensor data, even if it is used for the same process step. Since most algorithms can only work with complete data sets, it is important to replace or sort out these non-existent data. A combination of different methods is used for this, so that neither the number of wafers nor the number of parameters is minimised severely. Otherwise, relevant parameters can easily be removed[7]. After data pre-processing, 323 parameters were obtained for a first process step. To further reduce the number of parameters, only the relevant ones determined by a Wrapper Feature Selection (WFS) analysis are considered.

### A. Wrapper Feature Selection (WFS)

We needed an all-relevant feature selection wrapper algorithm. Therefore, the "Boruta" package in R was used [8]. Boruta can work with any classification method. However, the random forest is used by default. The only requirement is that the method provides a variable importance measure. The aim of Boruta is to assign importance to parameters.

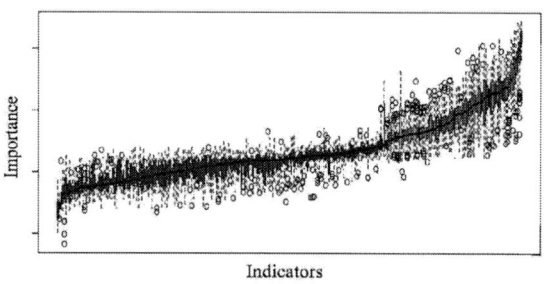

Fig. 1. WFS-analysis results: The green and some of the yellow parameters are important for the prediction

The green and some of the yellow parameters are used for further exploration. To do this, Boruta iteratively compares the importance of attributes with the importance of shadow attributes. These are created by shuffling the original attributes. If the attributes have a much lower importance than the shadow attributes, they are gradually removed. If they are much better than the shadow attributes, they are allowed as significant. The results are shown in green in Fig. 1. New shadows are created with each run. There are two possibilities to stop the algorithm. The first is that only attributes confirmed as relevant are left. The second is that the maximum number of runs has been reached. The number of the runs can be set manually. However, there are still attributes for which no decision has been made. These are shown in yellow in Fig. 1. To reduce the number, one could increase the number of maximum runs. However, their importance is indeterminate that they cannot be clearly assigned to red or green.

In our first investigations we set the maximum runs to 500. Because of too many unspecified parameters we increased the maximum runs to 1000. The runtime extended for a larger number of maximum runs too much. So, a larger maximum runs value was not used. This analysis narrowed down the number of parameters from 323 to 191 of the most influential parameters. The other process steps considered were treated the same way and the WFS analysis also reduced their number

of parameters. The number of parameters for these were reduced from approx. 200 to well below 100 in each case.

### B. Applied ML Algorithms

#### 1) The coefficent of determination (R squared)

To be able to compare the performance of the individual machine learning methods with the different hyperparameters, comparable key figures are required. These hyperparameters are the nuts and bolts for fine-tuning the ML. To compare them, we used the well-known metric of R squared. It is also called the coefficient of determination (R squared) [9]. R squared is a measure of the goodness of fit of a model. In regression, the R squared coefficient of determination is a statistical measure of how well the regression predictions approximate the real data points. This describes the proportion of the scatter in the output variable that can be explained by linear regression from the input variable. The R squared value is calculated by

$$R^2 \equiv \frac{SQE}{SQT} = \frac{\sum(\hat{y}_i - \overline{y})^2}{\sum(y_i - \overline{y})^2} = 1 - \frac{SQR}{SQT} = 1 - \frac{\sum(y_i - \hat{y}_i)^2}{\sum(y_i - \overline{y})^2}$$

The variables are defined as follows: SQE means Sum of Squares Explained. SQR is Sum of Squares Residuals. SQT represents Sum of Squared Totals. $y_i$ denotes the observed values. $\hat{y}_i$ indicates the predictive values. $\overline{y}$ represents the mean of the values. The aim of such constructive works is to identify models with higher R squared for the optimal selection of hyperparameters. For a given model, an interval of hyperparameters is specified and the model is trained on the data for each possible combination of hyperparameters [10]. For each data set, the power of the specified samples is calculated. Then the mean value and the standard deviation for each combination is calculated. The combination that has the optimal resampling statistics is selected as the optimal model. The entire training set is then used to fit a final model. We use regression methods and, depending on the method, different hyperparameters can be adjusted.

#### 2) Random Forest

The Random Forest [11] is designed to form an ensemble of weak, unbiased classifiers that combine their results in the final classification of each object. The individual classifiers are constructed as classification trees. Each tree is built using different bootstrap samples from the training set. Each bootstrap sample is the result of drawing the same number of objects as in the original training set. This results in that about 1/3 of the objects are not used to create a tree, but to perform an Out-of-Bag error estimation and importance measurement.

At each step of the tree construction, a different subset of attributes is randomly selected. The partitioning is performed based on the attribute that leads to the best distribution of data between the nodes of the tree. This procedure is carried out until the entire tree is constructed. The constructed tree is used to classify its Out-of-Bag objects, and the result is used to find approximation values for the classification error and to perform the calculation of confusion matrices. New objects are classified from all trees in the forest, and the final decision is made by simple voting. With Random Forest, the hyperparameter can be learned that specifies how many branches should leave a node.

#### 3) Neural Networks

Neural networks have the task of mapping the given input data to the corresponding output data with a minimum error. During learning, the network sends certain information via weighted connections between the neurons. Through continuous learning based on training examples, the network improves its performance by gradually adjusting the connection weights. We used the neural network called "nnet" in R. With this neural network [12], two hyperparameters are learned. One specifies the number of hidden nodes of the neural network and the other specifies the intervals at which the weighting of the hidden nodes is adjusted when training the neural network.

### 4) K-Nearest-Neighbours

The k-nearest-neighbours algorithm determines the k nearest neighbours using the Euclidean distance, where k is a freely selectable number. The value for k is generally chosen as the square root of the number of observations. With k-nearest-neighbours, the hyperparameter is learned that specifies how many neighbours are considered in the k-nearest-neighbours procedure.

### 5) Gerneralised Linear Model

The generalised linear model is a generalisation of ordinary linear regression that allows for response variables with error distribution models other than the normal distribution (e. g. Gaussian distribution). With generalised linear model, there are no hyperparameters.

### 6) Boosted Gerneralised Linear Model

The boosted generalised linear model is a generalisation of ordinary linear regression that allows for response variables with error distribution models other than the normal distribution. A boosted generalised linear model is fitted using a boosting algorithm based on component-wise univariate linear models. The fit or the regression coefficients, can be interpreted in the usual way.

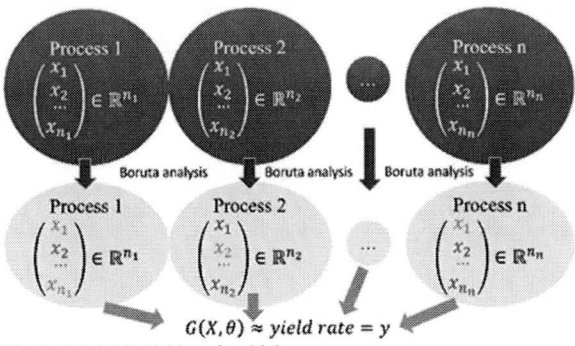

Fig. 2. Model for linking of multiple processes

### C. Model

With this reduced number of parameters, we then applied different machine learning methods to identify the best suited algorithm. To the first process step, we applied Random Forrest[13], Neural Networks, K-Nearest-Neighbours, generalised linear models, and boosted generalised linear model methods to predict the yield on the reduced number of 191 parameters. To avoid adverse data selection, we use 10-fold cross-validation and learned the optimal ML hyperparameters to optimise the R-squared value. Subsequently, we selected four other process steps with a high impact on the yield as advised by the experienced process engineers. The resulting multistep model is shown in Fig. 2. We also applied the different ML techniques to these four process steps. To ensure comparability of the results of the different techniques, we also used same metric of $R^2$ value.

## III. RESULTS

First, we are looking only at one process. It can be observed that our approach allows to correctly identify 99.37% of the good wafers and 97.98% of the bad wafers, which suits our purpose. Even, considering only one process step, the achieved R squared value is (0.389). The predicted yield versus the real yield is shown in Fig. 3.

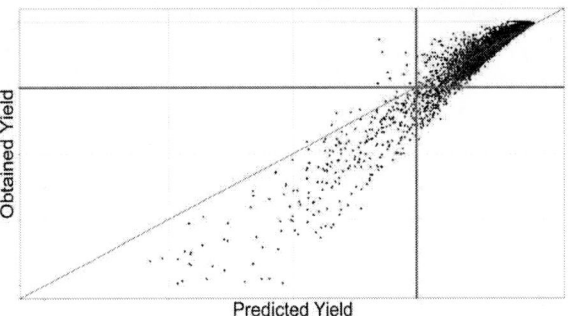

Fig. 3. Yield vs. with Random Forest predicted data of one process step after WFS analysis

The fat red lines indicate the yield value limits. For the values one the left side, a poor yield was predicted. For the ones below the big red line a poor yield was obtained. The thin red line shows the points where the predicted values are equal to the obtained ones. The result of our R squared is expected, as we are trying to predict an entire technology with many process steps, just considering one process. However, once a wafer has been declared to be faulty or with low yield, knowing the exact value of yield is of little value to a foundry.

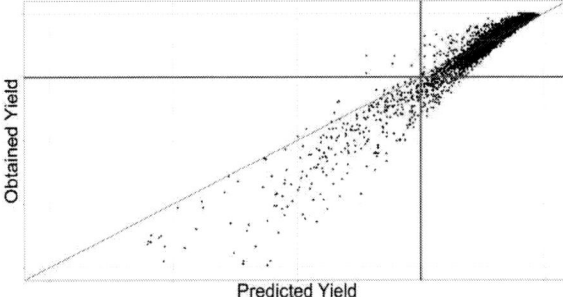

Fig. 4. Actual vs. Predicted Yield, when using four process steps to contribute to Random Forest together with WFS

Increasing the number of processes used for prediction to four leads to an improvement of the prediction, as expected. Now, 99.38% of good and 98.78% of bad are predicted correctly. The R-squared values are also increased to 0.54. We only used four process steps, so it is remarkable that we can still identify good and bad wafers with high accuracy. Especially, when we predict a poor yield value, the actual yield is much worse. This can be seen from the fact that they are nearly all below the thin red line. However, it is not important how poor the yield exactly is. We should now decide it is better to repair or throw it away.

## IV. PARAMETER LIMITS

Revisiting the work of Lee and co-authors [4], one may recall that in the metric the used first an attempt is made to find the optimal process parameters to increase the yield. A critical problem with this approach is the range of process parameter combinations in the search for the optimal values in real environments. Therefore, we will not search for the optimal values, nor ever will we try to find better parameter limits from existing data or validate known limits.

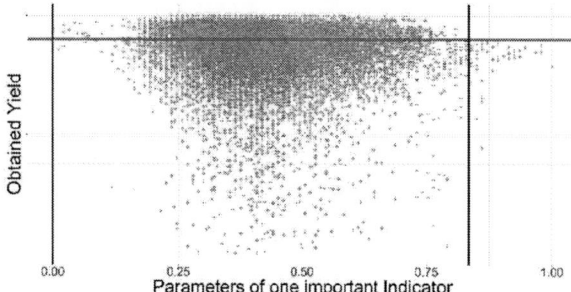

Fig. 5. Parameter vs. yield: better results with limitation the parameters to the green interval

One approach is not to find the values at which the yield becomes optimal, but to find the parameter limits, outside which the yield becomes worse. Since these limits are often only based on empirical values, one can adjust them. The process can be stabilised, and the yield improved by adjusting these limits. Fig. 5 shows one such example. As we identify the relevant parameters, it is possible to narrow down the number of parameters which need to be monitored. If the values of these parameters are plotted against the yield, we see that there is no single optimal value at which the yield becomes a maximum, but we can set limits (black lines). If the parameter value is outside these limits, it cannot exceed the limit we set vertically (red line) for the yield value. To achieve the necessary yield value, the limits of the parameters must not be outside the limits. Using these limits derived from the yield analysis, yield critical parameter value derivations can be identified, preventing a poor yield during production.

## V. SUMMARY AND CONCLUSIONS

We have shown that yield prediction is possible. We predicted good and bad wafer with a high accuracy, for using only one process with an accuracy of 99.37% and 99.38% for using four processes. The predicted bad values have an accuracy of 97.98 for one process and 98.78 for four processes. Exact yield predictions require significantly more work. The Random Forest algorithm gives the best results with an R squared value of 0.39 for one process and 0.54 for four respectively. In future studies, we will try to distinguish between the outliers and the good yield values. We hope to improve the accuracy of the yield predictions even further.

In our approach, we have applied ML to individual process steps, deriving the yield prediction based on summarizing the individual yield prediction values. In further investigations, we will evaluate if direct multi-stage ML methods produce even better results. Furthermore, we have shown an effective approach to identify parameter settings in production for the considered process steps.

### ACKNOWLEDGMENT

We thank Elmos Semiconductor SE for making their data available for our investigations. This research was part of the iDev40 project. The iDev40 project has received funding from the ECSEL Joint Undertaking (JU) under grant No 783163.

### REFERENCES

[1] F. Thuselt, „Physik der Halbleiterbauelemente", Berlin: Springer Berlin, vol. 3., pages 373-376 ,2018.

[2] J. Jang, B. W. Min, and C. O. Kim, "Denoised Residual Trace Analysis for Monitoring Semiconductor Process Faults," IEEE Transactions on Semiconductor Manufacturing, vol. 32, no. 3, pp. 293–301, 2019.

[3] L.-T. Chen, D. Lin, D. Muuniz, and C.-J. Wang,"Wafer Yield Estimation Using Support Vector Machines," in Advances in Neural Networks - ISNN 2006, Springer Berlin, pp. 1053–1058, 2006.

[4] Chang-Ho Lee , Dong-Hee Lee, Young-Mok Bae and Kwang-Jae Kim, "Determining golden process routes in semiconductor manufacturing process for yield management," IEEE International Conference on Industrial Engineering and Engineering Management (IEEM), pp. 2366–2370, 2017.

[5] Jie Zhang, Junliang Wang and Wei Qin, "Artificial Neural Networks in Production Scheduling and Yield Prediction of Semiconductor Wafer Fabrication System" , Artificial Neural Networks-Models and Applications, pp.355–387, 2016.

[6] Peter Czerner, " Big Data im Nanobereich: Halbleiter ICs mit Oracle, R & Co., " DOAG 2018 Konferenz, 2018.

[7] F. Chollet and J. J. Allaire, "Deep Learning mit R und Keras", 1.Auflage. Frechen: mitp, 2018.

[8] M. B. Kursa and W. R. Rudnicki, "Feature Selection with the Boruta Package," J. Stat. Soft., vol. 36, no. 11, 2010.

[9] L. Fahrmeir, C. Heumann, R. Künstler, I. Pigeot, and G. Tutz, „Statistik, Der Weg zur Datenanalyse", Springer Spektrum, pp. 121-155, 2016.

[10] M. Kuhn, "Building Predictive Models in R Using the caret Package", J. Stat. Soft,vol. 28, no. 5, 2008.

[11] Miron B. Kursa, Aleksander Jankowski and Witold R. Rudnicki. "Boruta A System for Feature Selection", In: *Fundamenta Informaticae* , p. 271-285, 2010.

[12] W. N. Venables and B. D. Ripley," Modern Applied Statistics with S", vol. 4, Springer, 2002.

[13] L. Breiman, "Random Forest" , vol. 45, pp 5-32, Machine Learning, 2001.

**Rebecca Busch** received her Dipl.- Wirt.- Math. at the University of Siegen in 2005. After familytime she came back to the University of Siegen as a researcher and Phd. student in 2016. Now she is working in the field of big data analysis and yield prediction in semiconductor manufacturing.

**Peter Czerner** received his diploma in physics at the University GH Essen 1996. He has been working for Elmos SE since 1997, first in IT as software developer, database designer and administrator. Since 2012 in production department as project manager FDC and since 2022 as head of Production Intelligence responsible for Data Governance, Digitalization and Data Science.

**Michael Wahl** received his academic degrees from University of Siegen. He is Head of the Digital Integrated Systems group at Siegen University and Distinguished Visiting Professor at Kerala University of Digital Sciences, Innovation and Technology. His research interest comprises design & test of integrated systems, test standards, semiconductor manufacturing, as well as economics & obsolescence of digital systems.

**Bhaskar Choubey** (Senior Member, IEEE) received the B.Tech. from Regional Engineering College, Warangal, India and the D.Phil. from the University of Oxford, Oxford. He is currently the Chair of Analog Circuits and Image Sensors, University of Siegen in Germany and an Associate Editor of the IEEE Sensors Journal.

# Positive/Negative Decision via Outlier Detection Towards Automatic Performance Evaluation for Defect Detector

Toshinori Yamauchi
*System Development Dept.1*
*Hitachi High-Tech Corporation*
Chuo-ku, Japan
toshinori.yamauchi.wd@hitachi-hightech.com

Kentaro Ohira
*Metrology Systems Solution*
*Development Dept.*
*Hitachi High-Tech Corporation*
Hitachinaka-shi, Japan
kentaro.ohira.yp@hitachi-hightech.com

Takefumi Kakinuma
*System Development Dept.1*
*Hitachi High-Tech Corporation*
Chuo-ku, Japan
takefumi.kakinuma.rk@hitachi-hightech.com

*Abstract— In the field of semiconductor defect inspection, it has been possible to detect defects with high accuracy thanks to the object detection model (defect detector) composed of the deep learning model. The performance of the deep learning model depends highly on training data; therefore, during the operational phase at the customer site, we need to frequently evaluate the model's performance to deal with shifts of appearance for defects. However, frequently executing general evaluation methods is difficult at the customer site; hence, we need a method to automatically evaluate performance. In this study, for the purpose of automatically evaluating the performance of the defect detector, we propose the Positive/Negative Decision via Outlier Detection (PNDOD). PNDOD decides on positive/negative for detection results based on comparing features corresponding to the detected result with statistics computed from training data. By using this method, we can calculate the estimated precision from the ratio of the estimated number of positive detections to the number of total detections, and we can evaluate the model performance automatically based on this estimated precision. In experiments using SiC wafer images, we confirmed that PNDOD can decide on positive/negative with high accuracy, and we can precisely evaluate the model's performance.*

*Keywords—Deep Learning, Object detection, Outlier detection, Semiconductor defect inspection, Defect detector.*

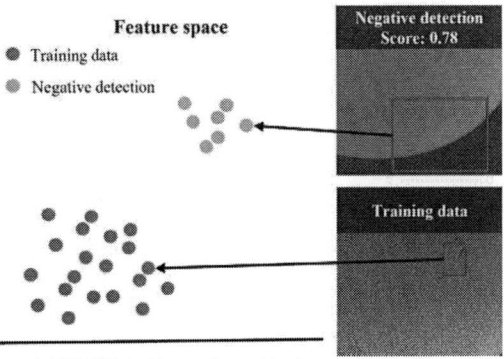

Figure 1: PNDOD decides on the positive/negative based on comparing features under the assumption that features corresponding to negative detections are distributed in different regions from features corresponding to training data. In this figure, the example image of negative detection is the case that the defect detector mis-detects the boundary between the wafer and the stage as the defect of Latent Scratch. You can see that the model mis-detects this region with high score.

## I. INTRODUCTION

In recent years, image processing based on deep learning has progressed remarkably and achieved significant results in a variety of tasks ([1], [2], [3], [4], [5], [6]). In terms of object detection models, which predict classes and positions of instances, various methods have been proposed ([2], [3], [4], [5]), and in the field of semiconductor defect inspection, it is possible to detect defects with high accuracy by applying such deep learning models.

On the other hand, the performance of the deep learning model highly depends on training data [7]. Therefore, during the operational phase at the customer site, there is a possibility of performance degradation for defects that have different features from training data, which occur because of changes in the manufacturing processes. In this case, we need to retrain using the above data to improve performance, and we need to frequently evaluate the model's performance to determine whether retraining is needed or not. However, general performance evaluations made by using labeled data or confirming detection results by humans frequently is difficult to make at the customer site. Hence, it is necessary to automatically evaluate the model performance with non-labeled data.

In this study, for the purpose of evaluating the performance of the defect detector composed of the object detection model automatically, we propose a method to decide on positive or negative for detection results named Positive/Negative Decision via Outlier Detection (PNDOD). By using PNDOD, we can automatically evaluate the model performance based on the precision, which is one of the evaluation indexes for the object detection model. The precision is defined as the ratio of the number of positive detections to the number of total detections. Therefore, if we decide on the positive/negative for all detections and identify detections that may be negative, we can determine the estimated number of positive detections and calculate the estimated precision.

In general, we can assume a detected result with a low score as negative, but as shown in Fig. 1, there is a case of negative detection with high scores. Therefore, it is difficult to decide on the positive/negative detections by adjusting the score threshold [7]. PNDOD decides on the positive/negative based on comparing features. Generally, the model may mis-detect for data that does not appear in the training data. PNDOD decides based on comparing features under the assumption that features corresponding to negative detections deviate from features corresponding to training data.

978-1-6654-7134-3/22 $31.00 © 2022 IEEE

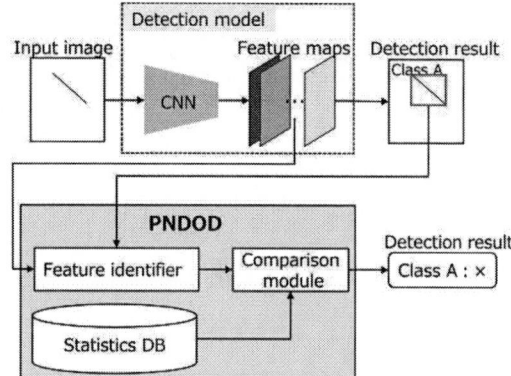

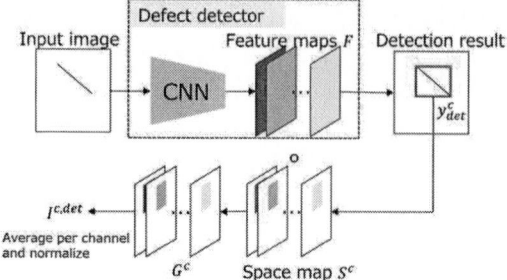

Figure 3: Overview of processes executed in the feature identifier. These processes are executed individually for each detected result.

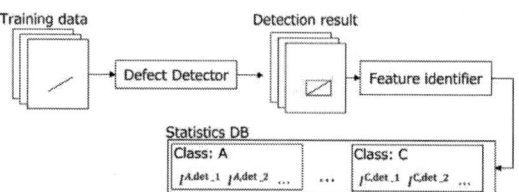

Figure 4: Overview of saving processes in the statistics database. We store features computed by applying the feature identifier to detection results in training data. These features are stored separately by class.

Figure 2: Overview of PNDOD. PNDOD is composed of three modules, the feature identifier, the statistics database, and the comparison module. PNDOD decides on the positive/negative by the following steps: 1) for each detected result, the feature identifier identifies features contributing to the detected result from feature maps, 2) the comparison module compares identified features with the statistics of features computed from training data, which are stored in the statistics database, 3) positive/negative decision is made by determining the detected result with features deviating from the statistics as negative and otherwise as positive.

Specifically, PNDOD compares features corresponding to the detected result with the statistics computed from training data and decides as negative for deviating detections and otherwise as positive.

Through experiments using silicon carbide (SiC) wafer images, we confirmed that PNDOD could decide on the positive/negative for detected results with high accuracy. Therefore, by using this method, we can calculate the estimated precision precisely and automatically evaluate the model performance with non-labeled data.

## II. PROPOSED METHOD

Fig. 2 shows an overview of PNDOD. In this study, we apply PNDOD to one of the well-known object detectors called the Single Shot Multibox Detector (SSD) [2]. For each detection result, PNDOD first identifies features that contribute to the detected result. Then, it compares identified features with statistics computed from training data. Finally, it decides as negative for detection results having features that deviate from the statistics and others as positive. As shown in Fig. 2, PNDOD is composed of a feature identifier, a statistics database, and a comparison module. We describe the details of processing in each module below.

### A. Feature identifier

The defect detector detects defects separately. Hence, we need to identify features that contributing to each detection result. Fig. 3 shows the processes in the feature identifier. The feature identifier first computes element-wise products for the model's extracted feature map (height $h$, width $w$, and channel $k$) and the space map having spatial information for the detected result.

$$G_k^c = F_k \circ S_k^c, \qquad (1)$$

where $c$ is the predicted class, $F_k \in \mathbb{R}^{h \times w}$ is the feature map of channel $k$, and $S_k^c \in \mathbb{R}^{h \times w}$ is the space map of channel $k$ represented as follows:

$$S_k^c = \frac{\left| \frac{\partial y_{det}^c}{\partial F_k} \right|}{max \left( \left| \frac{\partial y_{det}^c}{\partial F_k} \right| \right)}, \qquad (2)$$

where $y_{det}^c$ is the score for class $c$ in the detected bounding box and $max(\cdot)$ is a function to find the maximum value. The space map $S_k^c \in [0, 1]$ calculated by (2) indicates the relative numerical magnitude in the absolute value of gradients in the feature map $F_k$. Therefore, it represents the spatial importance of each neuron in the feature map $F_k$ for the detection result of $y_{det}^c$ [8], and we can use it as the index indicating spatial information for the detected result.

We can substitute masks with detected regions set to be 1 and others set to be 0 as space maps. Such masks only focus on the features in the detected regions, on the other hand, the space map computed by (2) can focus on not only the detected regions but also peripheral regions of the detected regions [8]. This is useful when the defect detector detects defects by paying attention to the peripheral regions. In this case, we need to identify features not only in the detected regions but also in their peripheral regions. For this reason, we apply the space map computed by (2) in this study.

We generate the feature map $G^c$ that identifies features contributing to the detected result from (1). We get $I^{c,det} \in \mathbb{R}^K$ as features corresponding to the detected result by calculating the average for each channel in $G^c$ and normalizing. Hence, each element in $I^{c,det}$ is represented by the following equation:

$$I_k^{c,det} = \frac{ave\left(TopN(G_k^c)\right)}{\sum_i^K ave\left(TopN(G_i^c)\right)}, \qquad (3)$$

where $K$ is the number of channels in the feature map, $TopN(\cdot)$ is a function to select the top N values (set to $N = 20$ in this study), $ave(\cdot)$ is a function to calculate the average.

## B. Statistics database

Fig. 4 shows flows of saving to the statistics database. We store features obtained from training data to the statistics database. Here, we identify features corresponding to each detection result in training data by using the feature identifier described in II-A and store those features separately by class. In this study, we identify features for detection results which have Intersection over Union (IoU) larger than 0.6 for the annotated bounding box and store those features.

## C. Comparison module

The comparison module compares features identified by the feature identifier with the statistics computed from samples corresponding to the predicted class of the model that is stored in the statistics database. In this study, we investigate L1 norm, L2 norm, Mahalanobis distance (MN), and one class support vector machine (OCSVM) as comparison methods. L1 norm, L2 norm and MN are represented as follows:

$$d_{L_1} = \sum_i^C |I_i^{c,det} - \hat{I}_i^c|, \tag{4}$$

$$d_{L_2} = \sqrt{\sum_i^C (I_i^{c,det} - \hat{I}_i^c)^2}, \tag{5}$$

$$d_{MN} = \sqrt{\left( (I^{c,det} - \hat{I}^c)^T \Sigma_c^{-1} (I^{c,det} - \hat{I}^c) \right)}, \tag{6}$$

where $\hat{I}^c \in \mathbb{R}^K$ is statistics for class $c$ computed from samples stored in the statistics database, and it is represented as follows:

$$\hat{I}^c = \frac{1}{N_c} \sum_i^{N_c} I^{c,i}, \tag{7}$$

where $N_c$ indicates the number of samples for class $c$ stored in the statistics database and $\Sigma_c^{-1} \in \mathbb{R}^{K \times K}$ in (6) is the inverse of covariance matrix calculated from those samples. In comparisons using L1, L2, and MN, each calculates the distance between features $I^{c,det}$ corresponding to the detected result and the statistics $\hat{I}^c$ corresponding to the predicted class. Then, if the distance is more than threshold $D$, it decides the detected result is negative because it is an outlier from the statistics, otherwise it decides it is positive.

OCSVM maps data into the different feature spaces using the kernel function, and those trained as positive data are mapped far from the origin in that space. In this study, we train OCSVM for each class separately using samples stored in the statistics database and decide on the positive/negative

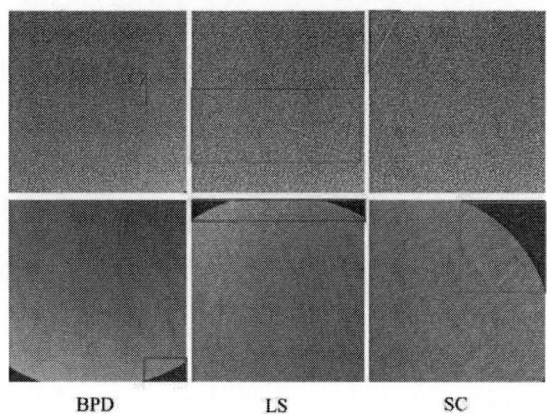

Figure 3: Examples of each defect (BPD, LS, SC). Top and bottom figures shows positive detections and negative detections, respectively.

Table 1: The number of samples stored in the statistics database ("Stored samples") and the number of detection results used for the evaluation ("Evaluation samples"). In the row of "Evaluation samples", "Positive" and "Negative" indicate the number of positive detections and negative detections, respectively. In the "Negative" row, (.) indicates the number of detected results that the model predicts with scores over 0.5. All evaluation samples are detected results from the model with score threshold is 0.3.

| Class | | BPD | LS | SC |
|---|---|---|---|---|
| Stored samples | | 2124 | 1826 | 1489 |
| Evaluation samples | Positive | 110 | 107 | 103 |
| | Negative | 22 (3) | 67 (52) | 19 (6) |

for features corresponding to the detected result by using OCSVM corresponding to the predicted class.

## III. RESULTS AND DISCUSSION

In this study, as the evaluation metrics for PNDOD, we use the true positive rate (TPR), which is the ratio of deciding on positive detections as the positive, and the true negative rate (TNR), which is the ratio of deciding on negative detections as negative. We also use the false positive rate (FPR), which is the ratio of deciding on negative detections as positive, that is $1 - $ TNR, in Fig. 7. We apply silicon carbide (SiC) wafer images as experimental data and evaluate positive and negative detection results of three types of defects: Basal Plane Dislocations (BPD), Latent Scratch (LS), and Scratch (SC), which all exist on the SiC wafer. Fig. 5 shows examples of positive and negative detections for each defect, and Table 1 indicates the number of samples stored in the statistics database and the number of detection results used for the evaluation. As shown in Table. 1, there are many cases of negative detections with high scores.

Table 2 shows the average of TPR and TNR. In each comparison method (L1, L2, MN, and OCSVM) of the PNDOD shown in Table 2, we list the best accuracy in the case of changing the threshold for evaluation samples. (In OCSVM, we execute multi-objective optimization using Optune [9] and obtain the best accuracy.) As shown in Table 2, PNDOD with the comparison method of MN or OCSVM can decide on the positive/negative with higher accuracy than the high score threshold and the PNDOD with comparison

Table 2: Average results of TPR and TNR for each method. In the high score threshold, we set the model score threshold to 0.5 and compute the average of TPR and TNR by comparing the detecton results when the model's score threshold is 0.3. We highlight in blur the best accuracy in each defect.

|  | BPD | LS | SC |
|---|---|---|---|
| High score threshold | 0.93 | 0.61 | 0.81 |
| PNDOD (L1) | 0.96 | 0.76 | 0.94 |
| PNDOD (L2) | 0.78 | 0.64 | 0.80 |
| PNDOD (MN) | 1.0 | 0.91 | 1.0 |
| PNDOD (OCSVM) | 1.0 | 1.0 | 1.0 |

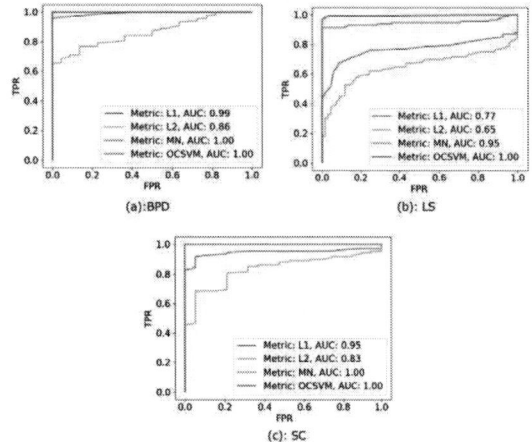

Figure 6: ROC curve and its AUC of PNDOD with each comparison method for each defect. (a), (b), and (c) are results for BPD, LS, and SC, respectively. In each figure, x and y axis indicate the FPR and the TPR, respectively.

methods of L1 and L2. In particular, PNDOD with OCSVM make decisions with 100 % accuracy for all defects.

Fig. 6 shows the Receiver Operating Characteristic curve (ROC) and its Area Under Curve (AUC) for the PNDOD with each comparison method in the case of changing the threshold. As shown in Fig. 6, the comparison method of MN and OCSVM have an AUC larger than 0.95 for all defects that indicate these two methods decide with high accuracy.

One possible reason for the higher accuracy of MN compared to L1 and L2 is that it is able to consider correlations of the data. In general, the Mahalanobis distance evaluates the influence strongly in the direction of small variance and weakly in the direction of large variance. Therefore, because it can consider such correlations of the data, it has a higher accuracy compared to other methods. In addition, because OCSVM can decide with high accuracy, we can say that it can correctly compute the decision boundary between the positive and the negative for the data.

Fig. 7 shows the two-dimensional space mapping results of features $I^{c,det}$ by using t-Distributed Stochastic Neighbor Embedding (t-SNE). Fig. 7 visualizes features corresponding to the detected results of the LS class (yellow: training data, blue: positive detections for inference data, purple: negative detections for inference data). As shown in Fig. 7, features corresponding to negative detections are mapped into different regions from features corresponding to training data and positive detections. Therefore, PNDOD, which decided on the positive/negative based on comparing features, is able

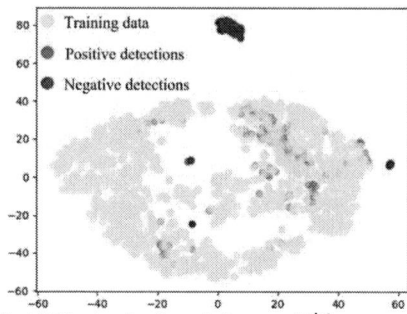

Figure 7: t-SNE visualization of features $I^{c,det}$ corresponding to detected result of LS class. In this figure, yellow, blue, purple dots indicate features corresponding to detected results for training data, positive detections for inference data, and negative detections for inference data, respectively

to distinguish between such positive detections and negative detections with high accuracy.

## IV. CONCLUSION

We proposed PNDOD, which decides on the positive/negative for detected results based on comparing features. Through experiments using SiC wafer images, we confirmed that PNDOD decide on the positive/negative with high accuracy. By using this method, we can calculate the estimated precision precisely, and we can evaluate the model performance automatically based on this estimated precision.

## REFERENCES

[1] Kaiming He, Xiangyu Zhang, Shaoqing Ren, and Jian Sun, "Deep residual learning for image recognition," in CVPR, 2016, pp. 770–778.

[2] Wei Liu, Dragomir Anguelov, Dumitru Erhan, Christian Szegedy, Scott E. Reed, Cheng-Yang Fu, and Alexander C. Berg, "SSD: Single shot multibox detector.," in ECCV. 2016, vol. 9905 of Lecture Notes in Computer Science, pp. 21–37, Springer.

[3] Joseph Redmon and Ali Farhadi, "YOLOv3: An Incremental Improvement," arXiv.org, pp. 1–6, Apr. 2018.

[4] Shaoqing Ren, Kaiming He, Ross B. Girshick, and Jian Sun, "Faster r-cnn: Towards real-time object detection with region proposal networks.," in NIPS, 2015, pp. 91–99.

[5] Tsung-Yi Lin, Piotr Dollr, Ross Girshick, Kaiming He, Bharath Hariharan, and Serge Belongie, "Feature pyramid networks for object detection," in CVPR, 2017, pp.936–944.

[6] Olaf Ronneberger, Philipp Fischer, and Thomas Brox,"U-net: Convolutional networks for biomedical image segmentation," in MICCAI. 2015, vol. 9351 of LNCS, pp. 234–241, Springer.

[7] A. R. Dhamija, M. Günther, J. Ventura and T. E. Boult, "The Overlooked Elephant of Object Detection: Open Set," 2020 IEEE Winter Conference on Applications of Computer Vision (WACV), 2020, pp. 1010-1019.

[8] T. Yamauchi and M. Ishikawa, "Spatial Sensitive GRAD-CAM: Visual Explanations for Object Detection by Incorporating Spatial Sensitivity," 2022 IEEE International Conference on Image Processing (ICIP), 2022, pp. 256-260

[9] Takuya Akiba, Shotaro Sano, Toshihiko Yanase, Takeru Ohta, and Masanori Koyama. 2019. Optuna: A Next-generation Hyperparameter Optimization Framework. In KDD.

## AUTHOR BIOGRAPY

The author is an engineer of Hitachi-High-Tech Corporation. His work focuses on developing AI models for semiconductor inspection. Especially in his most recent job is developing systems that enable to operate AI models at customer sites.

*** Formatting Issue - Best Available Paper/Graphic ***

YD-15

2022 International Symposium on Semiconductor Manufacturing (ISSM).
December 12–13, 2022

# A Study on Detection Method Using 2-Class Classifiers for Defective Wafer Maps

Seima Sakaguchi
*Mie University*
Tsu, Mie, Japan
422M223@m.mie-u.ac.jp

Yasushi Arimura
*KIOXIA Corporation*
Yokkaichi, Mie, Japan
yasushi1.arimura@kioxia.com

Takayuki Yamauchi
*KIOXIA Corporation*
Yokkaichi, Mie, Japan
takayuki2.yamauchi@kioxia.com

Yuichi Tokuyama
*KIOXIA Corporation*
Yokkaichi, Mie, Japan
yuichi1.tokuyama@kioxia.com

Tomoya Kawai
*KIOXIA Corporation*
Yokkaichi, Mie, Japan
tomoya.kawai@kioxia.com

Hidetaka Eguchi
*KIOXIA Corporation*
Yokkaichi, Mie, Japan
hidetaka.eguchi@kioxia.com

Hiroyuki Morinaga
*KIOXIA Corporation*
Yokkaichi, Mie, Japan
hiroyuki.morinaga@kioxia.com

Hiroharu Kawanaka
*Mie University*
Tsu, Mie, Japan
kawanaka@elec.mie-u.ac.jp

Tetsushi Wakabayashi
*Mie University*
Tsu, Mie, Japan
waka@hi.info.mie-u.ac.jp

*Abstract*—In semiconductor manufacturing, a pattern of chips with electrical failures in the wafer is usually used to identify failure factors. Wafers with similar in-plane trends are likely to have the same defect factors, and clustering techniques are often used to identify defect factors. It is, however, difficult for clustering approaches to make a cluster of infrequent unknown patterns. As a result, it will occur missing defect patterns. We discussed the method to detect infrequent unknown patterns and accurately classify frequent known defect patterns. We tried to make the proposed scheme with three strategies. As the first approach, VGG16 and SVM were used as the feature extractor and a classifier, respectively. The second approach is Convolutional Auto Encoder (CAE). We constructed CAEs for each known class, and the CAEs were trained to reconstruct the input images. When we take the above strategy, the constructed CAE cannot reconstruct the same image when the input image does not belong to the same class. It will be helpful to judge whether the given image belongs to the same class. The third approach uses the difference degree between the given image and typical images. The calculated value of the difference is used for distinguishing using thresholds. The experimental results show that the classification accuracy of known classes is 75.9%, the detection rate of unknown classes is 62.5%, and unknown clusters containing 35.7% of unknown classes are successfully created. To improve yield, it is essential to detect unknown defects at an early stage. If we can generate clusters that are mostly composed of unknown classes, it will be possible to recognize the occurrence of unknown defects. Since the proposed method was able to generate clusters in which unknown classes account for about 35%, we believe that it is sufficient to detect the occurrence of unknown defects.

*Keywords*—*SVM, Convolutional Auto Encoder, Wafer Map Classification, Intelligent System, Image Processing*

## I. Introduction

Yield analysis in semiconductor manufacturing improves productivity by identifying the causes of defects based on product quality inspection results and the processing history of each process. By using a large amount of data, more detailed yield analysis is possible, and significant productivity improvements are expected. However, the data has become huge and complex, so it has become difficult for technicians

to get useful information that can lead to improved productivity. One of the yield analysis methods is to identify the cause of defects from wafer maps. A wafer map is a map that shows the location information of defective chips identified by the inspection of each chip on the wafer. When a problem occurs in the manufacturing process, a characteristic defect pattern appears on the wafer. When multiple similar defective wafer maps occur, we assume that there is a common cause in the manufacturing process and investigate the cause. Defective wafer maps are various by-products, and new defect maps may appear that have not existed in the past. The patterns are various, but engineers need to find the critical defects early and identify the cause of the defects. That is one of the major challenges to improving productivity. In this study, we developed a system that automatically classifies patterns in defective wafer maps using machine learning to solve this problem. There are many cases of machine learning being used for wafer map classification[1][2][3]. In yield analysis based on wafer map patterns, it is important to accurately classify them. In addition to this, in actual semiconductor manufacturing sites, there is the problem of unknown wafer maps that may occur infrequently and be misclassified into clusters of similar known wafer maps. Misclassification of unknown wafer maps may lead to delays in trouble detection, resulting in significant damage. Therefore, in actual semiconductor manufacturing, in addition to accurately classifying a wide variety of wafer maps, it is essential to detect with high accuracy unknown wafer maps that may occur because of unexpected troubles.

## II. Experimental Material

In this study, we used open data published on Kaggle[4] as experimental material. These data were treated as a binary image. The areas with no chips and good chips are black, and the areas with defective chips are white. In addition, seven types of labels are pre-labeled based on the pattern of the wafer map, i.e., Center, Edge-Loc, Edge-Ring, Loc, Random, Scratch, and Donut(Fig. 1). We also checked the validity of the labels assigned to a total of 5,284 wafer map images randomly selected from the WM-811K wafer map dataset. After a careful examination, labels that were judged to be

978-1-6654-7134-3/22 $31.00 © 2022 IEEE

125

**\*\*\* Formatting Issue - Best Available Paper/Graphic \*\*\***

YD-15

2022 International Symposium on Semiconductor Manufacturing (ISSM),
December 12-13, 2022

(a) Center    (b) Edge-Loc   (c) Edge-Ring

(d) Loc      (e) Random     (f) Scratch     (g) Donut

Fig.1 Example of Wafer Map for each Label

unsuitable were changed properly. In addition, because some images could not be classified into any of the classes defined in WM-811K dataset, they were assigned the newly created label "Others". Center, Edge-Loc, Edge-Ring, Loc, Random, and Scratch class were treated as known defective wafer maps, and Donut and Other classes were treated as unknown defective wafer maps with infrequent occurrences. Each of the six types of defective wafer maps has about 500 images, the Donut class has 64 images, and Other class has 80 images. In this study, we aimed to classify the six known defective wafer maps and detect the Donut and Other classes as unknown defective wafer maps.

## III. CASCADE MODEL FOR WAFER MAP ANALYSIS

### A. Concept of Proposed Method

Create a 2-class classifier that classifies a specific class and the other classes for the six known defective wafer maps described above. For example, classifier1 classifies whether it is class1 or not class1. Similarly, classifier2 classifies whether it is class2 or not class2. In this way, a 2-class classifier is created for each of the six known classes and combined as shown in the Classification Flowchart(Fig. 2). Then, each class is classified as a filtering unit. When a known class is entered, it is detected by the specific classifier and a cluster is generated. On the other hand, if an unknown class is an input, it is not detected by any classifier and remains until the end, so an unknown cluster is generated with the remaining objects. As such, we proposed a method that can classify known classes and detect unknown classes of defective wafer maps at the same time.

### B. VGG16&SVM Based Approach

This is a method to create a binary classifier with SVM(Support Vector Machine)[6] using features extracted by VGG16, which is a kind of CNN. Fig. 3 is the structure of VGG16 used for feature extraction. The weights for VGG16 were pre-trained by ImageNet. In addition, only the 5th block layer is fine-tuned using the training data. The feature vector extracted by VGG16 is 7x7x512, and GlobalAveragePooling is performed to convert it into 512-dimensional features. These features are used for training SVM. The SVM is trained to judge whether the input data belongs to a specific class or others.

### C. Convolutional Auto Encoder Based Approach

We used the convolutional autoencoder[7] with thirteen layers for both the encoder and decoder layers. Fig. 4 is the structure of CAE. This convolutional autoencoder learns a

Fig.2 Classification Flowchart

Fig.3 Structure of VGG16

(a)Encoder

(b)Decoder

Fig.4 Structure of Convolutional Auto Encoder[8]

specific class and creates one that can restore only a specific class. When other classes are entered, the autoencoder created to restore a specific class will not properly restore the image. Using this characteristic, classification is performed by taking the difference between the input image and the restored image and setting a threshold value for the difference.

### D. Simple Similarity Based Method

A typical image contains features that represent its patterns and trends. In this paper, we carefully checked the given dataset and selected typical images. A typical image allows multiple images to be selected. In that case, if any one of several images exceeds the threshold value, it is judged as belonging to the specified class. The similarity is defined as the sum of the number of pixels of difference between the given image(s) and the typical image(s).

978-1-6654-7134-3/22 $31.00 © 2022 IEEE

### E. Concatenation of Each Classifier

This is an explanation of how to select the most accurate classifier from each of the classifiers created by the three methods. When selecting the most accurate classifier, the selection is based on the average of the known classification accuracy and the unknown detection rate. The higher the average is, the better the classifier. The known classification accuracy described here is the percentage of correctly classified specific labels and other labels. The unknown detection rate is the percentage of the input Donut class defined as an unknown label that was correctly classified into an unknown class. For example, to evaluate classifier1, six labeled known classes are input and the percentage of correctly classified class1 and the other classes is calculated as the known classification accuracy, and the percentage of the input Donut classes correctly classified as not class1 is calculated as the unknown detection rate. The next step is an explanation of the combining order of each selected classifier. The combining order is important for classifying known classes. For example, the reason for this is that once a class is classified as Class1, it does not flow to the next classifier, so if a faulty classifier is connected first, other classes will often be classified in the wrong cluster. Therefore, we combined the classifiers in order of highest Precision value, that is, the classifiers that can accurately extract only the appropriate labels.

## IV. EXPERIMENT

### A. Preparation of Dataset

We prepare three data sets (Dataset-1, Dataset-2, and Dataset-3). 80% of the known class data is training data, including validation data, and we used the remaining 20% of the data and 64 images of the Donut class, and 80 images of Other class data for evaluation. Dataset-1 and Dataset-2 have 2056 known class data for training and 514 known class data and 32 unknown class data(Donut) for testing. Dataset-1 was used to evaluate the classification accuracy of the known class and the detection rate of the unknown class for each classifier. The detection rate of an unknown class means how the classifier can accurately classify the input unknown data into an unknown class. Dataset-2 was used to evaluate the performance of the concatenated system. In Dataset 3, we evaluate the accuracy when the Donut and Other classes are defined as unknown classes, in order to evaluate the accuracy in more practical situations.

### B. Experimental Results and Discussion

Tables 1 and 2 show the experimental results. Table 1 shows the classification accuracy of the known class, and Table 2 shows the detection rate of the unknown class (Donut). Bold numbers in the table indicate the best approach in classifier creation, considering both known classification accuracy and unknown detection rate for each class. For example, for the Scratch and Random classes, the VGG16&SVM approach has the highest average values of known classification accuracy and unknown detection rate. Therefore, for the Scratch and Random classes, the 2-class classifier created by the VGG16&SVM approach is considered to be the best. In this approach, the least accurate classifier has a highly known classification accuracy of 83.3%. Therefore, it can be said that the known classification accuracy is sufficient. On the other hand, in the detection of unknown classes, the Loc class was particularly inaccurate at 31%.

Table 1 Classification Accuracy for Known Patterns (%)

|  | C | EL | ER | L | S | R |
|---|---|---|---|---|---|---|
| VGG16&SVM | 93.8 | 91.4 | 97.5 | 83.3 | **89.9** | **95.6** |
| CAE | 93.8 | **92.1** | 62.2 | **81.9** | 14.5 | 21.2 |
| Simple Similarity | **95.2** | 48.4 | **98.5** | 44.3 | 15.0 | 29.2 |

( C: Center, EL: Edge-Loc, ER: Edge-Ring, L: Loc, S: Scratch, R: Random)

Table 2 Detection Rate for Unknown Patterns (%)

|  | C | EL | ER | L | S | R |
|---|---|---|---|---|---|---|
| VGG16&SVM | 65.6 | 100 | 100 | 31.3 | **93.8** | **68.8** |
| CAE | 100 | **100** | 71.9 | **84.3** | 0 | 0 |
| Simple Similarity | **100** | 15.6 | **100** | 18.8 | 0 | 0 |

( C: Center, EL: Edge-Loc, ER: Edge-Ring, L: Loc, S: Scratch, R: Random)

Because their classes are similar to the Donut class, the Donut class is misclassified into Loc class. Results showed that the VGG16&SVM approach was able to classify all classes with high accuracy in classifying known classes, but it was unable to detect unknown classes well in classifiers that are similar to the donut defined as an unknown class. In the CAE approach, there was a significant difference in accuracy depending on the type of class. In particular, for the Scratch and Random classes, the classification accuracy for known classes were around 20% and the detection rate for unknown classes was 0%. This is largely influenced by whether the convolutional autoencoder was able to successfully restore the input image. Scratch and Random classes with small failure areas are not able to restore the input image well, therefore, the accuracy was poor. On the other hand, the Center, Loc, and Edge-Loc classes, which have relatively large failure areas, were able to restore well. So the classification accuracy of known classes and the detection rate of unknown classes were all able to obtain more than 80%. The approach of using simple similarity also varied in accuracy depending on the class. The reason for this is that a simple comparison cannot cover the case where the position of the failure chip area changes. However, for the Center and Edge-Ring classes, where the position of the failure chip area does not change, the average values of known classification accuracy and unknown detection rate were high. As mentioned above, we determined the most suitable method and the best classifier for each class.

The Precision and Recall for each determined classifier are shown in Table 3. In the proposed method, each classifier is combined in order of highest Precision, i.e., Scratch → Edge Ring → Edge Loc → Random → Center → Loc. Table 4 shows the overall known classification accuracy after combining each classifier, the unknown detection rate, and the percentage of unknown classes in the generated unknown clusters. Also, Table 4 compares the results and their respective accuracies when each classifier is created using only the VGG16+SVM approach, only the CAE approach, and only the Simple Similarity approach. In the proposed method, the known classification accuracy is 75.9%, the unknown detection rate is 62.5%, and the percentage of unknown classes in the generated unknown clusters is 35.7%. The proposed method that combines multiple approaches is the best in terms of all accuracies. In particular, it was found to be effective in detecting unknown classes and generating unknown cluster.

**\*\*\* Formatting Issue - Best Available Paper/Graphic \*\*\***

Table 3 Precision and Recall of Each Classifier (%)

|  | Classifier | | | | | |
|---|---|---|---|---|---|---|
|  | C | EL | ER | L | S | R |
| Precision | 62.2 | 70.6 | 86.8 | 46.2 | 90.9 | 67.6 |
| Recall | 92.6 | 83.3 | 94.2 | 70.7 | 84.3 | 84.5 |

( C: Center, EL: Edge-Loc, ER: Edge-Ring, L: Loc, S: Scratch, R: Random)

Table 4 Classification Accuracy (%)

|  | Approach | | | |
|---|---|---|---|---|
|  | VGG16&SVM only | CAE only | Simple Similarity only | Proposed Method |
| Classification Rate For Known Classes | 74.1 | 48.5 | 43.0 | 75.9 |
| Detection Rate For Unknown Classes | 9.3 | 0.0 | 0.0 | 62.5 |
| Accuracy Of Generated Unknown Clusters | 10.0 | 0.0 | 0.0 | 35.7 |

Table5 Confusion Matrix

|  |  | Predicted Label | | | | | | |
|---|---|---|---|---|---|---|---|---|
|  |  | C | EL | ER | L | S | R | U |
| True Label | C | 63 | 0 | 0 | 0 | 0 | 2 | 3 |
|  | EL | 0 | 70 | 3 | 1 | 3 | 5 | 2 |
|  | ER | 1 | 2 | 65 | 0 | 0 | 0 | 1 |
|  | L | 3 | 16 | 1 | 30 | 3 | 14 | 15 |
|  | S | 0 | 1 | 0 | 3 | 70 | 2 | 7 |
|  | R | 16 | 7 | 0 | 1 | 1 | 83 | 8 |
|  | D | 0 | 0 | 0 | 1 | 2 | 9 | 20 |

( C: Center, EL: Edge-Loc, ER: Edge-Ring, L: Loc, S: Scratch, R: Random, D:Donut, U:Unknown)

Table 5 shows the Confusion Matrix of the classification results using the proposed method. From Table 5, there are 15 cases where Loc class is misclassified as an unknown class, which decreases the percentage of unknown classes in the unknown cluster. The reason for this is the low recall of the created Loc classifier. Further improvement of the Loc classifier's accuracy is needed in the future.

### C. Further Experiment

To evaluate the accuracy in more practical situations using Dataset-3, we evaluated the accuracy when the Donut and Other classes are defined as an unknown classes. Table 6 shows the known classification accuracy, the unknown detection rate, and the percentage of unknown classes in the generated unknown clusters by using Dataset-3. Table 6 shows that the known classification accuracy, the unknown detection rate, and the percentage of unknown classes in the generated unknown clusters are 75.9%, 44.6%, and 58.1%, respectively. The unknown detection rate was about 20% lower when more practical data were used. On the other hand, the percentage of unknown classes in the generated clusters

improved by about 20%. This rate is the most important factor in the manufacturing process. We were able to create a cluster with 58% unknown classes, which means that it is effective in detecting unknown defects even in practical settings.

Table 6 Classification Accuracy [Dataset-3] (%)

| Classification Rate For Known Classes | 75.9 |
|---|---|
| Detection Rate For Unknown Classes | 44.6 |
| Accuracy Of Generated Unknown Clusters | 58.1 |

## V. CONCLUSION

### A. Conclusion and Future Tasks

In this study, we examined a method that simultaneously classifies known wafer maps and detects unknown wafer maps. For each class, 2-class classifiers were created by multiple methods and combined the best classifiers. The experimental results show that the classification accuracy of known classes is 75.9%, the detection rate of unknown classes is 62.5%, and unknown clusters containing 35.7% of unknown classes are successfully created. To improve yield, it is essential to detect unknown defects at an early stage. If we can generate clusters that are mostly composed of unknown classes, it will be possible to recognize the occurrence of unknown defects. Since the proposed method was able to generate clusters in which unknown classes account for about 35%, we believe that it is sufficient to detect the occurrence of unknown defects.

### REFERENCES

[1] T. Nakazawa and K. V. Deepak, Wafer map defect pattern classification and image retrieval using convolutional neural network, IEEE Transactions on Semiconductor Manufacturing, Vol.31, No.2, pp. 309-314, 2018

[2] S. Kang, Rotation-invariant wafer map pattern classification with convolutional neural networks, IEEE Access, 8, pp. 170650-170658, 2020

[3] H. Zheng et al., A Deep Convolutional Neural Network-Based Multi-Class Image Classification for Automatic Wafer Map Failure Recognition in Semiconductor Manufacturing, Applied Sciences, Vol.11, No.20, 9769, 2021

[4] WM-811K wafer map | kaggle, https://www.kaggle.com/qingyi/wm 811k-wafer-map

[5] K. Simonyan \& A. Zisserman, Very deep convolutional networks for large-scale image recognition, Proc. of ICLR, pp. 1-14, 2015

[6] A. J. Smola, and B. Schölkopf, A tutorial on Support Vector Regression, Statistics and Computing, vol.14, no.3, pp. 199-222, 2004.

[7] D. Bank, N. Koenigstein, R. Giryes, Autoencoders, ArXiv, 2003.05991, 2020

[8] LeNail, NN-SVG: Publication-Ready Neural Network Architecture Schematics. Journal of Open Source Software, 4(33), 747, 2019 (https://doi.org/10.21105/joss.00747)

YD-20

# Influence of High Temperature $N_2$ Annealing on Photoluminescence of SiC and Si Quantum Dots in $SiO_2$ Layer

Kohki Murakawa
*Dept. Science*
*Kanagawa University*
Hiratsuka, Japan
r201703321fu@jindai.jp

Norihito Mayama
*Physical Analysis Technology Center*
*Toshiba Nanoanalysis Corporation*
Yokohama, Japan
norihito.mayama@nanoanalysis.co.jp

Tomohisa Mizuno
*Dept. Science*
*Kanagawa University*
Hiratsuka, Japan
mizuno@kanagawa-u.ac.jp

*Abstract*—We experimentally studied the influence of high temperature $N_2$ annealing on the photoluminescence (PL) of SiC and Si quantum-dots (QDs) in $SiO_2$ layer fabricated by hot ion implantation technique. We demonstrated the increase of PL intensity of SiC- and Si-QDs after $N_2$ annealing, compared with that after Ar annealing, which is probably attributable to the reduction of dangling bond density at $SiO_2$/QD interface terminated by N atom trapping.

*Keywords— quantum dot, IV semiconductor, hot ion implantation, $N_2$ annealing, photoluminescence*

## I. INTRODUCTION

Recently, we experimentally developed group IV semiconductor quantum-dots (IV-QDs) of SiGe-, Si-, SiC-, and C-QDs in $SiO_2$ layer (OX) for photonic devices with diameter $\Phi$ of 2–4 nm and QD density of approximately $2\times10^{12}$ cm$^{-2}$, using very simple processes of hot ion implantation into OX layer and post $N_2$ annealing [1]–[4]. The IV-QDs were fabricated by the clustering effects of implanted ions in OX layer [1]–[4]. We also demonstrated strong photoluminescence (PL) from IV-QDs in a wide wavelength from near-UV to near-IR (370–940 nm) which is owing to the quantum mechanical effects of generated electrons in QDs, although bulk IV-semiconductors have indirect bandgap structures [1]–[4]. The PL intensity $I_{PL}$ from IV-QDs drastically increased with increasing the post $N_2$ annealing time $t_N$. In addition, the PL peak energy $E_{PH}$, which is equal to the bandgap $E_G$ of IV-QD, revealed the expanded $E_G$ values in the case of Si- and SiGe-QDs which is attributable to the quantum confinement of electrons in QD, that is, $E_G \propto \Phi^{-2}$ [1]–[4], [6]–[8].

In this work, we experimentally revealed the influence of high-temperature $N_2$ annealing on the $I_{PL}$ increase of SiC- and Si-QDs fabricated by hot ion implantation technique in OX layer.

## II. EXPERIMENTAL PROCEDURE

Figure 1 shows fabrication steps for IV-QDs, where Table 1 shows the process conditions for IV-QDs. SiC- and Si-QDs were fabricated by double $Si^+/C^+$ and single $Si^+$ hot ion implantation into OX layer, respectively, as shown in Fig. 1(b), after thermal dry-oxidation of Si substrate at 1000 °C shown in Fig. 1(a). Figure 1(c) shows that post $N_2$ annealing was carried out at annealing temperature $T_N$ for annealing time $t_N$ to recover the QD crystal quality. Post Ar (an insert gas) annealing was also performed at annealing temperature $T_A$ instead of $N_2$ annealing, as a reference. Both $T_N$ and $T_A$ were varied from 900 to 1100 °C. Process conditions for Si dose ($D_S$), C dose ($D_C$), and hot ion implantation temperature ($T$) are summarized in Table I.

Material structures for IV-QDs were analyzed by Atom Probe Tomography (APT), Secondary Ion Mass Spectrometry (SIMS), High-angle Annular Dark Field (HAADF) Scanning Transmission Electron Microscopy (STEM), and Corrector-Spherical Aberration TEM (CsTEM). PL properties of SiC- and Si-QDs were measured at room temperature, using excitation 325-nm (He-Cd) and 532-nm (green) lasers, respectively, where the laser beam diameter was 1 μm.

## III. RESULTS AND DISCUSSION

### A. Material Structures of QD

Figures 2 revealed the clear lattice images of the cross sections of hexagonal SiC-QD, respectively, which indicates that single crystalized SiC-QD were successfully formed, where average diameter ($\Phi \approx 2$ nm) and QD density ($N_{QD} \approx 2\times10^{12}$ cm$^{-2}$) of QD [2]. In addition, Figs. 3(a) and 3(b) show the CsTEM images of cross section of Si-QD before and after $N_2$ annealing, respectively, and $\Phi \approx 2$–3 nm after $N_2$ annealing. The lattice patterns of Si-QDs after $N_2$ annealing in Fig. 3(b) revealed single-crystal, although the lattice patterns before $N_2$ annealing in Fig. 3(a) were imperfect like amorphous. Thus, the crystal quality of Si-QD could be improved by high temperature $N_2$ annealing.

In addition, Fig. 4(a) show the APT results of plane view distributions of Si (blue dots), C (yellow dots), and N atoms (red dots) of SiC-QD at the projection range depth of ions (80 nm depth from the OX surface). It is noted that N atoms in SiC-QD could be observed after $N_2$ annealing without $N^+$ ion implantation. N atoms also clustered in OX layer, similar to Si (blue dots) and C atoms (yellow dots). Figures 4(b) and 4(c) show the atomic concentration contour-maps of N and Si atoms at the same data of Fig. 4(a), respectively. High density N atoms ($> 3\times10^{20}$ cm$^{-3}$) were observed and the cluster position of N atoms of Fig. 4(b) with the cluster size of approximately 3 nm nearly agreed with that of Si-atoms in Fig. 4(c), which suggests that N atom clustering are correlated with SiC-QD formation.

We also verified the N atoms evaluated by SIMS. Figure 5(a) shows the lateral concentration distribution of N atoms at $Y = 0$ in Fig. 4(b) evaluated by APT, which indicates that the N atom concentration was varied from $1.5 \times 10^{20}$ cm$^{-3}$ to 3.5

978-1-6654-7134-3/22 $31.00 © 2022 IEEE

$\times 10^{20}$ cm$^{-3}$ of N atom cluster region. In addition, Fig. 5(b) shows the SIMS analysis for SiC-QD, which revealed high density N atoms (> $3\times10^{20}$ cm$^{-3}$) after N$_2$ annealing (solid line) in OX layer after only Si$^+$/C$^+$ hot ion implantation into OX, and the peak N atom concentration was nearly equal to that by APT in Fig. 5(a). However, the Fig. 5(b) shows the dashed line without SiC-QD formation owing to Si$^+$/C$^+$ hot ion implantation revealed no N atom within the SIMS detection limit even after N$_2$ annealing, where the dotted line shows the Si-atom profile of SiC-QD. Thus, Fig. 5(b) suggests that N atoms are trapped near and inside SiC-QD.

Thus, we introduce a N atom trap model as shown as a schematic cross section of IV-QD in SiO$_2$ in Fig. 6, and an imperfect crystallization of hot ion implanted atoms in QDs causes vacancies which creates dangling bonds (shown as crosses) inside QD and at the OX/QD interface [3]. Thus, during high temperature N$_2$ annealing, some parts of the defects or dangling bonds at the OX/QD interface are terminated by the trapping of N atoms [5], which leads to the crystal quality improvement of QDs.

*B. PL Intensity Increase after N$_2$ Annealing*

Figures 7(a) and 7(b) show the annealing process dependence of PL spectra of SiC-and Si-QDs, respectively. The $I_{PL}$ of both SiC- and Si-QDs increased after N$_2$ (solid line) and Ar (dashed line) annealing, compared with that before annealing (dotted line). In addition, the $I_{PL}$ after N$_2$ annealing ($I_{PL}^N$) was much larger than that after Ar annealing ($I_{PL}^A$) even at the same conditions of annealing temperature $T$ and time $t$. Here, the maximum $I_{PL}^N$ ($I_{MAX}^N$) increase factor compared with the $I_{MAX}^A$ of peak $I_{PL}^A$ is defined by Eq. (1).

$$\Delta I_{MAX} \equiv \frac{I_{MAX}^N}{I_{MAX}^A} \tag{1}$$

The $\Delta I_{MAX}$ of Si-QD reached approximately 2.11, and had much larger than $\Delta I_{MAX}$ of SiC-QD ($\approx 1.13$).

Hereafter, we mainly discuss the N$_2$ annealing effects on the PL properties of Si-QD with larger $\Delta I_{MAX}$. Figures 8 shows the annealing time dependence of $I_{MAX}$ of Si-QD after N$_2$ (circles) and Ar (triangles) annealing at $T = 1000$ °C, respectively. Both $I_{MAX}^N$ and $I_{MAX}^A$ continued increasing with increasing annealing time, but $I_{MAX}^N > I_{MAX}^A$ under $t > 60$ min. As a result, $\Delta I_{MAX}$ (squares) under $t > 60$ min was greater than 1.5 and the maximum $\Delta I_{MAX}$ reached $\sim 2.1$. Thus, the $I_{PL}$ increase of IV-QDs after N$_2$ annealing may be attributable to the QD quality improvement owing to the trapped N atom-induced reduction of dangling bond density at QD/OX interface and inside QD [3]-[5], as discussed as high trapped density of N atoms shown in Figs. 3−5, and a schematic model for N atom terminating dangling bonds in Fig. 6. On the other hand, the $I_{PL}$ increase of IV-QDs after high temperature Ar annealing is possibly due to the recovery of QD crystal quality.

In addition, Fig. 9 shows the annealing time dependence of PL peak energy $E_{PH}$ of Si-QD after N$_2$ (circles) and Ar (triangles) annealing at $T = 1000$ °C, respectively, and revealed that the $E_{PH}$ (1.55−1.75 eV) was demonstrated to be much higher than the $E_G$ of bulk-Si (1.1 eV), which is attributable to the $E_G$ expanding in QD owing to $E_G \propto \Phi^{-2}$ [6]−[8]. The $E_{PH}$ was nearly independent of annealing time $t$ under $t \geq 5$min and equals to the expanding $E_G$ value at $\Phi \approx 3$ nm ever reported shown as the dashed arrow in Fig. 9 [7]. On the other hand, high $E_{PH}$ before annealing shown in Fig. 3(a)

is possibly attributable to large $E_G$ of amorphous Si ($\approx 1.6$ eV) [8]. In addition, the $E_{PH}$ after N$_2$ annealing was 0.02 eV higher than that after Ar annealing, whose physical mechanism is not understood now.

To evaluate the physical mechanism for N$_2$ annealing on PL properties of QDs, we carried out the annealing temperature $T$ dependence of PL of Si-QD, as shown as the Arrhenius plot for $I_{MAX}$ vs $1000/T$ at a fixed annealing time, as shown in Fig. 10. Both N$_2$ (Fig. 10(a)) and Ar (Fig. 10(b)) annealing data revealed that the $I_{MAX}$ significantly increased with increasing $T$. In addition, the $I_{MAX}$ could be explained by the following Arrhenius plot of $I_{MAX}$ in Eq. (2):

$$I_{MAX} \propto \exp\left(-\frac{E_A}{kT}\right) \tag{2}$$

where $E_A$ is an activating energy and $k$ is Boltzmann constant. The $E_A$ of N$_2$ data was 2.4 eV, which was nearly equal to the activation energy (2.3 eV) for the regrowth of amorphous Si (a-Si) by solid phase epitaxy (SPE) [9]. Thus, during high-$T$ N$_2$ annealing, the crystal quality of Si-QD was improved by the singly-crystallizing of a-Si before N$_2$ annealing, which leads to the long lifetime ($\tau$) of generated electrons in IV-QD. As a result, the PL intensity increases by the PL quantum coefficient ($\eta$) improvement, because $\eta \propto \tau$ [8]. On the other hand, the $E_A$ of Ar data (1.3 eV) was lower the $E_A$ of SPE, and thus, the crystal quality improvement during Ar annealing was imperfect to regrowth of a-Si, whereas Figs. 7 and 8 showed the $I_{PL}$ improvement even in Ar annealing. Thus, the $\Delta I_{MAX}$ of Eq. (1) increased with increasing $T$, and reached to approximately 5 at $T = 1100$ °C

## IV. CONCLUDION

In this study, we investigated the influence of high temperature N$_2$ annealing on the PL properties of IV-QDs fabricated by hot ion implantation into OX layer. Using APT and SIMS analyses, we experimentally demonstrated the N atom trapping in IV-QDs after high temperature N$_2$ annealing. Large PL intensity of IV-QDs was achieved by the QD quality improvement owing to N$_2$ annealing induced reduction of dangling bond density of IV-QDs. Consequently, it is necessary to optimize the N$_2$ annealing process, such as N$_2$ annealing temperature, to improve the PL intensity of IV-QDs.

### ACKNOWLEDGMENT

We would like to thank Dr. A. Shimazaki of Toshiba Nanoanalysis Corporation for continuous support of this work.

### REFERENCES

[1] T. Mizuno, R. Kanazawa, T. Aoki, and T. Sameshima, " SiC Quantum dot formation in SiO$_2$ layer using double hot-Si$^+$/C$^+$-ion implantation technique", *Jpn. J. Appl. Phys.*, vol. 59, SGGH02, 2020.

[2] T. Mizuno, R. Kanazawa, K. Yamamoto, K. Murakawa, K. Yoshimizu, M. Tanaka, T. Aoki, and T. Sameshima, " Group-IV-semiconductor quantum-dots in thermal SiO$_2$ layer fabricated by hot-ion implantation technique: different wavelength photon emissions", *Jpn. J. Appl. Phys.*, vol. 60, SBBK08, 2021.

[3] T. Mizuno, K. Murakawa, K. Yoshimizu, T. Aoki, and T. Sameshima, " Physical mechanism for photon emissions from group-IV-semiconductor quantum-dots in quartz-glass and thermal-oxide layers ", *Jpn. J. Appl. Phys.*, 61, SC1014, 2022.

[4] T. Mizuno, K. Murakawa, H. Ban, T. Aoki, and T. Sameshima, "Near-IR and UV Photon Emissions from SiGe- and C-Quantum-Dots Fabricated by Hot-Ion Implantation into Si-Oxide Layers", in *Extended Abst. of International Conference on Solid State Devices and Materials*, p.605, 2022.

[5] Keita Tachiki, Mitsuaki Kaneko , Takuma Kobayashi , and Tsunenobu Kimoto, "Formation of high-quality SiC(0001)/SiO₂ structures by excluding oxidation process with H₂ etching before SiO₂ deposition and high-temperature N₂ annealing", *Appl. Phys. Express*, vol. 13, 121002, 2020.

[6] S. Takeoka, K. Toshikiyo, M. Fujii, S. Hayashi, and K. Yamamoto, "Photoluminescence from $Si_{1-x}Ge_x$ alloy nanocrystals", *Phys. Rev. B*, vol.61, 15988 (2000).

[7] S. Takeoka, M. Fujii, and S. Hayashi, "Size-dependent photoluminescence from surface-oxidized Si nanocrystals in a weak confinement regime", *Phys. Rev. B*, vol.61, 16820 (2000).

[8] S. M. Sze and K. K. Ng, Physics of Semiconductor Devices, Wiley, New York, 2007.

[9] S. M. Sze, VLSI Technology, McGraw-Hill, New York, 1988.

TABLE I.     PROCESS CONDITIONS FOR SiC-QD AND Si-QD

| QD | $D_S$ (x10¹⁶cm⁻²) | $D_C$ (x10¹⁶cm⁻²) | $T$ (°C) | Anneal (°C) |
|---|---|---|---|---|
| SiC | 6 | 4 | 200 – 400 | 900 – 1100 |
| Si | 3 | | 600 | |

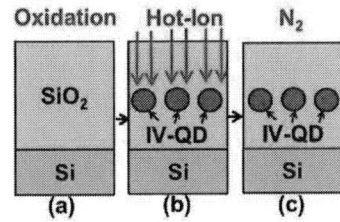

Fig. 1. Fabrication steps for IV-QD in SiO₂ layer, using hot ion implantation technique. (a) Oxidation of Si substrate. (b) Hot ion implantation into SiO₂. (c) Post N₂ annealing at 1000°C.

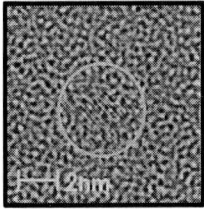

Fig. 2. Clear lattice images of hexagonal SiC-QD encircled by yellow circle observed by CsTEM after N₂ annealing, where $T_N$ = 1000 °C and $t_N$=60 min.

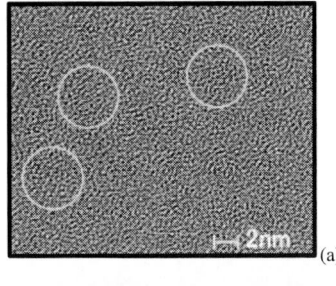

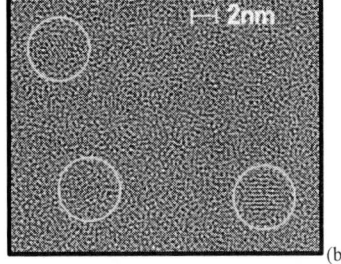

Fig. 3. CsTEM images of Si-QD encircled by yellow circles (a) before and (b) after N₂ annealing at $T_N$ = 1000 °C and $t_N$=150 min.

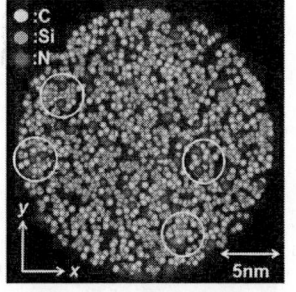

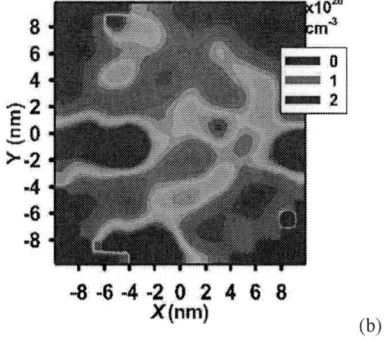

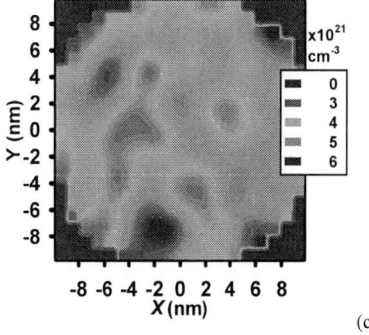

Fig. 4. APT analyses for SiC-QD at the peak atomic density of 80 nm depth from the OX surface of (a) plane-distribution of Si (blue dots), C (yellow dots), and N atoms (red dots), and atomic concentration contour-maps of (b) N atom and (c) Si atom, where the depth $z$=80nm from OX surface. White circles in Fig. 4(a) show the clustering areas of Si, C, and N atoms.

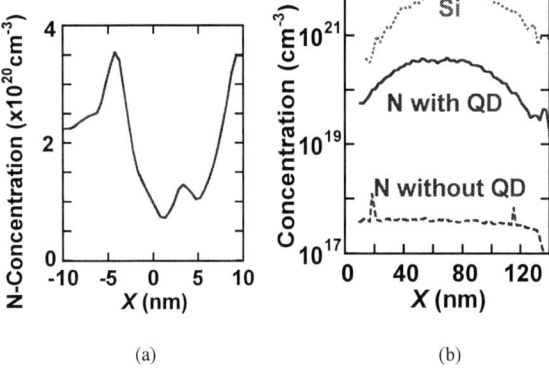

Fig. 5. (a) Lateral concentration distribution of N atom evaluated by APT in Fig. 4(b). (b) SIMS analysis for depth profiles of N atom after post N₂ annealing at $T_N$ = 1000 °C and $t_N$=150 min in OX layer with (solid line) and without (dashed line) SiC-QD formation. Dotted line shows implanted Si profile.

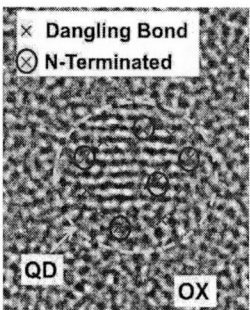

Fig. 6. N atom termination model for the density reduction of dangling bonds at OX/QD interface and inside QD (yellow circle), where crosses show the dangling bonds and encircled crosses indicate N atom terminated dangling bonds.

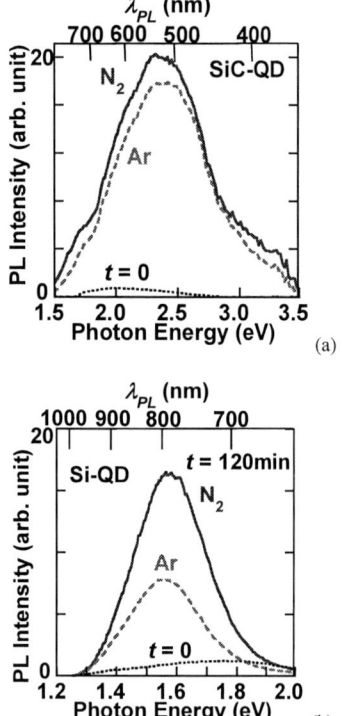

Fig. 7. PL spectra of (a) SiC-QD at $t = 5$ min and (b) Si-QD at $t = 120$ min after N$_2$ (solid line) or Ar annealing (dashed line), where $T = 1000$ °C. Dotted lines show the data before annealing.

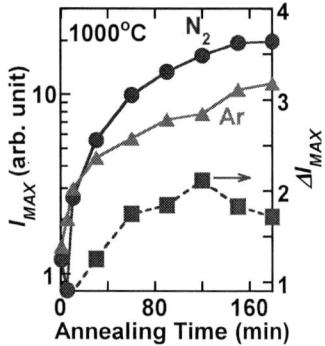

Fig. 8. Annealing time dependence of $I_{MAX}$ after N$_2$ (circles) and Ar (triangles) annealing of Si-QD at $T = 1000$ °C. Squares show $\Delta I_{MAX}$ of Eq. (1).

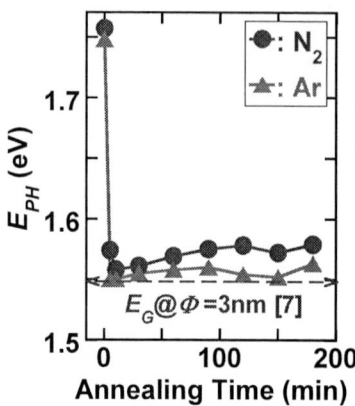

Fig. 9. Annealing time dependence of $E_{PH}$ after N$_2$ (circles) and Ar (triangles) annealing of Si-QD at $T = 1000$ °C. Dashed arrow shows the expanding $E_G$ value of Si-QD at $\Phi \approx 3$ nm owing to quantum confinement effects of electrons in QD [7].

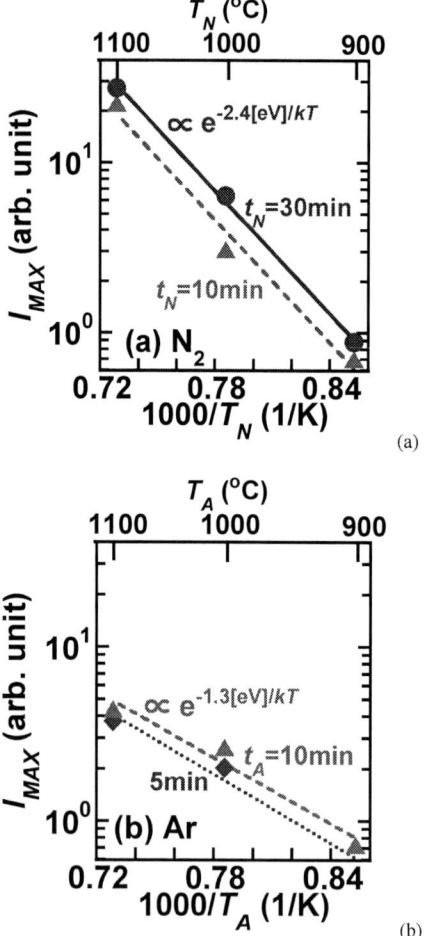

Fig. 10. Arrhenius plot for $I_{MAX}$ vs $1000/T$ of Si-QD after (a) N$_2$ and (b) Ar annealing, where $t_N$=5min (rhombi), 10min (triangles), and 30min (circles). Data can be well fitted by approximation lines of $I_{MAX} \propto \exp(-E_A/kT)$ of Eq. (2) with the correlation coefficient of approximately 1. Solid, dashed, and dotted lines show approximate exponent functions at $t$=30min, and 10min, and 5min, respectively. The activation energies of Si-QD after N$_2$ and Ar were approximately 2.4 and 1.3 eV, respectively.

# Noise Reduction in SEM Images using Deep Learning

Yuki Sato
Tokyo Electron Ltd.
Sapporo, Japan
yuki.sato@tel.com

Masato Kazui
Tokyo Electron Ltd.
Sapporo, Japan
masato.kazui@tel.com

Shinji Kobayashi
Tokyo Electron Ltd.
Koshi, Japan
shinji1.kobayashi@tel.com

*Abstract*— Measurement of patterns formed on wafers is required for defect inspection in mass production and for pattern quality evaluation in research and development. Scanning electron microscope (SEM) images are used for pattern measurement. The number of SEM scans must be reduced because of the incidents such as reduced throughput and damage to the resist. However, frame average images from fewer SEM images are noisy, and the noise makes it difficult to measure the pattern. In our proposed method, a deep learning was trained to perform noise reduction to measure patterns from noisy SEM images. Denoised images using the proposed method were evaluated with a 256-frame average image as a pseudo-correction image. The evaluation was made with PSNR and SSIM image quality evaluation, and with RMSE and power spectral density (PSD) of edge positions estimated using the tool. The results of noise reduction of single-frame image with proposed method were PSNR 32dB, SSIM 0.91, and RMSE 0.43nm, and showed high image quality and high accuracy in edge position estimation. With proposed method, an unbiased PSD-like graph with no noise floor was obtained. In addition, there is no significant difference between PSD graphs using single-frame images and 16-frame average images. These results indicate that proposed method can effectively remove noise from a few-frame average images, and that the denoised images can be used for pattern measurement and roughness evaluation using PSD.

*Keywords—measurement, SEM, noise reduction, denoise, machine learning, deep learning, PSD*

## I. INTRODUCTION

In order to improve the performance of semiconductors, miniaturization of patterning is required. In recent years, EUV lithography has been used to mass produce fine patterns. However, for fine patterns of 17 nm or less, the line edge roughness (LER) is relatively large compared to the line width, which significantly affects transistor performance. Therefore, high-precision pattern measurement is becoming more important to improve the yield of semiconductors.

A scanning electron microscope (SEM) is used for pattern measurement. SEM irradiates an imaging target with a beam of electrons and images the shape of it based on the electron dose returned from it. Since SEM images are noisy, frame average (FA) images are created by accumulating images of the same location. The more frames are averaged, the less noise of in-frame average image. However, increasing number of scans causes problems such as a decrease in throughput, deformation of the resist, and electrification. As patterns become finer, the effect of resist deformation increases. Therefore, it is required to measure patterns with high accuracy and fewer in number of imaging times. However, frame average images created from a small number of frames have a large amount of noise, and a large amount of noise reduces the accuracy of pattern measurement. Noise must be removed for accurate pattern measurement.

Noise reduction (NR) is used to stabilize evaluations using SEM images. Conventional NR methods are rule-based and include filtering methods such as Box filter and Gaussian filter, and non-local methods such as non-local means (NLM) and Block-matching and 3D (BM3D). However, these conventional methods have a problem of distorting edge roughness as the same time as NR, so they are unsuitable for pattern measurement.

In recent years, many deep learning (DL) methods have been proposed, and NR using DL (DL-NR) can achieve higher performance than conventional NR methods[1]. General DL-NR model is trained using noisy images as source images and clean images without noise as target images. However, since SEM cannot acquire clean images, general DL-NR cannot be trained with SEM images. On the other hand, Noise2Noise (N2N) [2], which trains DL-NR using only noisy images, is a method when clean images cannot be acquired. When training DL-NR based on Noise2Noise, two images that differ only in noise distribution are used as the source and target images in pairs. Since SEM images the same location multiple times, it is possible to acquire images that can be assumed to differ only in noise distribution.

The prior art, NR of SEM images by Noise2Noise for the purpose of NR of logic and line/space patterns in one net-work, achieved NR of both patterns by training to mix them [3]. SEM image quality varies significantly depending on imaging conditions such as electron beam scanning speed, resolution, and number of integrations, but no research has been conducted to train Noise2Noise considering these conditions.

## II. METHOD

The purpose of this study is to train a DL network model for NR considering multiple conditions during SEM imaging. The network of Noise2Noise-based learning method was used and SEM images were used as the training dataset.

### A. Implementation

The network architecture is a u-net and the various network parameters are as follows. The optimizer is Adam with $\beta 1$ as 0.9, $\beta 2$ as 0.99, and $\varepsilon$ as 10E-8. The learning rate is 0.001. The minibatch size is 16. The input images are cropped images of 128x128. The training epochs is set up to 500 and uses early-stopping with patience value 100.

### B. Datasets

The training dataset used SEM images mixed with multiple conditions during SEM imaging, including scanning speed, image resolution, and number of frames to be averaged. The imaging conditions were as follows: the electron beam scanning speeds, which had a trade-off relationship with image quality, were 1x, 2x, and 4x, and the image resolutions were 512x512, 1024x1024, and 2048x2048. The maximum number of cumulative images was 32, which was commonly

978-1-6654-7134-3/22 $31.00 © 2022 IEEE

used, and set to 1, 2, 4, 8, 16, 32 so that it could correspond to a smaller number of images.

### C. Correction of image shift

To improve the accuracy of network training, we corrected the SEM device-derived imaging position shift that occurred in SEM images. Even if the same location is scanned by SEM, electron beam control can be shifted by about 10 nm. As a result, SEM images taken at the same location have a slightly misalignment. Since Noise2Noise assumes training with identical images except for the noise distribution, the training accuracy can be improved by correcting for image shifts. Methods to correct for image shift include pattern matching and phase-only correlation [4], and in this study, phase-only correlation was used to do it.

### D. Selection of training image pairs

Fig.1 shows how to select image pairs for training. Training image pairs are selected from the same imaging conditions. For example, if there are data sets such as Data Set 1 (Line/space, 1x speed, 512x512, 1 frame images) and Data Set 2 (Hole, 2x speed, 1024x1024, 2-frame average images), a training image pair is selected from Data Set 1 or Data Set 2, without crossing data sets. In addition, an image taken at the next timing of the source image is selected as the target image to reduce the effect of resist deformation between paired images.

### III. RESULTS

To show how the DL-NR algorithm contributes to improving the accuracy of pattern measurement, NR images using proposed method were compared with pseudo-correct images. The 256-frame average images were used as pseudo-correct images. For the NR image quality evaluation, peak signal to noise ratio (PSNR) and structural similarity index (SSIM) [5] were used. For quantitative evaluation, LER (3σ) and RMSE were used, which was calculated from edge positions estimated by use of a tool.

### A. Noise reduction results

Fig.2 shows the NR images. It indicates that the proposed method was able to effectively reduce noise without disturbing line/space and hole edges while suppressing the loss of LER, even when the input images were not averaged. In addition, when the input images were large number of frame average image, such as 16-frame average image, edges could be represented with higher accuracy. On the other hand, the conventional method was unable to achieve both NR and reproduction of pattern edges in NR images of 1-frame image or 4-frame average image. When the input images were 16-frame average image, the noise were reduced without disturbing edges.

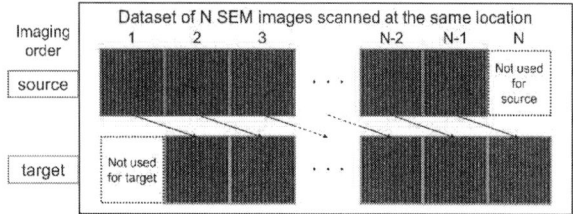

Fig. 1. How to select source and target images in pairs. Two images that are consecutive in time are considered as a pair.

### B. Evaluation

Fig.3(a)-(d) show PSNR and SSIM. In line/space pattern, the proposed method were PSNR=31.95 and SSIM=0.909 even when the input was single-frame image, PSNR=36.25 and SSIM=0.919 when the input was 4-frame average image, and PSNR=36.40 and SSIM=0.922 when the input wad 16-frame average image. On the other hand, the conventional method, BM3D, showed PSNR=31.05 and SSIM=0.850 when the input was single-frame image, PSNR=34.21 and SSIM=0.890 when the input was 4-frame average image, and PSNR=36.25 and SSIM=0.902 when the input was 16-frame average image.

Table 1 shows LER and RMSE of edge positions. The proposed method were LER=1.18 and RMSE=0.43 even when the inputs was single-frame image, LER=1.19 and RMSE=0.34 when the input was 4-frame average image, and LER=1.18 and RMSE=0.27 when the input was 16-frame average image. The conventional method BM3D showed LER=2.24 and RMSE=0.79 when the input was single-frame image, LER=1.54 and RMSE=0.44 when the input was 4-frame average image, and LER=1.32 and RMSE=0.28 when the input was 16-frame average image. Incidentally, the correct image had LER=1.35.

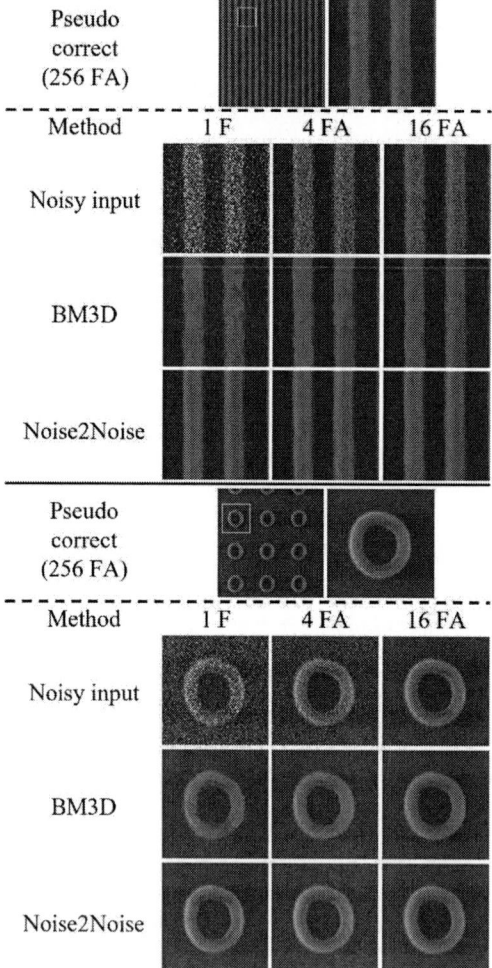

Fig. 2. Noise reduction results. Noisy input and NR images of 1-frame image, 4-frame average image, and 16-frame averaged image, and the pseudo-correct images are shown.

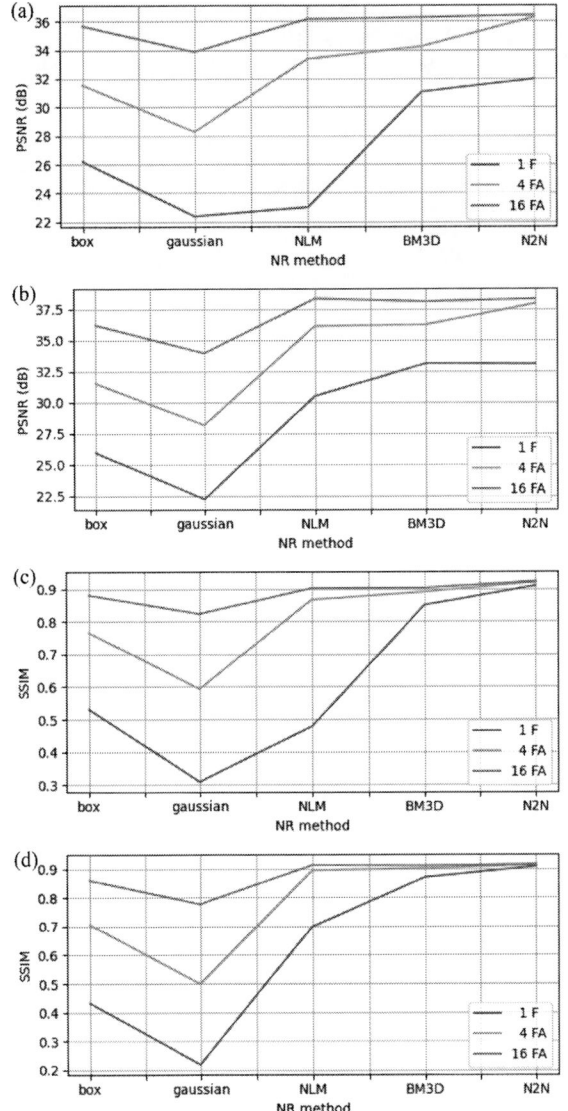

Fig. 3. PSNR and SSIM, (a) PSNR for line/space, (b) PSNR for hole, (c) SSIM for line/space, (d) SSIM for hole.

## C. Power spectral density

Fig.4(a)-(b) show the PSD graphs. The PSD of LER was calculated from NR images using the proposed method or from noisy images. When the proposed method was used, the graph does not have a flat area such as noise floor, and is less disturbed than when the conventional method was used. Regarding graph shape, the graph from the NR images varies little with the number of images for frame averaging, while the unbiased PSD graph calculated from noisy images shows variation in the high frequency area depending on the number of images for frame averaging. Spikes that do not occur in the other methods appear in the graphs with the proposed method.

## IV. DISCUSSION

The comparison results of the NR images show that the proposed method outperforms the conventional methods. The proposed method was able to remove noise without edge distortion regardless of the number of frame average in the

TABLE I.    LER AND RMSE OF EDGE POSITIONS, SHOWN AS LER / RMSE.

| NR method | Number of input frames | | |
| --- | --- | --- | --- |
| | *1 frame* | *4-frame average* | *16-frame average* |
| Proposed | 1.181 / 0.431 | 1.187 / 0.338 | 1.182 / 0.265 |
| BM3D | 2.244 / 0.789 | 1.538 / 0.439 | 1.322 / 0.276 |

NR input image. The image quality evaluation using PSNR and SSIM also shows that the proposed method performs better. The image quality evaluation of the proposed method using single-frame images as input was equal to or better than that of the conventional method using 16-frame average images as input. These results indicate that the proposed method can be used for visualized evaluation such as defect inspection with the NR images of a few-frame average images.

The LER and RMSE of edge positions show that the proposed method can reconstruct edges with higher accuracy than conventional methods. The RMSE of the proposed method is 0.43 nm even when using a single-frame image as input, so that means a very small error with the pseudo-correct image. In addition, the LER is 1.18 regardless of the average number of frames, and this indicates that edge restoration is stable. The PSD graph also has almost the same shape regardless of the average number of frames. These results as mentioned show the possibility of pattern quality evaluation by PSD of NR images using the small number of frame averaged images. However, the cause of the peak appearing in the PSD graph by the proposed method has not been identified, and further investigation is required.

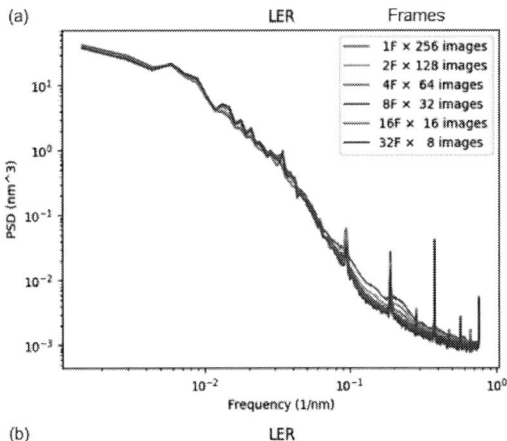

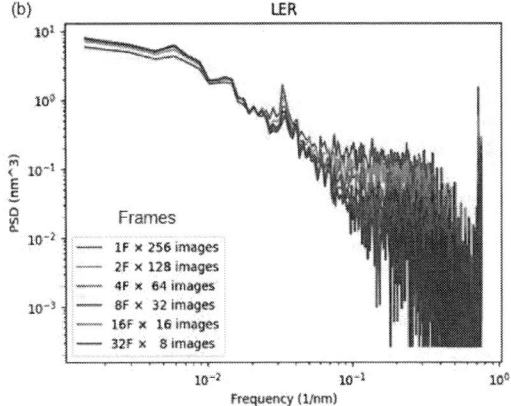

Fig. 4. PSD graphs of LER, (a) PSD graph of NR images used the proposed method, (b) unbiased PSD graph from noisy image.

## V. CONCLUSION

We trained the noise reduction deep learning model that is for variations in patterns such as line and hole, and for variations in SEM imaging conditions such as scanning speed, image resolution, and the number of frame average. The network was the Noise2Noise learning method, and it was trained using SEM images with a mixture of the pattern variations and imaging condition variations. The images used for training were corrected for the shift of the image position derived from the SEM. Using 256-frame average images as the correct image, we compared the noise reduction effect of the proposed method with that of the conventional method. The proposed method was able to obtain higher image quality evaluation values and better noise reduction than the conventional method. By using noise reduction images with the proposed method, unbiased PSD-like graphs without noise floors were obtained. In addition, there was no difference in the obtained PSD graphs between the single-frame images and the 16-frame average images. These results indicate that the proposed method can improve pattern measurement accuracy and throughput by denoising a few-frame average images and enabling pattern measurement and roughness evaluation using PSD from noise reduction images.

## REFERENCES

[1] K.Zhang, W.Zuo, Y.Chen, D.Meng, and L.Zhang, "Beyond a Gaussian Denoiser: Residual Learning of Deep CNN for Image Denoising," IEEE Trans Image Process, vol.26, pp.3142–3155, 2017.

[2] J.Lehtinen, J.Munkberg, J.Hasselgren, S.Laine, T.Karras, M.Aittala, and T.Aila, "Noise2Noise: Learning Image Restoration without Clean Data," ICML J Mach Learn Res, pp.2965–2974, 2018.

[3] L.Hairong, T.Cho, Y.Liangjiang, F.Gino, P.Lingling, and F.Wei, "Denoising sample-limited SEM images without clean data," Proc. SPIE Advanced Lithography 116111A, 2021.

[4] K.Kobayashi, H.Nakajima, T.Aoki, M.Kawamata, and T.Higuchi, "Principles of Phase Only Correlation and Its Applications," ITE TechnicalReport, Vol.20, no.41, pp.1–6, 1996.

[5] Z.Wang, A.Bovik, H.Sheikh, and E.Simoncelli, "Image Quality Assessment : From Error Visibility to Structural Similarity," IEEE Transactions on Information Theory, vol.13, no.4, pp.600–612, 2004.

# Automatic classification of C-SAM voids for root cause identification of bonding yield degradation

Julien Baderot
Technology Department
Pollen Metrology
38430 Moirans, France
julien.baderot@pollen-metrology.com

Solange Garrais
Technology Department
Pollen Metrology
38430 Moirans, France
solange.garrais@pollen-metrology.com

Sergio Martinez
Technology Department
Pollen Metrology
38430 Moirans, France
sergio.martinez@pollen-metrology.com

Johann Foucher
CEO
Pollen Metrology
38430 Moirans, France
johann@pollen-metrology.com

Ryuji Eto
Process Technology Center
Tower Partners
Semiconductor Co., Ltd.
800 Higashiyama, Uozu
City, Toyama 937-8585,
Japan
eto.ryuji@tpsemico.com

Kazumasa Tanida
Process Technology Center
Tower Partners
Semiconductor Co., Ltd.
800 Higashiyama, Uozu
City, Toyama 937-8585,
Japan
kazumasa.tanida@tpsemico.com

Takatoshi Yasui
Process Technology Center
Tower Partners
Semiconductor Co., Ltd.
800 Higashiyama, Uozu
City, Toyama 937-8585,
Japan
yasui.takatoshi@tpsemico.com

Tomoya Tanaka
Process Technology
Development Group
Tower Partners
Semiconductor Co., Ltd.
Uozu City, Toyama 937-8585, Japan
tanaka.tomoya@tpsemico.com

*Abstract*—**Wafer-level direct bonding technology is a key process for the production of backside illuminated (BSI) CMOS image sensor (CIS). Usually, constant-depth mode scanning acoustic microscope (C-SAM) 300mm wafer images are acquired and defect size distribution is provided to monitor defects that degrade bonding yield. Current solutions are not providing information detailed enough to identify the root cause of this degradation. In this paper, we propose a rule-based method for the classification of the defects and automatic segmentation of the defects to extract precise measurements depending on the type of defect. All these information will allow to reduce the time to analyze the images and improve the precision and consistency of the analysis.**

*Keywords—defects, classification, segmentation, voids*

## I. INTRODUCTION

Recently, wide variety of backside illuminated (BSI) CMOS image sensor (CIS) has been developed and in mass production [1]. One of the key processes for BSI is wafer-level direct bonding technology. It requires no defects between the two bonded wafers, such as voids, particles, and so on. In order to detect these defects, constant-depth mode scanning acoustic microscope (C-SAM), which has been often used for bump and under fill of chip-on-chip technology (COC), is also used. Usually, C-SAM inspector provides 300 mm wafer image and defect size distribution on inspected wafer. However, only this information is insufficient to find suspicious root cause and/or to judge if the inspected wafer can be proceeded to the next process step. That is because some large defects have little impacts on yield and vice versa. It requires very careful inspection and classification of defect properties such as shape, contrast, and so on, to detect all of defects and to find suspicious root cause of defects. These inspections and its investigations are very problematic due to time consuming and human error variation when it is done manually. In this work, we describe the development of high-performance classifier for C-SAM wafer images.

## II. METHODOLOGY

The first part of this work aims to reproduce the manual work done by an engineer with the automatic classification of the defect to reduce the time of analysis and improve the repeatability. Once the defects are correctly detected and classified, we propose to push further the metrological analysis by introducing new measurements that are either not possible or repeatable manually.

### A. Classification method

For this application, the analysis of defect is performed on a complete 300mm wafer image. A previous manual analysis paired with the expertise of the process revealed several types defect with various impact on the yield. The current classification contains only two main types of defects concerning the yield: the killer ones and the others. Nonetheless, the definition of a killer defect can be caused by the size of the defect, their spatial organization or their number. So, it is better to relate on the defect about their origin or appearance. To start with the most critical ones, defects that are killer because of their location and size, we have scratch, edge and delamination defects. Scratch are scarce defects that appear on the full wafer in the form of an arc of a circle and is caused by the chemical mechanical polishing (CMP) process abnormality producing a groove on the wafer [2]. The second one, delamination, which is rare too, appears as a blob at the boundary of the wafer caused by a defect such as a particle or an abnormality of wafer edge shape. Finally, edge defects which are a bit more frequent are a grouping of defect present in a circular band at the periphery of the wafer caused by hybrid bonding surface formation abnormality. These defects can be observed in figure 1. For the less critical defects, we have random defects which are circular defects caused by particle on bonding surface [3] and pattern defects caused by insufficient planarization of bonding surface [3]. These two types or defects are becoming critical when their number or individual size is too high. Finally, two other types of defects can be detected. The small defects that were not monitored when the procedure was manual. Up to now, they are not know

978-1-6654-7134-3/22 $31.00 © 2022 IEEE

as becoming killers. And the nuisance defects that are artifacts of the acquisition condition which can raise false alarms for automatic analysis. These less critical defects are displayed in figure 2.

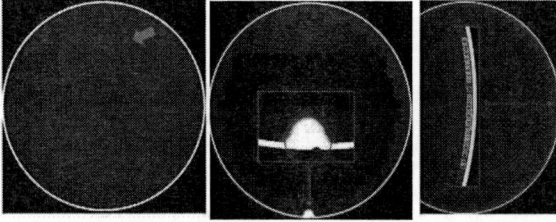

Fig. 1. Examples of scratch (left), delamination (middle) and edge (right) defect.

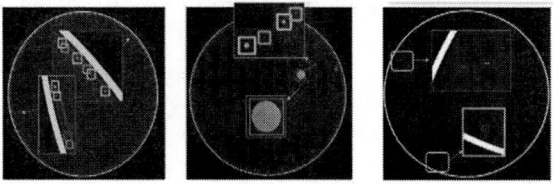

Fig. 2. Examples of defects and their classification. Random in green, pattern in blue, noise in red and small in orange.

Based on the image quality and homogeneity, a thresholding method was robust enough to detect each defect on the wafer with a strong accuracy. Otsu's thresholding, a classical algorithm is used to account for the small variability of the images [4]. Nonetheless, for some defects types, a refinement of the segmentation is performed posterior to the classification to improve the quality of the segmentation thanks to the prior knowledge about shape and position of the defect given its class. Once each defects detected, we need to classify them. One important feature of these defects is that each individual instance of a defect is not necessarily considered as unique and a defect can be formed by a group of nearby defects.

State of the art methods for classification are most of the time relying on deep learning or AI technologies [5], [6], but to achieve such performances, the quality and quantity of data for the training of these models are of tremendous importance. About the quality, as the criteria for the different classes are really precise, a human annotator is not be able to accurately and repeatably classify defects. As for quantity, for the simple case of deep learning training, a minimum of 10 or so is needed to get the final result, but some defects have less than 5 examples in the dataset. For these reasons, we decided that using a strategy of hierarchical clustering and boosting strategies are more adapted [7], [8] as described in figure 3. As rules for the classification are really precise and can be decomposed into several steps, these kinds of approach seem more suitable for this challenge. In particular, in our case, as all the defects are available after the thresholding, we can apply successive steps of classification to identify the different type one by one.

Defects can be separated into two main categories for the classification aspect: defects that can be classified by their position or by their shape. Small defects are set aside as the criteria is based on the area and are defect that were not taken into account in the original dataset because of the difficulty for a human annotator to detect them. This classification of defects types can be observed in figure 3. Each defect category is inspected to find defects matching with the requirements then passing to the next class until all classes are investigated, thus all defects classified. More precisely, scratch is a grouping of defects that are lying on a circle projection. This step is handled by sampling several times random defects on the wafer and try to fit a parametric model of a circle. If enough defects are taken into account and the circle model good enough, we consider the existence of a scratch defect and attach all defects lying on the computed circle to this defect. The edge defect is also a grouping of defects defined by defects lying on a circular band at the boundary of the wafer. After filtering defects based on their location, they are clustered by distance between each other to identify the groups of defects that are close together. A group long and numerous enough define an edge defect. Finally, delamination are defects touching the edge of the wafer, the challenge being measuring them correctly as the precise boundary detection with thresholding is challenging. Then, for the geometrical defects, shape of individual remaining defects is considered. Nuisance type is represented by defects that are thin, long and horizontal. They depend of the microscope used. Criteria about the aspect ratio and the size of the defect allows the classification. This defect, if long enough can be mistaken with scratch as a two- or three-pixels width line can be approximated by a circle with a really large diameter, thus particular care is taken at the scratch step is avoid this misclassification.

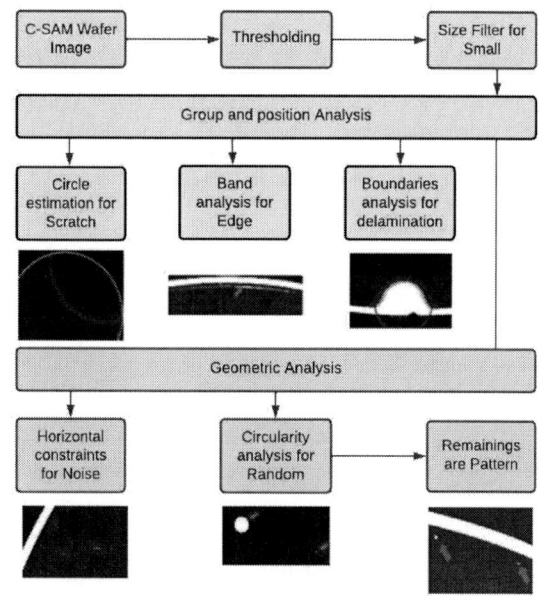

Fig. 3. Pipeline of classification of the defects. Blue boxes (first row) are image processing steps for the detection of the defects and the filtering of the small defect. Orange boxes (second and third rows) are related to the classification based on the position of the defects regarding the wafer or the other defects. Green boxes (last two tows) are related to geometric defects that are classified in a second time.

Finally, for random and pattern defects the discrimination is based on their roundness, round defects being randoms and more squared ones being patterns. The main difficulty for this part is to adapt the threshold on the roundness depending on the size of the defect as the analyzed data is lying on a square grid, small defects tend to have a lower roundness no matter their shape. The roundness criterion is described in part B, as it is also used for the measurement.

### B. Measurement strategies

Measurements are a key element to be able to identify the root cause of the degradation of bonding yield. Indeed, they are the only way a defect type can be analyzed in large quantity to establish correlations and conclusions. In addition, based on the variety of defects and their characteristics, we have to provide different sets of measurements in order to be accurate in the description of each defect. The complete description about how each type of defect is measured is provided in Tab. I.

TABLE I.    MEASUREMENT IN FUNCTION OF DEFECT TYPE

| Measurement | Defect Type | | | |
|---|---|---|---|---|
| | Scratch | Edge | Delamination | Geometric |
| Area | X | X | | X |
| Arc Length | X | X | | |
| Max Width | X | X | X | |
| Centroid | | X | | X |
| Length | | X | | X |
| Geometric | | | | X |

The first measurement, common to almost all the classes is the area that the defect is occupying on the wafer. Then, the arc length of a defect is the curved distance between the two most far points of the same defect as displayed in figure 4. For scratch, a circle is already fitted for the classification part so it is the same that is used for the computation of the arc length. Edge is using the center of the wafer and a fixed diameter, as the circular band for edge defects classification is fixed and thin. For scratch and edge, it is the maximum width of each individual defect forming the scratch or edge as described in figure 4. For delamination, the max width corresponds to the largest dimension of the smallest rotated rectangle containing the whole defect as described in figure 5.

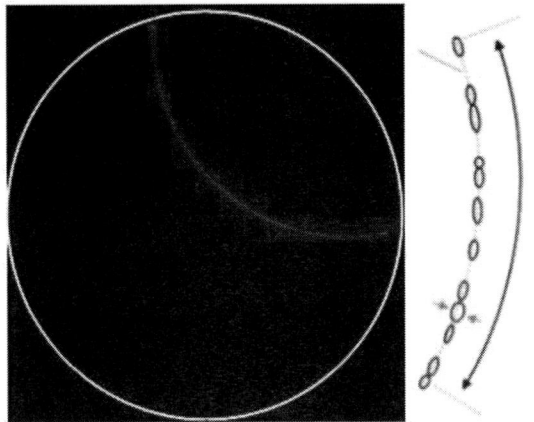

Fig. 4.    Illustration of a scratch defect in blue on the left. Small pink dots are visible to show the extreme points for the arc length. On the right a more synthetic illustration explain with the curved arrow the arc length measurement and blue arrows the max width.

The length is defined by the diameter of the minimum enclosing circle of the defect as illustrated in figure 5. Finally, for the geometric defects, we provide the centroid, the distance of the centroid of the defect from the center of the wafer, the maximum Feret diameter, the diameter of the equivalent circle in term of area and the circularity as defined in (1). The centroid is the center of gravity of the defect. The maximum Feret diameter correspond to the largest distance between two parallel lines tangent to the boundaries of the defect, and is also referred as caliper diameter [9], [10].

Fig. 5.    Illustration of max width and length on a delamination defect. The largest dimension of the rectangle is used for the max width. The diamaeter of the circle correspond to the length.

$$circularity = \frac{4\pi Area}{Perimeter^2} \ (1)$$

These measurements aim to provide an accurate quantification of each defect class individually to try to identify the root cause of the yield degradation. The variety in the measurement is representative of the variety present in the defect classes appearance to better characterize their individual particularities. If the classification of the defects can already be assessed on current process and discussed below, analysis of the measurements is still undergoing to identify the source of the degradation.

### III. RESULTS

The main objective of this work is to provide accurate and automatic classification of the defect to be able to precisely identify the root cause of defects that are degrading yield of bonding as manual analysis is difficult and can be biased by the engineer. The total dataset, at the development time, is 151 wafer images with a third of them being defect free. The method is able to correctly not detect defects on these defect free images. A second important point is that the detection step is more accurate than the human as some of the defects have really low contrast and are really small regarding the size of the image. The detection step is validated in comparison of an engineer who first annotated the excepted defects and in a second time validated the additionally found defects. For each detected defect, the engineer is providing a class which is then compared to the automatic classification. Tab. II is compiling all the results of the classification, as 100% of the defects are detected and segmented. In the analysis of the root cause of bonding yield degradation, edge defect is a special type of defect that is categorized both

TABLE II. CONFUSION MATRIX OF THE PROPOSED METHOD. "ADC" MEANS AUTOMATIC DEFECT CLASSIFICATION BY THE TOOL, AND "MANUAL CLASS" MEANS CLASSIFICATION BY AN ENGINEER

| | Count | ADC Class | | | | | | | |
| | | Delamination | Noise | Pattern | Random | Scratch | Small | Total | Row accuracy % |
|---|---|---|---|---|---|---|---|---|---|
| Manual class | Delamination | 3 | 0 | 0 | 0 | 0 | 0 | 3 | 100.0% |
| | Noise | 0 | 38 | 0 | 0 | 0 | 0 | 38 | 100.0% |
| | Pattern | 0 | 1 | 198 | 9 | 0 | 0 | 208 | 95.2% |
| | Random | 0 | 4 | 67 | 79 | 0 | 0 | 150 | 52.7% |
| | Scratch | 0 | 0 | 0 | 0 | 3 | 0 | 3 | 100.0% |
| | Small | 0 | 0 | 0 | 0 | 0 | 3552 | 3552 | 100.0% |
| | Total | 3 | 43 | 265 | 88 | 3 | 3552 | 3954 | |
| | Col accuracy % | 100.0% | 88.4% | 74.7% | 89.8% | 100.0% | 100.0% | | 98.0% |

globally based on the position as edge defect and individually by considering each geometrical defect forming the edge. To simplify the analysis, we decided to consider on the geometrical aspect.

From these results we can see that the algorithm reaches 100% accuracy on scratch and delamination defects which are critical. It is important to avoid sending defective wafers to the next step. Nonetheless, as these classes of defects are in small number, concerning a total of six wafers, these results should be used with caution as the current dataset is not big enough to allow the validation of the classification generalization. A second important result is the small defects which is represent 3552 defects that were not detected by the engineer because of their size and contrast. This discovery could be an important element in the identification of root cause of bonding yield degradation. Finally, geometric defects are the harder to distinguish automatically but we are still able to reach 88.4% accuracy for the noise defects, 89.8% accuracy for the random defect and 74.7% accuracy for the pattern defect. From the results the most significant error is random defect being classified automatically as pattern and concern 67 defects in the dataset. Two reasons can explain this difference. The first one is that for small defects it is difficult to estimate the circularity as the pixels are square. As an extreme example, a circle of one pixel of diameter has a circularity of 0.78 while we are expecting 1. In the proposed method, two sets of thresholds are provided to account this artifact depending on the size of the defect and the rectangular grid on which defects are lying on. A second possibility is that the different thresholds for the classification of pattern and random is not defined precisely enough to be representative of the reality of the process. This second hypothesis could be validated with further analysis of the classification and measurements. Additionally, the automation of the whole process will allow to validate and refine this criterion on a larger set of representative images.

## IV. CONCLUSION

In this work, we demonstrated the possibility to automate the detection and classification of defects in C-SAM wafer images. The first contribution is the detection step that allowed to precisely quantify a large class of defect that was not possible to handle manually. We hope that this more extended analysis will allow to better understand bonding process. A second contribution of the work is the accurate

automatic classification of the defects. The automation allows to process a larger quantity of wafer which will provide more statistical information about the defects. In addition to a larger quantity, the repeatability of the classification is increased reducing the bias, or at least fixing it constant when the classification is wrong. Finally, the large panel of proposed measurements are more accurately quantifying each defect. In addition, some measurements are difficult to do manually and others are even not possible. It allows to obtain at the same time new measurements with a better quality.

Thanks to this new approach, further analysis is undergoing to more accurately identify the root cause of the degradation of bonding yield. As the quantity of wafer analyzed increased with the automation of the classification, new types of defects are identified and will be added to this first classification to extend the capabilities of the algorithm and refine the final analysis.

## REFERENCES

[1] Y. Kagawa et al., Novel Stacked CMOS Image Sensor with Advanced Cu2Cu Hybrid Bonding, Proc. of IEEE Int. Electron Devices Meeting (IEDM), 2016, pp.208-211.

[2] A. Castex et al., Mechanism of Edge Bonding Void Formation in Hydrophiric Direct Wafer Bonding, 2013 ECS Solid State Lett. 2 P47.

[3] C. Cavaco et al., On the Fabrication of Backside Illuminated Image Sensors: Bonding Oxide, Edge Trimming and CMP Rework Routes, 2015 ECS Trans. 64 123.

[4] OTSU, Nobuyuki. A threshold selection method from gray-level histograms. IEEE transactions on systems, man, and cybernetics, 1979, vol. 9, no 1, p. 62-66.

[5] LIU, Ze, LIN, Yutong, CAO, Yue, et al. Swin transformer: Hierarchical vision transformer using shifted windows. In : Proceedings of the IEEE/CVF International Conference on Computer Vision. 2021. p. 10012-10022.

[6] LI, Shutao, SONG, Weiwei, FANG, Leyuan, et al. Deep learning for hyperspectral image classification: An overview. IEEE Transactions on Geoscience and Remote Sensing, 2019, vol. 57, no 9, p. 6690-6709.

[7] SILLA, Carlos N. et FREITAS, Alex A. A survey of hierarchical classification across different application domains. Data Mining and Knowledge Discovery, 2011, vol. 22, no 1, p. 31-72.

[8] WANG, Ruihu. AdaBoost for feature selection, classification and its relation with SVM, a review. Physics Procedia, 2012, vol. 25, p. 800-807.

[9] MERKUS, Henk G. Particle size measurements: fundamentals, practice, quality. Springer Science & Business Media, 2009, pp15.

[10] HODOROABA, Vasile-Dan et al. 17NRM04 nPSize, "Improvedtraceability chain of nanoparticle size measurements" European Project 2021 https://www.euramet.org/research-innovation/search-research-projects/details/project/improved-traceability-chain-of-nanoparticle-size-measurements

978-1-6654-7134-3/22 $31.00 © 2022 IEEE

# AUTHOR INDEX

Al Dujaili, H. .......... 59
Alpysbayeva, Balaussa Ye. .......... 93
Arima, Sumika .......... 1, 41
Arimura, Yasushi .......... 125
Baderot, Julien .......... 137
Bode, Christopher .......... 5
Britrun, Nikolay .......... 66
Busch, Rebecca .......... 117
Chalvin, Florian .......... 16
Chang, Shi-Chung .......... 26
Chang, Stephanie Y .......... 54
Chen, Chieh-Yu .......... 26
Chen, Po-Chih .......... 23
Choubey, Bhaskar .......... 117
Choudhary, Sumit .......... 97
Colard, M. .......... 59
Czerner, Peter .......... 117
Dewolf, T. .......... 59
Dupre, C. .......... 59
Eguchi, Hidetaka .......... 125
Estrada, Teresa .......... 105
Eto, Ryuji .......... 137
Faugier-Tovar, J. .......... 59
Fengjie, Guo .......... 114
Foucher, Johann .......... 137
Fujimori, Toru .......... 45
Funk, Hannes S. .......... 97
Gabdullin, Maratbek T. .......... 93
Garcia, S. .......... 59
Garnica, Sergio .......... 30
Garrais, Solange .......... 137
Goto, Tetsuya .......... 69
Hamaya, Aoi .......... 69
Hanafusa, Hiroaki .......... 90
Harashima, Yosuke .......... 34, 79
Hasunuma, Ryu .......... 34, 79
Higashi, Seiichiro .......... 90
Hirose, Ryo .......... 110
Hori, Masaru .......... 66
Hsiao, Shih-Nan .......... 66, 83
Hsieh, Yun-Che .......... 26
Huynh, Thinh .......... 73
Imai, Yusuke .......... 66
Inada, Takafumi .......... 69
Ishibashi, Shoji .......... 79
Ishikawa, Kenji .......... 66
Ito, Shiho .......... 9
Ito, Takumi .......... 12

Ito, Toshiki .......... 105
Ito, Yuto .......... 105
Iuti, Takashi .......... 9
Kadono, Takeshi .......... 110
Kakinuma, Takefumi .......... 121
Kasashima, Yuji .......... 51
Kataoka, Yuki .......... 63
Kataoka, Yuto .......... 41
Kato, Hibiki .......... 90
Katouda, Michio .......... 34
Kawai, Tomoya .......... 125
Kawanaka, Hiroharu .......... 125
Kawata, Isao .......... 105
Kazui, Masato .......... 133
Kim, Young-Bok .......... 73
Kitsunezuka, Masaki .......... 63
Kobayashi, Dai .......... 63
Kobayashi, Daisuke .......... 9
Kobayashi, Koji .......... 110
Kobayashi, Shinji .......... 133
Koga, Hiroaki .......... 34
Koga, Yoshihiro .......... 110
Kubo, Atsushi .......... 79
Kui, Ma .......... 101, 114
Kuniie, Shinji .......... 51
Kuo, Chia-Cheng .......... 23
Kurita, Kazunari .......... 110
Kuroda, Rihito .......... 69
Lai, Yang .......... 101
Le Tiec, R. .......... 59
Lee, Dong-Hun .......... 73
Lee, Y. .......... 59
Levi, S. .......... 59
Liao, Da-Yin .......... 26
Lin, Jia .......... 1, 41
Liu, Weijun .......... 105
Luh, Peter B. .......... 26
Luu, Lam .......... 54
Mao, Chending .......... 1, 41
Martinez, Sergio .......... 137
Masada, Ayumi .......... 110
Matsui, Hidefumi .......... 34, 79
Mayama, Norihito .......... 129
Meynard, B. .......... 59
Michishio, Koji .......... 79
Midoh, Yoshihiro .......... 47
Millard, K. .......... 59
Miyamae, Yoshinori .......... 16

| | |
|---|---|
| Mizuno, Shinsuke | 59 |
| Mizuno, Tomohisa | 129 |
| Morie, Sho | 41 |
| Morimoto, Tatsuo | 69 |
| Morinaga, Hiroyuki | 125 |
| Moriya, Tsuyoshi | 34, 79 |
| Murakawa, Kohki | 129 |
| Nagane, Kouhei | 105 |
| Nagashima, Kazutaka | 12 |
| Nakaegawa, Tatsuhiro | 83 |
| Nakahara, Ken | 16 |
| Nemkayeva, Renata R. | 93 |
| Nguyen, Hung H | 19 |
| Ni, Zeyuan | 34, 79 |
| Notake, Akira | 34, 79 |
| Ohira, Kentaro | 121 |
| Ohmi, Shun-Ichiro | 86 |
| Oikawa, Tetsu | 69 |
| Okamoto, Kazuya | 47 |
| Oku, Yoshiaki | 16 |
| Okuyama, Ryosuke | 110 |
| Oomuro, Yasuhisa | 12 |
| Oshima, Nagayasu | 79 |
| Ramazanov, Tlekkabul S. | 93 |
| Sakaguchi, Seima | 125 |
| Sakai, Yushi | 69 |
| Sasaki, Yu | 41 |
| Sato, Takuma | 90 |
| Sato, Yuki | 133 |
| Sayama, Toshiyuki | 51 |
| Schulze, Jöorg | 97 |
| Schwarz, Daniel | 97 |
| Sekine, Makoto | 66, 83 |
| Sharma, Kumar Palit | 97 |
| Sharma, Satinder K. | 97 |
| Shiba, Yoshinobu | 69 |
| Shigematsu, Satoshi | 110 |
| Shigeta, Yasuteru | 34, 79 |
| Shin, Joong-Won | 86 |
| Shinagawa, Jun | 63 |
| Shirai, Yasuyuki | 69 |
| Shun, Yang Fa | 101, 114 |
| Shuo, Wang | 114 |
| Si, Mrinal Kanti | 34 |
| Smith, Holland | 37 |
| Stachowiak, Timothy | 105 |
| Sugawa, Shigetoshi | 69 |
| Sutoh, Akihito | 69 |
| Suwa, Tomoyuki | 69 |
| Suzuki, Akihiro | 110 |
| Tabaru, Tatsuo | 51 |
| Takagi, Shigeyuki | 83 |

| | |
|---|---|
| Takahashi, Naomichi | 79 |
| Tanaka, Tomoya | 137 |
| Tanida, Kazumasa | 137 |
| Tanuma, Masakazu | 86 |
| Tiku, Shiban | 54 |
| Tokuyama, Yuichi | 125 |
| Tseng, Chang-Tsun | 23 |
| Tsuchiyama, Hirofumi | 37 |
| Tsutsumi, Takayoshi | 66 |
| Uedono, Akira | 34, 79 |
| Ueyama, Ken-Ichi | 105 |
| Vannuffel, C. | 59 |
| Wahl, Michael | 117 |
| Wakabayashi, Tetsushi | 125 |
| Watahiki, Kosuke | 47 |
| Wieland, Robert | 30 |
| Xu, Wang | 101 |
| Xueting, Wang | 12 |
| Yamauchi, Takayuki | 125 |
| Yamauchi, Toshinori | 121 |
| Yang, Ran Jing | 101, 114 |
| Yasuda, Shunsaku | 9 |
| Yasui, Takatoshi | 137 |
| Yerlanuly, Yerassyl | 93 |
| Yonehara, Takehiro | 34 |
| Zhang, Wei | 105 |
| Zhumadilov, Rakhymzhan | 93 |

**IEEE**
445 Hoes Lane
Piscataway, NJ 08854-4141

ISBN 978-1-6654-7134-3